# THE BIOLOGY AND ECOLOGY OF GIANT KELP FORESTS

# THE BIOLOGY AND ECOLOGY OF GIANT KELP FORESTS

David R. Schiel and Michael S. Foster

UNIVERSITY OF CALIFORNIA PRESS

University of California Press, one of the most distinguished university presses in the United States, enriches lives around the world by advancing scholarship in the humanities, social sciences, and natural sciences. Its activities are supported by the UC Press Foundation and by philanthropic contributions from individuals and institutions. For more information, visit www.ucpress.edu.

University of California Press
Oakland, California

Library of Congress Cataloging-in-Publication Data
Schiel, David R., author.
   The biology and ecology of giant kelp forests /
David R. Schiel & Michael S. Foster.
      pages   cm
   Includes bibliographical references and index.
   ISBN 978-0-520-27886-8 (cloth)
   ISBN 978-0-520-96109-8 (ebook)
   1. Giant kelp.   I. Foster, Michael S., author.   II. Title.
   QK569.I53S35 2015
   577.3-dc23                        2014041839

Manufactured in the United States of America
22   21   20   19   18   17   16   15
10   9   8   7   6   5   4   3   2   1

The paper used in this publication meets the minimum requirements of ANSI / NISO Z39.48-1992 (R 2002) (*Permanence of Paper*).

Front cover image: Giant kelp forest, San Clemente Island, southern California; photo courtesy Ron J. McPeak, UCSB Library Digital Collection. Back cover image: Portion of a giant kelp (*Macrocystis pyrifera*) frond; photo by John Heine.

*To the pioneers of giant kelp forest research, especially
Conrad Limbaugh, Wheeler North, and Michael Neushul,
whose love of scuba diving, observations, insights, and publications
helped in opening new frontiers.*

# CONTENTS

# PREFACE

So many interesting points are connected with the *Macrocystis*, that a
book might be instructively filled with its history, anatomy, physiology and
distribution; whilst its economy, its relation to other vegetables and to the
myriads of living creatures which depend on it for food, attachment, shelter and
means of transport, constitute so extensive a field of research that the mind of
a philosopher might shrink from the task of describing them.

—Hooker (1847)

Giant kelp, *Macrocystis pyrifera*, dominates nearshore rocky reefs of many temperate
zone coastlines of both hemispheres. Perhaps no other single species is so important
in providing biogenic habitat in which thousands of species can live and interact. From
historical to contemporary times, giant kelp forests have been valued for their provision
of numerous food species of fish and invertebrates, biomass production, industrial uses
such as extraction of potash and alginates, and their intrinsic beauty. Darwin (1839)
compared their importance to that of terrestrial forests and seemed to have been awe-
struck by the diversity of life within giant kelp forests and how so many species, including native human populations, depended on them. Giant kelp forests have therefore
long stood as important components of coastal ecosystems. The reduction in kelp forest
cover in some areas, especially along heavily populated coastlines such as southern California that are subjected to multiple stressors, has spawned considerable debate about
the status of kelp forests, causative mechanisms for their sometimes great variation in
abundance, and ways to restore them and their associated communities to full functioning. With an increasing awareness of the need to maintain the integrity of communities
and ecosystems, it is timely to have a comprehensive review of one of the most important
marine ecosystems in temperate waters and the giant kelp that defines it.

Since our last book on giant kelp communities was published (Foster and Schiel
1985), there has been a considerable increase in kelp forest research and understanding,
from microscopic to ecosystem-level effects. This includes a much greater knowledge
about dispersal mechanisms, genetic relationships, long-term coast-wide variation in

distribution and abundance of kelp forests, and their large-scale linkages with the ocean climate. At the same time there has been increasing emphasis on management and conservation, especially through marine protected areas and reduction in fishing pressure on species that can be strong interactors in food web dynamics. As in many areas where societal values, general awareness of the imperative need for better management of natural resources, strong advocacy, and science intersect, the debates can be heated about underlying mechanisms and how best to proceed.

We address these issues here with an up-to-date presentation of what is known scientifically about giant kelp, their forests worldwide, and how they function. We include a review of the literature, both from peer-reviewed publications in scientific journals and the "gray" literature of reports to state and federal agencies, which are often not reviewed externally and can be difficult to obtain. We were as comprehensive as possible in using the literature available to August 2014. We also point readers to several published reviews, which give details of various aspects of giant kelp biology and kelp forest ecology (Dayton 1985a, Schiel and Foster 1986, 2006, Graham, Vásquez et al. 2007, Foster and Schiel 2010, Foster et al. 2013). We have endeavored to provide the currently accepted scientific names for genera and species, based on listings in AlgaeBase, WoRMS, and FishBase. The pace of name changes continues to increase, however, so we encourage readers wishing to know the current names and their history to consult the appropriate databases and associated literature.

Finally, we emphasize the continuing need for empirical, field-based studies, not just new models and meta-analyses, to provide the necessary underpinnings for advancing knowledge and informed management. We hope our book proves to be a useful resource for students, scientists, managers, and the general public who share our passion and concern for these iconic parts of the marine ecosystem.

# ACKNOWLEDGMENTS

We thank the many colleagues who have provided advice, critique, information, discussion, photographs, and personal communications. We especially thank Dan Reed, John Pearse, and Mike Graham who took on the prodigious task of reviewing the entire manuscript and provided much valuable comment and criticism. We thank Genny Jo Schiel for her dedication and talents in producing the figures and artwork. Many others helped in various ways, and we gratefully acknowledge their contributions to this book: Stacie Lilley (for the massive task of checking and compiling all references, obtaining permissions, and proofreading); Kerry South (for pitching in when needed); Linda and Alan McCarter (for providing a quiet place for DRS to write); Bruce Menge and Jane Lubchenco who hosted DRS during a sabbatical at Oregon State University; and Joan Parker and the staff of the Moss Landing Marine Laboratory library for tracking down references. For personal communications, information, or reviews of parts of the book and photos, we thank Robert Anderson, Craig Barilotti, Jarrett Byrnes, Inka Bartsch, Roger Beattie, Ed Bierman, John Carter, Kyle Cavanaugh, Jiaxin Chen, Louis Druehl, Matt Edwards, Rebecca Flores Miller, Jon Geller, Mike Graham, Michael Guiry, Kamille Hammerstrom, John Heine, Gustavo Hernández-Carmona, Lawrence Honma, Joanna Kain, Brenda Konar, David Kushner, Chris McQuaid, Ron McPeak (via the University of California at Santa Barbara Library Digital Collection), Dieter Müller, Peter Neushul, Dan Reed, Susan Saupe, John Steinbeck, Feijiu Wang, James Watanabe, and Renado Westermeier. We gratefully acknowledge Moss Landing Marine Laboratories for literature and technical support of MSF, Canterbury University and the School of Biological Sciences for their continued support of DRS, and the funding bodies which supported this work, especially the New Zealand Ministry of Science and Innovation, the Royal Society of New Zealand Marsden Fund, and the Andrew W. Mellon Foundation of New York, which have generously supported the research and writing of DRS. The online resources of Google Scholar, Google Earth, AlgaeBase, WoRMS (World Register of Marine Species), FishBase, and Wikipedia were invaluable in accessing information. Finally, we thank UC Press, especially Chuck Crumly, who invited us to write this book, and our editor, Merrik Bush-Pirkle, who graciously guided it to fruition.

# INTRODUCTION

There is one marine production, which from its importance is worthy of a
particular history... I know few things more surprising than to see this plant
growing and flourishing amidst those great breakers of the western ocean,
which no mass of rock, let it be ever so hard, can long resist. The number of
living creatures of all orders, whose existence intimately depends on the kelp, is
wonderful. A great volume might be written, describing the inhabitants of one
of these beds of sea-weed... I can only compare these great aquatic forests of
the southern hemisphere with the terrestrial ones in the intertropical regions.
Yet if the latter should be destroyed in any country, I do not believe nearly so
many species of animals would perish, as, under similar circumstances, would
happen with the kelp. Amidst the leaves of this plant numerous species of
fish live, which nowhere else would find food or shelter; with their destruction
the many cormorants, divers, and other fishing birds, the otters, seals, and
porpoises, would soon perish also.

—Darwin (1839)

Giant kelp (*Macrocystis*) forests support some of the most species-rich communities on earth. With plants reported up to 60 m long growing from the seafloor and extending along the sea surface in lush canopies, these forests are true "biogenic engineers" that provide extensive vertical habitat in a largely two-dimensional seascape, alter the light environment, and dampen water motion. Well before Darwin (1839) published the first observations on their ecology, these forests provided food and other resources for human populations. Even with only limited observations from the sea surface and collections, Darwin was clearly fascinated by giant kelp and the diverse organisms associated with it, and made the first analogy between this community and terrestrial forests. There were a few subsequent reports on *Macrocystis* distribution, morphology, and growth, but it was around a hundred years after Darwin's observations in South America that the study of giant kelp forest ecology began. Andrews' (1945) pioneering research on the fauna of giant kelp holdfasts included some underwater observations made during surface-supplied, hard-hat diving, but this and other early studies were hampered by the lack of simple diving equipment. With the advent of scuba in the early 1950s, direct observations of kelp forests became relatively simple and, because of mounting concern over the effects of sewage discharges, loss of kelp habitat, and possible impacts of kelp harvesting, several kelp research programs were started.

The considerable amount of research and monitoring in the 69 years since Andrews' time has provided information from which to glean a more comprehensive view of the biology and ecology of giant kelp, and the structuring forces that determine its production and abundance patterns. This includes studies on plant structure, physiology, and growth, molecular studies on phylogenetic relationships, observations and experiments on structuring processes and interactions, aerial surveys of kelp canopy abundance, subtidal monitoring of plant and community structure, and climatic and oceanic processes that affect giant kelp. These studies encompass not only the pioneering work done along the coast of California but also, especially over the past 30 years, considerable work in several southern hemisphere countries where giant kelp also occurs. Common aims of research have been to understand the environmental drivers underlying the great spatial and temporal variation in kelp forests and the role of food web (trophic) dynamics in these fluctuations. Recent and historical overextraction of top predators such as sea otters, fishes, and lobsters, grazing by sea urchins, and numerous stressors such as sedimentation, increased coastal runoff of nutrients, alteration to coastal morphology, and huge increases in populations of coastal cities have contributed to kelp fluctuations and losses. It is clear that spatial and temporal variation in *Macrocystis* is controlled not by a single factor but by many, usually acting in concert.

Our aim is to describe the giant kelp forest environment and to discuss the sources of variability in the distribution and abundance of the organisms in it, especially that of *Macrocystis*. We do this with reference to the complete life history of the kelp, from spores through to the large sporophyte stages, and the biotic and abiotic environment in which they live. In Part II, we discuss worldwide community structure, productivity, competition, demography, grazing, and predation, and, where possible, their influence on different phases of the life cycle of *Macrocystis*. Part III addresses human uses, effects and management of giant kelp, and associated forest organisms. We end with Part IV on climate change and our conclusions, with suggestions for future research. We trust that this comprehensive treatment, along with its figures and extensive reference list, will aid others in studying, understanding, enjoying, and managing giant kelp communities.

# PART I

# THE BIOLOGY OF GIANT KELP

# INTRODUCTION TO GIANT
# KELP FORESTS WORLDWIDE

> After a very attentive examination of many hundreds of specimens, we have arrived at the conclusion that all the described species of this genus which have come under our notice may safely be referred to as *Macrocystis pyrifera*.
>
> —Hooker (1847)

> Numerous recent studies on *Macrocystis* interfertility, genetic relatedness, and morphological plasticity all suggest that the genus is monospecific. We propose that the genus be collapsed back into a single species, with nomenclatural priority given to *M. pyrifera*.
>
> —Demes et al. (2009)

## TAXONOMIC CLASSIFICATION

*Macrocystis*, commonly called giant kelp but also known as giant bladder kelp, string kelp (Australia), huiro (Chile), and sargasso gigante (Mexico), is a genus of brown algae, a group characterized by containing the accessory photosynthetic pigment fucoxanthin that gives them their characteristic color. Historically, brown algae were classified as plants in the Domain Eukaryota, Kingdom Plantae, and Phylum (Division) Phaeophyta. The Plantae contained most terrestrial plants and also included two other common algal phyla with multicellular species, the green (Chlorophyta) and red algae (Rhodophyta). Collectively, the large marine species in these three phyla are commonly called "seaweeds." It is now recognized through modern techniques, however, that brown and red seaweeds have characteristics so distinct that they are separate from true "plants." High-resolution microscopy has revealed striking differences in plastid and flagella morphology among many of the traditional plant phyla, or groups within them, as well as similarities in flagella morphology and other characteristics to some colorless flagellates. These more fundamental relationships have generally been supported by analyses of genetic similarities and a better understanding of the role of endosymbiosis in shaping the photosynthetic apparatus. This flood of new information and interpretation has led to fundamental taxonomic rearrangements and the creation of new kingdoms, but there is still debate about appropriate classification (e.g., Parfrey et al. 2006). Revisions will no doubt continue, but the basic classification scheme of Cavalier-Smith (2010) is currently accepted by most phycologists (e.g., Graham et al. 2009, Guiry and Guiry 2012). This scheme places the former Phaeophyta within the Kingdom Chromista, Phylum Ochrophyta, Class Phaeophyceae, with "kelp" being the term used to refer to members in the order Laminariales (figure 1.1). Diatoms are the other major group in the Ochrophyta. The kingdom name Heterokontophyta is preferred by some researchers, others use Stramenopila (discussion in Graham et al. 2009), while some prefer the supergroup designation Chromaveolata (e.g., Adl et al. 2005, Cock et al. 2010).

Within the order Laminariales, giant kelp is now usually placed in the family Laminariaceae (Lane et al. 2006, Guiry and Guiry 2012). Based on genetic similarities, Yoon et al. (2001) placed *Macrocystis* within the family Lessoniaceae but Lane et al. (2006) argued this lacked bootstrap support. Genetic analyses in both studies indicate that the kelp genera *Pelagophycus*, *Nereocystis*, and *Postelsia* are most closely related to *Macrocystis*, forming a clade or two closely related clades within the family (figure 1.1). There has been considerable taxonomic work on *Macrocystis* and the genus is now considered to be monospecific, the sole species worldwide being *Macrocystis pyrifera* (discussed below). *Pelagophycus*, *Nereocystis*, and *Postelsia* are also considered to be monospecific (Guiry and Guiry 2012) and are endemic to the Northeast Pacific. All but *Postelsia* form subtidal kelp forests (Abbott and Hollenberg 1976).

"Kelp" originally referred to the calcined ashes resulting from burning large brown algae. It is sometimes used as the common name for all large brown algae, but particularly species in the order Laminariales. In this book, we use "kelp" to mean only species in this order. Some argue that kelps should be called chromistans, ocrophytes, phaeophyceans, etc., and not plants, because they are no longer in the kingdom Plantae. Although taxonomically correct, these names are awkward in usage, so we will refer to kelps and other algae as seaweeds, algae, or plants.

*Macrocystis* and its putative species have undergone considerable taxonomic revision since originally being described in 1771 by Linnaeus, who included it with other brown algae under the name *Fucus pyriferus* (reviews in Womersley 1954, Neushul 1971a, Coyer et al. 2001, Demes et al. 2009). Agardh (1820) placed *F. pyriferus* into a separate genus, *Macrocystis*, and described three species based on differences in blade and float (pneumatocyst) morphology. The number of species based on these characters increased to 10 by the mid-1800s. Hooker's (1847) extensive field observations indicated these characters were highly variable and he argued there was only one species, *M. pyrifera*. Recognition of variability in these characteristics led to species revisions based primarily on holdfast morphology. This character was not considered by early investigators because they relied on specimens collected by others, specimens that generally did not include holdfasts (Womersley 1954). More recent investigators examined plants as they grew in the field, and used holdfast morphology as the primary character to distinguish species. This resulted in three commonly recognized species in both hemispheres:

**Kingdom Chromista** ("colored")
Members have a plastid bounded by four membranes and containing chlorophyll *a* and *c* and various xanthophylls. Also includes organisms that have lost these characteristics but are closely related, based on multigene trees.

**Phylum Ochrophyta** ("golden-brown")
Chromists with two flagella, one smooth and one with tripartite tubular hairs, or with modified versions of this structure.

**Class Phaeophyceae** ("dusky seaweed")
Brown algae. Ochrophytes with chromistan plastids, ochrophycean flagella (on spores and/or gametes) and multicellular thalli with alginate in the cell wall. Parenchymatous members have pores or plasmodesmata between cells.

**Order Laminariales** ("flat blade")
Kelps. Phaeophyceans that have an alternation of generations: gametophytes (haploid) are small, free-living, and filamentous with apical growth; sporophytes (diploid) are large, free-living, and parenchymatous with intercalary growth.

**Family Laminariaceae**
Members are currently based on genetic affinities. There are no common morphological features distinct from other families in this order.

**Clade** with *Macrocystis* and its closest relatives
(Illustrated below). The three kelps to the left can reach lengths of 10s of meters.

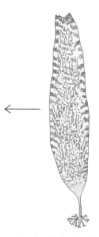

*Laminaria farlowii* illustrating a typical species of the type genus of the order

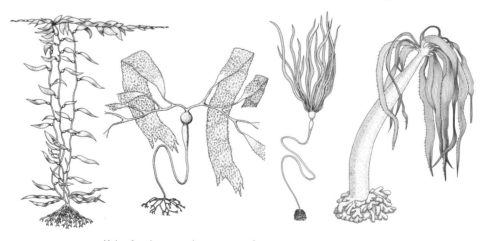

| Kelps forming canopies on sea surface | | | Smaller kelp (~0.25 m) |
|---|---|---|---|
| *Macrocystis pyrifera* | *Pelagophycus porra* | *Nereocystis luetkeana* | *Postelsia palmaeformis* |

FIGURE 1.1

*Macrocystis* classification based on genetic and morphological characteristics, the primary ones of which are listed. See Chapter 1 text for references.

*M. pyrifera, M. integrifolia,* and *M. angustifolia* (Womersley 1954, Neushul 1971a). Hay (1986) described a fourth species, *M. laevis,* from the subantarctic Marion Islands based on its unusually smooth blades (compared to rugose / corrugated blades in other recognized giant kelp species). Morphometric measurements and transplant experiments by Brostoff (1988) showed that the holdfast morphology of *M. angustifolia* populations described by Neushul (1971a) in California intergraded with that of *M. pyrifera. M. angustifolia* was subsequently considered to occur only in the southern hemisphere (e.g., Macaya and Zuccarello 2010a).

Holdfast morphology did not stand up to scrutiny as a species indicator, however, as more research was done. Field observations and transplant experiments have shown that holdfast morphology varies with environment and that blade smoothness also does not distinguish species well (review in Demes et al. 2009; figure 1.2). Genetic and molecular studies confirm that external morphological characters are not good discriminators of species. Lewis and Neushul (1994) showed that the three "species" (*M. 'pyrifera'* and *M. 'integrifolia'* from the Northeast Pacific, and *M. 'angustifolia'* from Australia) distinguished by differences in holdfast morphology could hybridize and produce normal sporophytes, and Westermeier et al. (2007) produced hybrids with normal sporophytes in crosses of *M. 'pyrifera'* and *M. 'integrifolia'* from Chile. Molecular taxonomic work comparing the similarity of noncoding rDNA internal transcribed spacer regions (ITS1 and ITS2) by Coyer et al. (2001), and DNA barcoding (Macaya and Zuccarello 2010a) of all four "species" in the southern hemisphere and *M. pyrifera* and *M. integrifolia* in the northern hemisphere indicate that *Macrocystis* is a monospecific genus, and this was confirmed by Astorga et al. (2012). All these findings indicate there is only one species of giant kelp, *M. pyrifera* (Linnaeus) C. Agardh, and this is currently accepted (Guiry and Guiry 2012). Unless otherwise noted, therefore, we refer to this species in the text as giant kelp or *Macrocystis.* However, when referring to the literature on *Macrocystis* it can be advantageous for clarity to refer to the former species names. Where necessary we designate these, including *M. 'pyrifera,'* as ecomorphs by enclosing the ecomorph name in quotes.

## EVOLUTION

The timing of origins of kelp and their relatives is problematic and not completely resolved, but significant progress has been made in the past 20 years. A variety of morphological, biochemical, and, most recently, genetic evidence indicates that species in the kingdom Chromista, as well as other photosynthetic eukaryotes, obtained their plastids via endosymbiosis with other organisms that took up residence inside cells (reviews in Yoon et al. 2004, Graham et al. 2009) early in the evolution of eukaryotic organisms. Primary endosymbiosis between prokaryotic cyanobacteria (blue-green algae) and eukaryotic protists resulted in the red and green algae (with green algae being the progenitor of "higher plants"). A secondary endosymbiosis between a uni-

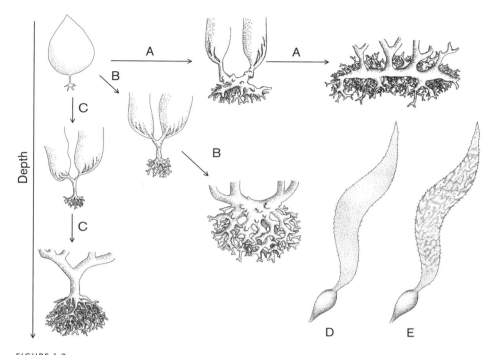

FIGURE 1.2

The development and holdfast morphology of *Macrocystis* ecomorphs, beginning as a small sporo-phyte (top left). (A) *M. 'integrifolia,'* (B) *M. 'augustifolia,'* and (C) *M. 'pyrifera.'* (D) The smooth blade morphology of *M. 'laevis'* compared to (E) more typical corrugated *Macrocystis* blades. 'Depth' indi-cates the relative depth distribution, from shallow to deep, of the holdfast ecomorphs.

SOURCE: Modified from Demes et al. (2009), reprinted with permission from Wiley Publishing.

cellular red alga and another eukaryotic protist resulted in the golden-brown algae, a lineage of which was the progenitor of kelps. Given the number of green chloroplast genes in the genome of the filamentous brown alga *Ectocarpus siliculosus*, the protist in the partnership that produced the brown algae may have been previously inhabited by a green chloroplast (Cock et al. 2010). Molecular clock methods indicate that red and green algae arose around 1500 Ma (Ma = SI unit for mega-annum or million years ago), and the secondary symbiosis that eventually led to the chromists occurred around 1300 Ma (Yoon et al. 2004) during the late Mesoproterzoic era, after the earth's tran-sition to a more highly oxygenated atmosphere with an ozone screen (Cloud 1976). Fossil evidence (Cloud 1976) is consistent with these gene-based estimates. Medlin et al. (1997) suggested the chromists originated between 275 and 175 Ma (in the Per-mian–Jurassic period), but Yoon et al. (2004) suggested a much earlier origin at about 1000 Ma at the Mesoproterozoic–Neoproterozoic boundary, with the Ochrophyta aris-ing soon afterward. An earlier study by Saunders and Druehl (1992) analyzed 5S rRNA similarities and concluded that the Phaeophyceae originated "within 200 Ma." Medlin

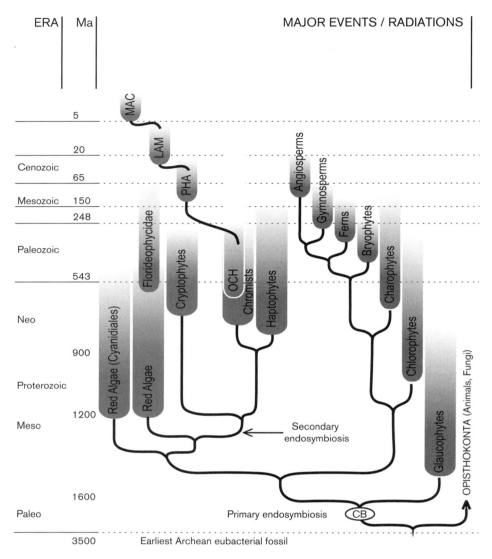

ERA | Ma | MAJOR EVENTS / RADIATIONS

FIGURE 1.3

Evolutionary timing and relationships leading to *Macrocystis*.

Ma = millions of years ago; CB = cyanobacteria; OCH = Ochrophytes; PHA = Phaeophytes;
LAM = Laminariales; MAC = *Macrocystis*.

SOURCE: Modified from Yoon et al. (2004), reprinted with permission from Oxford University Press.

et al. (1997) estimated this origin at 150–90 Ma (in the Jurassic–Cretaceous period) from a molecular clock analysis based on 18S rRNAs (figure 1.3).

The fossil record has not shed much light on the origins and radiation of kelp and other brown algae because, with the exception of a few lightly calcified species, they lack hard parts and do not fossilize well. There are no calcified Laminariales and the only

FIGURE 1.4

*Julescraneia grandicornis* from the upper Miocene of California. The fossil and inset reconstruction (with the fossil location indicated by the rectangle), pneumatocyst (P) is about 16 cm in diameter.

SOURCE: From Parker and Dawson (1965), reprinted with permission from Schweizerbart Science Publishing (www.schweizerbart.de).

fossil found so far that is generally accepted as a kelp is the extinct species *Julescraneia grandicornis* from the Miocene Monterey formation in southern California (Parker and Dawson 1965). This fossil consists of impressions on two rock fragments, one with a portion of a blade and another with "antler-like branches from a large pneumatocyst" (figure 1.4). Parker and Dawson (1965) interpreted *J. grandicornis* as having characteristics of modern *Pelagophycus* and *Nereocystis*, which are close relatives of *Macrocystis* (cf. figure 1.1). The deposit was dated at 13.5–7.5 Ma (upper Miocene). The molecular clock analyses of Saunders and Druehl (1992) and their review of similar analyses by others indicate kelps diverged from other brown algae 30–16 Ma (Oligocene–Miocene), and that morphologically similar kelp taxa radiated 3 – 6 Ma. These periods roughly correspond to those suggested by Lüning and Dieck (1990) based on temperature tolerances and paleo-oceanographic conditions (figure 1.3). Saunders and Druehl (1992) pointed out that "The extensive morphological variation observed among the kelp comes from genotypically similar plants." They reference other studies supporting these dates, including the fossil record of kelp-associated limpets, the number of monospecific kelp taxa, biogeographic distributions, and hybridization among kelps.

Various selective processes may have stimulated the radiation of kelp genera and species within the times discussed above. Evolutionary attention has been focused on processes in the North Pacific region because it contains the greatest diversity of kelps (review in Lane et al. 2006) and is therefore considered to be the likely area of origin and high diversification. Various processes have been hypothesized as being primary drivers of kelp evolution. Modern kelps occur in generally cool temperate waters and so the primary evolutionary processes are likely to be related to paleoenvironmental changes that caused North Pacific waters to cool. A series of ice ages between 23.7 and 5.3 Ma (Miocene) cooled the North Pacific, as did the growth of the Isthmus of Panama, both of which generally coincided with the appearance of kelps (review in Stanley 2009). Some argue that trophic dynamics were a factor. For example, Estes and Steinberg (1988) wrote that kelp evolution in the region from Canada to northern Japan was facilitated by predatory marine mammals like sea otters that primarily foraged at depths shallower than 30 m, a depth range that coincided with the high-light zone that most kelps occupy. Their proposed mechanism was that by eating sea urchins, which can be extensive grazers of kelp, these mammals removed a potential impediment to kelp evolution. Domning (1989), a noted marine mammal biologist, countered their argument by pointing out that Stellar's sea cows (*Hydrodamalis gigas*, large, surface-dwelling, herbivorous marine mammals related to dugongs and manatees, which were hunted to extinction around 1770) and their relatives were also present, and grazing by these mammals may have affected kelp evolution. He questioned the assumption that the strength of the modern sea otter / sea urchin / kelp interaction applies to the past, given that the dynamics of kelp stands may have been quite different then. He further postulated that kelp diversification may have occurred in habitats, including those in deep water, that were inaccessible to predatory and grazing mammals.

The degree to which the various processes affected kelp evolution remains unresolved, but there is some intriguing recent evidence that past notions of the causes of shallow-water diversification as well as modern distributions may have to be altered if kelp radiation occurred in deep water. Hypotheses about diversification are commonly based on attempted reconstructions of past physical and ecological processes operating in shallow, cold, surface waters (<30 m) where modern kelps are most abundant. However, the increasing use of deep SCUBA diving, remotely operated vehicles (ROVs), and remote sensing technologies has shown that kelp can be abundant in deep temperate waters (>30 m) if they are clear and transmit light to the depths (review in Spalding et al. 2003). For example, non-float-bearing kelps can be found to 45 m deep in central California (Spalding et al. 2003) and to 130 m in southern California (Lissner and Dorsey 1986). Perissinotto and McQuaid (1992) discovered stands of *Macrocystis* nearly 70 m deep at Prince Edward Island in the Southern Ocean (the fronds of this population did not reach the sea surface). Graham, Kinlan, et al. (2007) developed a model to predict kelp distribution in tropical waters using the photosynthetic compensation depth of kelp and the depth of the mixed layer below which nutrients are sufficient for kelp growth. The model predicted both the known distribution of kelp in the tropics and that kelp populations could occur to depths of 200 m on hard substrata when light, temperature, and nutrients are suitable. The recognition that kelps are common in clear, deep waters, including those in the tropics, should lead to a re-evaluation of the causes of kelp diversification and distribution.

Other hypotheses about processes shaping kelp evolution emphasize potential selection by the abiotic environment leading to the occupation of new habitats. For example, moving from the sea bottom to sea surface in the subtidal zone traverses diverse hydrodynamic conditions, from a benthic boundary layer to a surge zone to mid-water mixing, with orbital motion in breaking surf at the surface. Neushul (1972) argued that many developmental and morphological features of kelps may reflect the consequences of these changes because they affect phenomena such as nutrient availability and the ability of kelp to remain attached to the substratum. All kelp sporophytes have a similar morphology as juveniles, consisting of a small holdfast and a short stipe bearing a single blade (figure 1.1). It may be that this universal morphology, which involves juvenile kelp quickly getting blades elevated above the substratum on a stipe, is an adaptation that makes it more difficult for benthic grazers to remove the blade, the primary site of photosynthesis. Selection in response to abiotic factors such as water motion and gradients of light intensity and spectral composition between the surface and the bottom may have affected the morphology of float-bearing kelps like *Macrocystis* that occupy the full water column. These and numerous other habitat–kelp relationships indicate that the diversification of genera and species within the Laminariales inhabiting communities from deep water to the wave-swept rocky intertidal zone was most likely influenced by myriad biotic and abiotic variables. Sorting out these relationships and their relative influences is difficult in the present, let alone trying to reconstruct the past. Although the present often holds the key to the past, there are many possible keys that probably turned together. New molecular tools will clarify phylogenetic relationships, but the thorny problem of understanding the underpinning evolutionary forces that produced kelp diversification will most likely remain in the realm of plausible "just so" stories and speculation without a better fossil record and more detailed paleo-oceanographic information.

## BIOGEOGRAPHY

Assuming that *Macrocystis* evolved a few million years ago, what may have influenced its biogeographic distribution and differences among populations? Giant kelp occurs along the west coast of North America, but it is most widely distributed in the southern hemisphere where there are populations along the east and west coasts of South America, off southern South Africa, Tasmania and south Australia, central and southern New Zealand and its offshore islands, and the subantarctic islands. Recent work by Macaya and Zuccarello (2010b) clarifies the geographic occurrence of *Macrocystis* (figure 1.5) and highlights the broad distributional pattern of the species and ecomorphs. In the southern hemisphere, giant kelp occurs from as far north as Peru, at around 6° S latitude, and as far south as the subantarctic islands, at around 57° S. These latitudes span water temperatures that reach as low as a few degrees Celsius during parts of the year to as high as around 25°C in subtropical areas of the north. In the northern hemisphere, the latitudinal range was thought to be from around 27° N latitude in Baja California, Mexico, to around Sitka, Alaska, at 57° N (Druehl 1970, 1981), but the northern limit was recently extended as a few small stands were found in Alaska to 60° N at Icy Bay and southwest to Kodiak Island (Saupe 2011). Chapter 5 provides more detailed distributional information as well as descriptions of forests across biogeographic regions.

As for other species with anti-tropical distributions, it is generally thought that temperature is the chief barrier to geographic expansion into warmer waters (Hedgpeth 1957). However, despite the generalization that temperature can be a major factor limiting the distribution of seaweeds, acting on various life history stages (van den Hoek 1982), it seems clear that water temperature alone poses few biogeographic limits on giant kelp. For example, the general inverse relationship between temperature and nutrients in central and southern California, and Baja California, Mexico (Jackson 1977, Edwards and Estes 2006, Lucas et al. 2011), indicates that nutrients can affect expansion and distribution, and this has been shown by Graham, Kinlan et al. (2007) to affect kelp distribution in the tropics. Biogeographically, the southern latitudinal limit of giant kelp in the southern hemisphere coincides with the furthest landmasses in the cold temperate zone between the southern continents and Antarctica. On a wider geographic scale, wave force may well be a primary influence on the distribution of giant kelp. Our own personal observations, those of Hooker (1847), more recent algal studies (e.g., Smith and Smith 1998), and oceanographic studies all highlight the severe wave climate of the Southern Ocean. This region has consistently high wind speeds, little variation in wind direction, and no continental land barriers to impede wind and waves forces (Tomczak and Godfrey 2003). Nevertheless, the subantarctic islands have extensive giant kelp forests, but almost exclusively along lee shores and in protected inlets, despite being surrounded by some of the most tumultuous seas in the world.

It may well be the case that a combination of cold water and low light acting on the various life history stages of giant kelp prevents range expansion toward the poles (cf. van den Hoek 1982, Jackson 1987). However, Gaines and Lubchenco (1982) suggested that because herbivory increases inversely with latitude, the spread of kelp into warmer waters may also be limited by grazing. Their model, like some others for kelp diversification (see Evolution section above), is based on modern grazing interactions in shallow water. Whether these interactions are similar in deep water now or were in the past is unknown.

Early hypotheses about the origin and spread of *Macrocystis* were based primarily on its present distribution, the number of species in the genus thought to be present in the two hemispheres at the time, fossil evidence, and the distribution of other kelp species and genera (review in Coyer et al. 2001). Parker and Dawson (1965) and North (1971a) concluded, based on the more widespread distribution of the genus in the southern hemisphere, that there was a southern hemisphere origin of giant kelp with an expansion to the north (figure 1.5). They argued that a longer residence time in the southern hemisphere, indicated by the greater number of giant kelp species described there (at the time), also supported this hypothesis. Nicholson (1979) favored a northern hemisphere

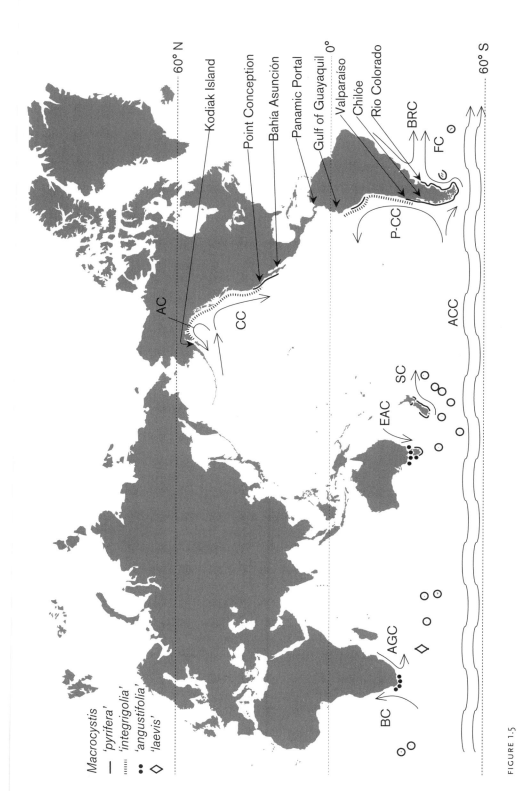

FIGURE 1.5

Worldwide distribution of *Macrocystis* and its ecomorphs, important biogeographic locations, and relationships to major currents.

Currents: AC = Alaskan; ACC = Antarctic Circumpolar also known as the West Wind Drift; AGC = Agulhas; BC = Bengula; BRC = Brazil; CC = California; EAC = East Australian; FC = Falkland; P-CC = Peru-Chile; SC = Southland.

NOTES: Open circles in the Southern Hemisphere indicate *Macrocystis* '*pyrifera*' around subantarctic islands, as does the diamond for *M.* '*laevis*'.

SOURCE: Modified from Macaya and Zuccarello (2010a), with permission from Wiley Publishing.

origin with spread to the south because more related kelp genera occur in the north and the fossil *Julescraneia* was found there. Chin et al. (1991) suggested a vicariant origin with *Macrocystis* evolving where it occurs now from a widespread ancestral complex. This hypothesis is not supported by the genetic analyses discussed below. The former hypotheses require explaining how, assuming conditions were somewhat similar to the present, a cold water species like *Macrocystis* crossed the equator, and how it moved north against the southward-flowing California Current or south against the northward-flowing Peru–Chile Current. Parker and Dawson (1965) suggested crossing the equator may have occurred during a period like the late Pliocene–Pleistocene (2–0.1 Ma) when temperatures were lower. Spread counter to the flow of major currents may have been via the transport of drifting reproductive plants in countercurrents and eddies (Nicholson 1979). Spread within the southern hemisphere could have been relatively rapid due to the cold, circular, west to east flow of the Antarctic Circumpolar Current (figure 1.5) where, as observed by Hooker (1847) and many others, vast amounts of drift *Macrocystis* can occur. With speeds up to 1.8 km hour$^{-1}$ (Fyfe and Saenko 2005), this current could disperse drifting *Macrocystis* as well as attached animals, such as the brooding pelecypod *Gaimardia trapesina*, between suitable habitats (Helmuth et al. 1994).

Early evidence and speculation about the origins of *Macrocystis* have been superseded to some extent by recent studies. A better understanding of genetic structure and relatedness, morphological plasticity, dispersal via surface drift, and depth distribution now point strongly to *Macrocystis* having originated in the northern hemisphere and then spreading south, with *Macrocystis* 'pyrifera' the likely northern source (Astorga et al. 2012). As discussed earlier (see the "Taxonomic Classification" section) genetic and morphological data indicate that *Macrocystis* is one species with several ecomorphs, eliminating the argument for a southern hemisphere origin based on having more species there. Moreover, genetic analyses by Coyer et al. (2001) of relationships between and within ecomorphs and within and between hemispheres show that populations in the southern hemisphere are more similar to each other than those in the northern hemisphere. This evidence combined with the greater diversity of the genetic sequences in individuals from the north are indications that *Macrocystis* has existed longer in the northern than in the southern hemisphere (Coyer et al. 2001), and therefore is of northern origin. High genetic similarity among southern hemisphere populations indicates either high gene flow or very recent dispersal (Coyer et al. 2001, Macaya and Zuccarello 2010b), most likely from western South America where the landmass is continuous south of the distribution of *Macrocystis* along the west coast of North America. Coyer et al. (2001) speculated that there were probably multiple crossings between 3.1 Ma and as recent as 0.01 Ma.

The argument that giant kelp spread across the equator has become less problematic with improved knowledge of opportunities and mechanisms. Lindberg (1991) reviewed the biological and paleo-oceanographic evidence for interchange across the equator and concluded there were two main periods when this occurred. One was during the Pliocene (5.3–1.6 Ma) when northern species moved south as cooling occurred because of the closure of the Panamic Portal and the other was in the early Pleistocene (about 1 Ma) when species moved in both directions as a result of glacial cooling and increased upwelling. The importance of upwelling is difficult to assess because glaciation cycles also caused changes in sea level that may have affected upwelling (Lindberg 1991) as well as the distribution and abundance of giant kelp (Graham et al. 2003). Because giant kelp can grow in deep water (e.g., Perissinotto and McQuaid 1992) and it is now known that other kelps occur in deep tropical waters (review in Graham, Kinlan, et al. 2007), the tropical thermal barrier appears to be less daunting and, historically, more permeable because the spread of kelp across the equator could potentially have occurred through submerged populations. The probability that spread across the equator could have occurred by surface drift of adult plants has also increased because we now know that rafts of giant kelp can be driven against currents by the wind, and can survive and produce viable spores for over 100 days and hundreds of kilometers if surface temperatures are suitable (Macaya et al. 2005, Hernández-Carmona et al. 2006, Rothäusler et al. 2011). Alberto et al. (2011) determined that current speed and direction, combined with spore production, explained more of the genetic differences in giant kelp populations than geographic distance and habitat continuity, at least on a regional scale. It remains to be determined, however, if spores from drift giant kelp can arrive in suitable habitats at great enough densities and with the genetic diversity likely to produce viable new populations. Drifters may not be important locally over short time scales but may be important regionally over long time scales, especially if very large masses of giant kelp strand on shore and release spores.

We have outlined where *Macrocystis* occurs and potential influences on its range limits but an intriguing question is, why doesn't it occur elsewhere, particularly along the coastlines of the western Pacific Ocean? This question is an old one, first posed in the literature by Setchell (1932). In discussing the worldwide distribution of *Macrocystis*, he noted that "the waters of the Bering Sea, of the Ochotsk, and the Kuriles are probably not colder than those of Cape Horn, South Georgia, the Falklands, and those of the Antarctic islands. Hence its absence presents the question: specific identities or seasonal or other variables?" Whatever the underlying mechanisms for the absence of giant kelp from the shores of the western Pacific, a combination of prevailing currents, cold temperatures, seasonal ice, low light, and competition with other kelps would have helped prevent the spread of *Macrocystis* westward. Furthermore, the ice bridge along the Bering Sea between Alaska and eastern Siberia may have prevented the spread of giant kelp over much of its history. During the Last Glacial Maximum (LGM), other algal taxa were not abundant. Caissie et al. (2010) pointed out that during the LGM through to the early deglacial period (23,000–17,000 years ago) there were more than 6 months of sea ice present each year, corresponding with low diatom concentrations. As ice-free conditions later prevailed, there was a transition from diatom assemblages dominated by sea-ice species to those dominated by species indicative of high productivity. Similarly, de Vernal et al. (2005) found that dinocyst concentrations were lower during the LGM, which they interpreted as being the result of limited biogenic production due to limited light because of permanent or quasi-permanent sea-ice cover. With respect to *Macrocystis*, both Graham, Vásquez, et al. (2007) and Macaya and Zuccarrello (2010a, 2010b) concluded that glaciations probably had a great effect on the distribution, abundance, and productivity of giant kelp. A more recent paper highlights this possibility for a similar large, drifting alga of the southern hemisphere. Fraser et al. (2009) used molecular techniques to assess the history of the southern bull kelp, *Durvillaea antarctica* (the largest fucoid alga, not a laminarian kelp), across the southern hemisphere landmasses. They found considerable genetic diversity among northern and southern Chilean sites, and around the coast of mainland New Zealand. Samples across the subantarctic islands, however, showed remarkable genetic homogeneity. Fraser et al. interpreted this similarity across the vast distances of the subantarctic islands as the result of subantarctic kelp populations being eliminated during the LGM due to ice scour, with recolonization occurring through a series of long-distance rafting events.

As in all areas where reconstructions, correlative evidence, and informed surmisal are involved, we will probably never know why giant kelp did not get to some regions of the Pacific basin and establish unless, of course, it suddenly appears there and we can determine how it did so. Nevertheless, many find it interesting to use the accumulating knowledge across a wide range of disciplines and continue the long tradition of speculation. The problems of origins, diversification, and spread will no doubt be areas of research and speculation in the future. As Druehl (1981, quoting Silva 1962) pointed out, however, "the field of phytogeography is vast, the literature is inexhaustible, its data capable of a variety of interpretation, extraordinary manipulation, and distortion. It has the fascination of a chess game. It is a valid, though treacherous, field of investigation."

## HISTORY OF RESEARCH (TO 1970)

If you see beds of weede, take heed of them and keep off from them.

— sailing directions from the mid-1500s cited by Hooker (1847)

The ocean is an inexhaustible treasury of varied wealth, but its riches are stored in so attenuated a form that we are powerless to gather them without the aid of natural processes that go on continually. Marine plants include annually certain portions of this wealth, and offer it for our acceptance like dividends due; should we refuse, it is returned to the treasury and, as time advances, offered us again and again.

— Balch (1909)

No level of species' protection or reserve status will be effective if water quality, coastal run-off, increased sedimentation, and contamination impact the ability of giant kelp to survive and thrive.

— Foster and Schiel (2010)

Giant kelp forests have provided food, medicine, materials, and probably aesthetic value since the earliest humans inhabited temperate coasts (Dillehay et al. 2008). Forests and drift plants first entered the written record in the logs of European explorers in the mid-1500s. Cabrillo, one of the earliest Spanish explorers in the Northeast Pacific, was apparently the first to mention giant kelp in this region, which he observed while anchored at Bahía San Pablo in Baja California in 1542 (Kelsey 1986). Other explorers noted giant kelp in the Southern Ocean at around the same time and commonly used drifting and attached plants to indicate proximity to land and the presence of shallow, submerged rocks (Hooker 1847). The presence of "porra," individuals or rafts of clubbed-shaped drifting kelp, was especially significant for the Spanish Manila-Acapulco merchant galleons active between 1565 and 1815. Porra indicated they were near the often fog-shrouded west coast of California / Baja California on the eastward leg of their route and needed to bear south to Acapulco (Schurz 1917). Setchell (1908) noted that porra could have been *Nereocystis luetkeana*, *Pelagophycus porra*, or *Macrocystis pyrifera*, based on voyage narratives and depending on where the galleons arrived along the coast. Distributional records and descriptions of new species and forms proliferated as increases in ocean commerce and voyages of exploration led to more collections and observations. The resulting taxonomic descriptions, however, were commonly based only on portions of plants that were often collected as drift. Moreover, the descriptions were usually not done by the collector but by botanists who saw only the collected portion and not entire plants growing in the field (Setchell 1932). The early efforts to describe *Macrocystis* formally are exemplified by Bauhin (1651) who produced the first written description and drawing, naming giant kelp BULBUS MARINUS CRINATUS (figure 1.6), and by Linnaeus' official description (review in Womersley 1954). Bauhin clearly did not see an entire plant growing in nature. An even more inaccurate illustration of the plant appeared in Darwin (1890). To Darwin's credit, however, he was not responsible for the illustrations in this 1890 edition of what has become known as the "Voyage of the Beagle"; most drawings, including that of giant kelp, were done by the illustrator R. T. Pritchett after Darwin died. Pritchett supposedly made the sketches "on the spot with Mr. Darwin's book by his side" (publisher's prefatory notice to the 1890 illustrated edition), but this is unlikely for the *Macrocystis* illustration, which looks like a "tribrid" of the kelps *Macrocystis* and *Lessonia*, and the fucalean *Marginariella* (figure 1.6).

The accuracy of descriptions was greatly improved by Hooker (1847) and Skottsberg (1907) (figures 1.6B and 1.6C) who spent several years in the Southern Ocean observing and collecting drift and attached plants. Both of these field botanists showed that much of the morphological variation previously used to describe species was a result of limited collections of parts of plants. They concluded that the variation previously used to discriminate between species could often be found as a result of changes in a single plant during development, within a mature plant, and among plants along environmental gradients. Skottsberg's (1907) research is particularly noteworthy as he also accurately described how giant kelp plants developed, and critically reviewed and added

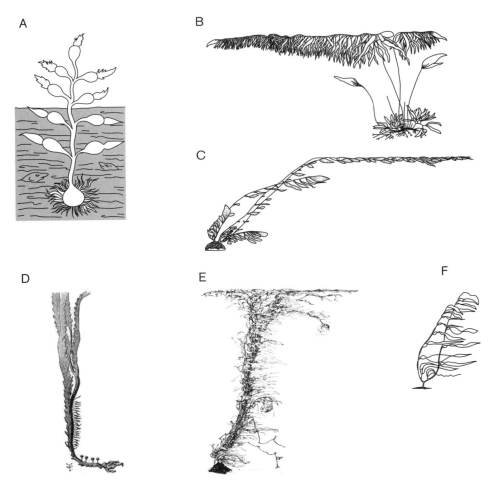

FIGURE 1.6
Evolution of giant kelp illustrations. The Skottsberg and Neushul illustrations are anatomically correct
as they show all fronds arise from dichotomous branching (obscured in more fully developed plants)
and that floats and blades occur from the base to the tip of vegetative fronds (see also figure 1.2).
SOURCE: Drawings from (A) redrawn from Bauhin (1651), (B) Hooker (1847) (from North 1971a and redrawn
by L.G. Jones), (C) Skottsberg (1907), (D) Darwin (1890) by R.T. Pritchett, (E) and (F) Neushul (adult and
juvenile forms) from Dawson et al. (1960). (B) is reprinted with permission from Schweizerbart Science
Publishing (www.schweizerbart.de).

to the understanding of its internal morphology. This early research was summarized
by Setchell (1932), Womersley (1954), and North (1971a). Setchell (1932) produced the
first world map showing the distribution of the genus. Womersley (1954) reviewed and
more completely described what were thought to be three species based on holdfast
morphology, and mapped the distribution of species worldwide, a map very similar to
the most recent one in figure 1.5.

As it has for many species, research and research funding increased significantly when giant kelp became economically important. With the outbreak of World War I, the U.S. government was concerned over the possible termination of potash supplies imported from Germany (Cameron 1915). Potash, composed of soluble salts of potassium, can be used as fertilizer as well as in the manufacture of gunpowder and numerous other materials. Potash was traditionally obtained by burning wood or other plant materials, extracting the soluble potash from the ashes with water and then evaporating the water. It became more readily available and cheaper with the discovery of large, subterranean deposits in Germany in 1852 (Neushul 1989). In 1912 and 1913 around 900,000 tons of potash worth about $13,000,000 (in US dollars at the time) were imported to the United States from German mines, primarily for use as fertilizer on east coast farms (Cameron 1915). In addition to growing concern over this loss of supply with the onset of war (imports from Germany were embargoed in 1916; Neushul 1989), there was also interest in new supplies for farms on the west coast, and Balch (1909) had shown that potash could be produced from giant and other large kelps in the west coast region. In response, the U.S. Department of Agriculture, Bureau of Soils, funded several studies of kelp, and particularly giant kelp, beginning in 1911. These studies included kelp distribution, abundance, and productivity, as well as harvesting and extraction methods (reviewed in Neushul 1989). Funding continued into the 1920s, stimulating the giant kelp harvesting and processing industry in southern California. The information on giant kelp biology appeared in government publications by Cameron (1915) and Brandt (1923) and was summarized along with other U.S. efforts to develop its own potash supplies by Turrentine (1926).

With the end of World War I and resumption of potash imports from German mines, it became uneconomical to produce potash and other war-related materials from giant kelp. Fortunately for the kelp harvesting industry, a relatively new product, alginate, could be extracted in quantity from giant kelp. Alginate (also referred to as algin or alginic acid) was discovered in kelp by Stanford (1883). It is a complex cell wall polysaccharide that absorbs water and forms a viscous gum that probably provides strength and flexibility to the thallus (review in Draget et al. 2005). Alginate in various chemically modified forms has numerous industrial uses, particularly in the food processing and pharmaceutical industries. Its extraction from fresh giant kelp sustained the California harvesting industry and stimulated considerable research through the 20th century. The potential for harvesting giant kelp as a source of alginate also stimulated growth, harvesting, and resource assessments in New Zealand (e.g., Moore 1943), Australia (Cribb 1954), and Canada (Scagel 1947) but given the large size of forests and their proximity to processing facilities, the alginate industry became centered in southern California. The California harvest for alginate ended in 2006 as dried kelp that was collected and processed in other countries became the major source.

The government-sponsored research as well as other studies in the early 1900s revealed the fundamentals of giant kelp biology and led to harvest management. Detailed maps were made of giant kelp distribution from Pacific Mexico to Puget Sound. Observations indicated that storms could greatly diminish stands and that they recovered in 2–3 years, and field experiments showed that fronds cut by harvesting degenerated but new fronds grew up from the base of plants to replace them (Crandall 1915). Brandt's (1923) extensive field and laboratory work is especially noteworthy. He followed the growth and survivorship of a cohort of juveniles in the low intertidal zone, confirming Skottsberg's (1907) observations on development. Brandt (1923) found it took 3 months for visible sporophytes to develop from spores in the laboratory, and an additional 7 months for juvenile sporophytes to develop into harvestable adults in the field. Like Crandall (1915), he noted that storms were a major source of adult mortality but fronds and plants also died during periods of warm, calm weather. In such conditions, portions of surface fronds turn black, a phenomenon called "black rot." The need to manage the California harvest became apparent as many harvesting companies increased the annual harvest to 400,000 wet tons by 1917 (Crandall 1918). The California Legislature gave management authority to the California Department of Fish and Wildlife (until January 1 2013, this was called the California Department of Fish and Game) which developed a numbering system for the various beds in southern California, required harvesters to purchase a license ($10.00), and collected a $0.015 fee per ton of wet kelp harvested. The method, location, and timing of harvests were also regulated to promote sustainable yields (Crandall 1918).

Other important early studies explored the relationship between giant kelp morphology and physiology and its ability to grow rapidly, and clarified its life history. With sufficient nutrients, average frond growth rates are in the order of 15 cm day$^{-1}$ (Zimmerman and Kremer 1984), and the holdfast can grow in the shade of the canopy. These observations coupled with the presence in the stipe of sieve elements, the specialized cells morphologically similar to sieve elements in the phloem of terrestrial plants, led early investigators to speculate that giant and other kelps with similar structures could actively transport (translocate) the products of photosynthesis within the thallus (e.g., Skottsberg 1907). Sargent and Lantrip (1952) showed that translocation was necessary to support the growth of rapidly growing blades, and Parker (1965) used dye and C$^{14}$ labeling to demonstrate that the products of photosynthesis were actively transported through the sieve elements. The life history of giant kelp was first completely determined by Levyns (1933) who showed it to be similar to other genera in the Laminariales with a large, parenchymatous sporophyte and small, filamentous, free-living male and female gametophytes. The sporophytes, the large visible structures that were the subject of almost all early investigations, are therefore only part of the plant if its entire life cycle is considered; knowledge of the biology and ecology of the gametophytes has become essential to understanding the dynamics of sporophyte populations.

The first insights into the community ecology of giant kelp forests were published by Darwin (1839) who was impressed by the diversity and abundance of associated species and by Skottsberg (1907) who observed that the holdfast "teemed with animal life." Population studies began with the research of Crandall (1915) and Brandt (1923) discussed above. These early observations and studies were necessarily made from the surface, on plants pulled from the bottom or on intertidal populations at low tide. Giant kelp forests are most luxuriously developed in the subtidal zone, however, and so progress in understanding them required a suitable air supply and a warm, flexible diving suit to enable sustained underwater work. The first in situ research in giant kelp forests was that of Andrews (1945) who used a diving helmet with air supplied through a hose from the surface to collect giant kelp holdfasts for identification and enumeration of their fauna. The difficulties of working easily, efficiently, and safely under water were essentially eliminated by the development of scuba by Cousteau and Gagnan in 1943 (Dugan 1965) and the neoprene wet suit for divers by Bradner in 1951 (Rainey 1998). A few marine science graduate students in southern California began using scuba soon after the first "Aqua-Lungs" were imported from France in 1949. The beginnings of the use of scuba to study kelp forests were largely due to the curiosity and efforts of Conrad Limbaugh (figure 1.7), a graduate student at Scripps Institution of Oceanography. He developed training procedures, offered informal training courses, and began the first institutional marine science diving program at Scripps in 1953 (Dugan 1965, Price 2008).

The use of this new technology in kelp forest ecological research was initially concentrated along the southern California mainland, stimulated by concerns over the effects of harvesting giant kelp on sport fishing and the effects of sewage discharges on the "health and sustenance" of kelp forests. Limbaugh (1955) was the first to investigate the effects of giant kelp harvesting on species other than kelp, providing information on habitat utilization and feeding habits of numerous kelp forest fishes, and the distribution of common invertebrates. His research included comparing fish assemblages in harvested and non-harvested stands, the first such field experiment in a kelp forest. Aleem (1956) used scuba to provide the first quantitative description of kelp forest zonation and the standing crop of algae and invertebrates. The studies of Limbaugh and Aleem were followed by those of Limbaugh's colleague Wheeler J. North (figure 1.7). North's investigations were focused on kelp harvesting and sewage discharge effects, and done primarily from the perspective of the effects on *Macrocystis* itself (North 1964, North and Hubbs 1968). North's early observations of sea urchin grazing in southern California kelp forests and his suggestion that this might be caused by overharvesting of sea otters (North and Pearse 1970) stimulated much additional research and controversy over the effects of sea urchins and their predators on kelp population dynamics (review in Foster and Schiel 2010). Michael Neushul (figure 1.7), collaborating with North as well as the macroalgal taxonomist E.Y. Dawson, expanded on Aleem's (1956) zonation studies with surveys of macroalgal distribution in many giant kelp forests in southern California and northern Baja California (e.g., Dawson et al. 1960). Neushul also did innovative field and laboratory experiments to examine giant kelp growth and reproduction (Neushul 1963, Neushul and Haxo 1963). These early studies provided most of the basic descriptive and life history information about kelp forests and their associated organisms, and the phenomena that affect their abundance and distribution in California (summarized in North 1971a).

FIGURE 1.7

Some pioneers of giant kelp research.

Top (left to right): Johann Bauhin, Joseph Dalton Hooker, and Carl Skottsberg.

Bottom (left to right): Conrad Limbaugh, Wheeler J. North, and Michael Neushul.

SOURCE: Credits: Bauhin (1651); Hooker (Henry Joseph Whitlock © National Portrait Gallery, London); Skottsberg (1911). Limbaugh and North photos are from Scripps Institution of Oceanography Archives, UC San Diego; Neushul photo by M. S. Foster.

In recent decades, *Macrocystis* research has expanded geographically and in scope. Studies in Canada (e.g., Druehl and Breen 1986), Chile (review in Graham, Vásquez, et al. 2007), New Zealand (Schiel 1990, Schiel et al. 1995, Schiel and Hickford 2001), the Subantarctic Islands (e.g., Beckley and Branch 1992), Australia (e.g., Ling et al. 2009), and Argentina (Barrales and Lobban 1975) have greatly improved our appreciation of the global differences in giant kelp communities. Furthermore, long-term programs, such as those of Paul Dayton, Mia Tegner, and colleagues around Point Loma (San Diego), which have been running since the 1970s, and the Santa Barbara Coastal LTER (long-term ecological research network of the U.S. National Science Foundation, begun in 2000) with the work of Principal Investigator Dan Reed and colleagues, have provided considerable new insights into temporal dynamics, production, dispersal, connectivity, and interactions with the physical environment of kelp forests.

# 2

# THE STRUCTURE, FUNCTION, AND ABIOTIC REQUIREMENTS OF GIANT KELP

The fact that frutification is produced only on the submerged young bladderless and small fronds, within a few inches of the very root, is highly remarkable. What then is the function of the floating mass of the plant?

—Hooker (1847)

*Macrocystis* may be unique among large, community-dominating plants because it is essentially a "weed." Within a wide range of environmental conditions and if spores and suitable rocky substrata are available, *Macrocystis* can quickly colonize surfaces, grow rapidly, and become reproductive in less than a year. It does not require facilitation by other species and, with some local exceptions, its population dynamics are largely driven by changes in the oceanographic environment. Its biological attributes are therefore particularly relevant to its ecology. To understand how *Macrocystis* functions within the environment, it is necessary to consider not only its life history traits but also the structures and internal processes that allow plants to survive and thrive.

Giant kelp sporophytes are the largest among the kelps and, combined with its relatively complex internal and external morphology, beg comparison to large land plants. In growth rates, kelps are most like bamboos, which are large grasses, and in ecological importance they are analogous to forest trees. Unlike its terrestrial analogues, however, giant kelp has an alternation of generations with free-living, 1N, microscopic gametophytes and 2N sporophytes (figure 2.1). Understanding the biology of giant kelp and its ecological relationships requires consideration of both of these life stages, which act almost like two different organisms because of their differing structures, development, and requirements, and the linkage between them provided by spores and gametes. The following information unless otherwise noted comes primarily from studies on *Macrocystis 'pyrifera,'* the ecomorph for which we have by far the most information.

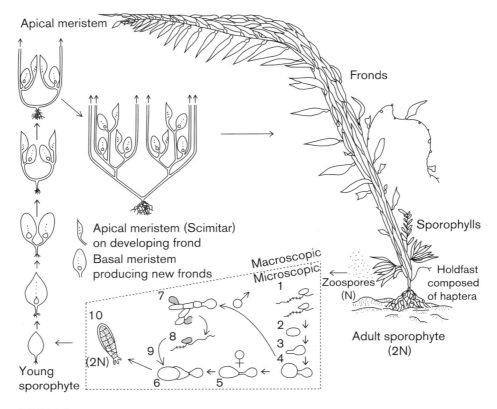

FIGURE 2.1

*Macrocystis* life history and development. The adult sporophyte (2N) illustrated is *M. 'pyrifera.'* Inset depicts microscopic stages of development from 1N zoospores to multicellular 2N sporophytes. Illustration at the left shows stylized sporophyte development, with dichotomous growth.

1. Zoospores
2. Settled spore
3. Developing germ tube
4. Germinated gametophyte
5. Female gametophyte with oogonium
6. Female gametophyte with extruded egg to left
7. Multicellular male gametophyte with antheridia
8. Antherozoid (sperm)
9. Fertilization
10. Embryonic sporophyte with rhizoids for attachment

SOURCE: Modified from North (1994).

## SPOROPHYTES

### GROWTH AND DEVELOPMENT

As a giant kelp sporophyte grows, it undergoes several developmental transitions (reviewed in North 1994; detailed descriptions of growth and development in the work of North [1971b], Parker [1971], and Lobban [1978]). Investigators have defined the results of these transitions in various ways for various reasons (e.g., Dean et al. 1989). We will distinguish between juveniles and adults, and define juveniles as plants that are visible to the naked eye but have not yet reached the sea surface. Fronds on adults may all be beneath the surface in deep water (Perissinotto and McQuaid 1992) or because of removal of long fronds by storms. These are distinguished from juveniles either by their high number of fronds or by their large holdfasts.

Soon after fertilization the embryonic sporophyte develops into a parenchymatous blade with basal cells that elongate to produce rhizoids. Continued growth and differentiation produces a juvenile sporophyte around 5–10 cm tall with a holdfast composed of haptera, a stipe (the primary stipe), and a blade (lamina), which are characteristics of all juvenile Laminariales (figures 1.1 and 2.1). A meristem at the stipe-blade juncture contributes to further elongation, and a meristoderm (also called meristematic epidermis) near the outer cortex (i.e., the surface) of all parts of the plant contributes to frond enlargement. As growth continues, the meristem and blade split longitudinally and form the first dichotomy, also referred to as the primary or basal dichotomy. Continued dichotomous branching produces more fronds, some of which develop rapidly and can eventually spread over the sea surface.

Vegetative fronds consist of a branch made up of a central stipe with attached floats (called pneumatocysts) and blades (lamina). The floats and blades develop as the apical blade, sometimes referred to as the scimitar blade because of its shape, splits into numerous blades that subsequently develop basal floats and short pedicels arranged alternately along the frond. Fully developed blades commonly develop undulations and marginal spines (figure 1.2) that may enhance blade movement, thereby reducing self-shading and enhancing water flow (Hurd and Pilditch 2011) and potentially facilitating nutrient and carbon uptake (Wheeler 1980a). Skottsberg (1907) suggested the undulations may function to stiffen the blade and thus reduce tearing. Fronds can grow very rapidly while forming a surface canopy. For example, Clendenning (1971a) found their maximum elongation rate to be "the fastest autotrophic elongation on record for any marine or land plant." He estimated a maximum growth rate of around 50 cm day$^{-1}$, with the closest contenders being bamboo and the flower stalks of the century plant, *Agave americana*, which came in second and third at 30 and 16 cm day$^{-1}$, respectively. However, there clearly is considerable variation in maximal frond elongation rates of *Macrocystis*. Zimmerman and Kremer (1984) measured average frond growth rates of 15 cm day$^{-1}$ at a site in southern California when nutrients were sufficient for growth. Stewart et al. (2009) examined frond growth for over a year inside and at the edge of a giant kelp forest near Santa Barbara. Growth rates ranged from 10 to 35 cm day$^{-1}$ with highest rates at the edge when the canopy was thick. Other fronds from the dichotomous divisions in the understory grow very slowly until larger fronds senesce or are otherwise lost, and still others produce the numerous reproductive blades that cluster near the base of a plant. These blades usually lack floats, divide longitudinally, and become sporophylls (Neushul 1963, North 1971b).

Haptera production, growth, and attachment to the substratum are critical to sporophyte survival in the often turbulent waters they occupy. As described by Chien (1972, 1973), haptera are first produced at the base of the developing primary stipe. As the plant grows, haptera continue to be produced up the primary stipe and from stipes of the basal dichotomy, although the exact process depends on plant age and ecomorph (figures 1.2 and 2.1). Elongation is presumably a result of very active meristoderm at the tips. Haptera branch dichotomously and are negatively phototactic, growing downward and covering older haptera. The ability of new haptera to attach to older haptera undoubtedly reinforces the integrity of the entire holdfast. Haptera continue growth and form an adhesion disc that produces sticky rhizoids when they contact the substratum. Plant attachment is therefore accomplished both chemically by this rhizoidal "glue" composed of secreted polysaccharides and perhaps phenolics (Chien 1973, Tovey and Moss 1978), and mechanically by haptera growing around objects on the bottom. North (1994) reported that holdfasts of old *Macrocystis* '*pyrifera*' plants grew up to 2 m tall and 3 m in diameter. Furthermore, plant anchoring shows adaptations to characteristics of sites. For example, holdfasts exceeding 5 m in diameter were found in stands of giant kelp growing on sand near Santa Barbara, California, prior to the 1983–1984 El Niño (Thompson 1959, North 1971a). These probably established initially on worm tubes (Neushul 1971a) or other hard substrata in the sand, and were maintained over the years by several plants growing together and several generations of plants using the original holdfast for recruitment and continued growth (e.g., Schiel and Foster 1992).

If not removed prematurely by storms or grazers, large vegetative fronds eventually produce a terminal blade, cease growing, and senesce; individual fronds of *M.* '*pyrifera*' usually live no more than 4–9 months (Rodriguez et al. 2013). Although it is commonly thought that high water motion and low nutrients / high temperatures are a major contributor to frond loss (see Chapter 3), Rodriguez et al. found that senescence and loss of fronds on plants in the Santa Barbara region are "age-dependent," controlled by intrinsic biological processes that result in progressive senescence. Around 50% of the variation in frond initiation and loss rates could be explained by frond age structure. This phenomenon is similar to that controlling the growth and loss of leaves in many terrestrial plants, and needs to be considered in studies evaluating the causes of frond dynamics.

Individual plants can live up to 9 years or so (review in Schiel and Foster 2006), but there is considerable variation in longevity. In southern Chile, for example, Buschmann et al. (2004) reported that plants in protected populations were annual. Long-term survivorship studies have not been done for the *M.* '*integrifolia*' ecomorph, but qualitative observations suggest it could live much longer than *M.* '*pyrifera*' because storms or other disturbances may remove only part of the stipe-holdfast (rhizome) and the remainder continues to grow (Brandt 1923). Chien (1973) estimated that an individual hapteron lived less than 6 months and as the holdfast ages it becomes a core of old, dead haptera with an outer ring of younger, living haptera (Ghelardi 1971). Taken as a whole, the persistence of the sporophyte therefore depends on the continual production of new vegetative fronds as well as new haptera. The quick turnover in biomass of *Macrocystis* is therefore quite unlike that of terrestrial trees, where biomass continues to accumulate in trunks and branches over time. North (1994) estimated a turnover rate of once or twice a year, whereas others have estimated rates in central and southern California of 6–7 times annually (Gerard 1976, Reed et al. 2008).

Giant kelp is a large, complex organism, and various metrics are used to describe its size. One metric is length which, like a fish caught with no witnesses or photographs, seems to have been exaggerated over time. North (1971a) found that despite rumors of plants with fronds over 150 m long, the most reliable records of fronds on attached plants indicate the longest are likely to be around 60 m. This is, of course, a labile characteristic of plants. For example, fronds of plants growing in shallow water are shorter than those on plants growing in deep water (e.g., North 1971a, Lobban 1978). The biomass of plants is important for estimates of productivity (Chapter 6) and, although labor intensive, can be estimated by combining measures of plant density, frond abundance and length, and frond length–weight relationships (Rassweiler et al. 2008). Acquiring these metrics is compounded by the sheer sizes of some plants. For example, the largest plant collected from the La Jolla kelp forest in southern California weighed over 400 kg (Neushul 1959). The most common metric used for a relative measure of plant size is the number of fronds per plant which, along with frond length and plant density (in numbers per area), can be used to estimate the amount of vegetation in the water column across a kelp forest. It is also indicative of the degree of shading below the canopy and carrying capacity (Tegner et al. 1997). North (1971a) estimated plant weights by taking the average wet weight of M. 'pyrifera' fronds on plants, which was around 1.25 kg, and multiplying by the number of fronds. The largest plant he sampled had 400 fronds and therefore would weigh 0.5 metric tons. Given the many thousands of *Macrocystis* plants that normally inhabit coastal kelp forests, this represents a considerable biomass or standing stock. This is seen somewhat dramatically in the large amounts of *Macrocystis* that wash up on shores after storms, which contribute to onshore and nearshore food webs and seed colonization of coastal vegetation (figure 2.2).

It is also noteworthy that there is considerable biogeographic variation in kelp frond and plant sizes, although the degree to which this represents specific site characteristics or true large-scale differences in growth, survival, and size is unknown. For example, in a southern New Zealand kelp forest, the maximum elongation rate of fronds was less than 40 mm day$^{-1}$, and the maximum life span of fronds was 7 months at one site and 15 months at another (Pirker 2002). These were sheltered sites, however, and affected by sediment and occasionally poor light quality. Nevertheless, even the largest plants in most southern New Zealand populations rarely have more than 10 fronds and those in perennial populations in southern Chile have only 3–4 fronds (Buschmann et al. 2004) compared to the hundreds of fronds occasionally seen in California populations.

## INTERNAL ANATOMY

The internal morphology of giant kelp rivals the complexity of its exterior and is among the most highly developed of all the algae. This is perhaps not surprising given the

FIGURE 2.2

*Macrocystis* wrack on California beaches. (A) Numerous piles, up to 0.5 m high, scattered along the shore after a moderate storm. (B) Huge piles, up to several meters deep, in front of the Santa Barbara Yacht Club after large swells during the 1982–1983 El Niño. (C) Blade on the left consumed by amphipods that then burrowed into the sand (apparently floats are not palatable); fresh blade placed on the right to illustrate how blade appeared prior to consumption. (D) Embryonic dune formed by a giant kelp holdfast and colonized by sea rocket (*Cakile maritima*).

SOURCE: Photo (B) courtesy of Ron J. McPeak, UCSB Library Digital Collection; others by M. S. Foster.

development, size, complexity, and turnover rates of the various plant parts which must function in an environment that can be highly variable over time and with depth. Sykes (1908) provided the first thorough anatomical description of adult plants, and Parker (1971) expanded on this (see also Lobban 1978, North 1994). A cross section of a mature stipe reveals the basic features that occur in a radially symmetrical pattern from edge to center (figure 2.3): a thin surface cuticle, a pigmented meristoderm, a cortex with mucilage ducts near its interface with the meristoderm, a zone of sieve elements, and

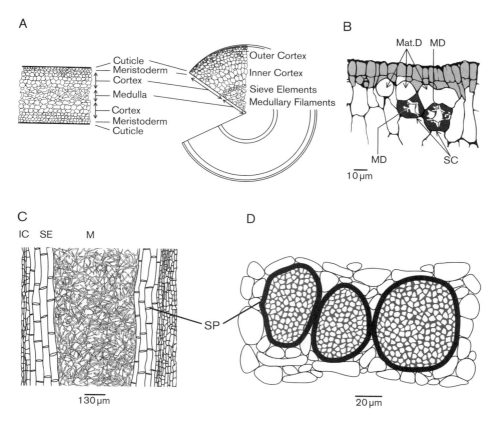

FIGURE 2.3

Internal morphology of *Macrocystis*. (A) Cross sections of a blade (top left; about 2 mm thick) and stipe (top center; about 1 cm diameter; modified from North 1994). (B) Cross section through the stipe meristoderm and outer cortex with developing mucilage ducts (MD) surrounded by secretory cells (SC), and mature ducts (Mat.D) above them (scale bar = 10 μm; from Grenville et al. 1982). (C) Photomicrograph of longitudinal section of stipe with inner cortex (IC), region of sieve elements (SE) showing sieve plates (SP) and medulla (M) (scale bar = 130 μm; drawn from Parker 1971). (D) Cross section through sieve elements showing sieve plates (scale bar = 20 μm; modified from Graham and Wilcox 2000).

SOURCE: (B) is reprinted with permission from the Phycological Society of America; (C) is reprinted with permission from Scweizerbart Science Publishing (www.schweizerbart.de).

the medulla. The meristoderm and outer cortex are composed of parenchymatous, isodiametric cells, with the meristoderm containing chloroplasts and serving as the primary photosynthetic tissue. Cells elongate in the inner cortex, and the innermost cells develop hyphal extensions that produce the filamentous medulla. The structure of the blade is like a flattened stipe but without mucilage ducts, and with sieve elements among the medullary filaments (Parker 1971; figure 2.3). Haptera have a meristoderm and cortex but lack a distinct filamentous medulla. Mucilage ducts are scattered in the cortex (Grenville et al. 1982) and sieve elements are mixed with medullary filaments in the inner cortex (Parker 1971).

Mucilage ducts are produced when secretory cells derived from the meristoderm separate to form a cavity, a process called schizogamy (figure 2.3). This occurs throughout the thallus and forms an extensive but sometimes discontinuous longitudinal network with side connections and cul-de-sacs (Grenville et al. 1982). The mucilage in the ducts is composed primarily of the sulfated polysaccharide fucoidan but also contains other polysaccharides and small amounts of fat, all presumably produced by the secretory cells. Various functions of the duct system and its contents have been proposed (review in Grenville et al. 1982), but the most likely in giant kelp seems to be for carbohydrate storage and perhaps osmoregulation. Mucilage may move through the meristoderm to the surface of the plant. If so, it could also serve to reduce fouling, decrease desiccation of surface fronds and perhaps reduce hydrodynamic drag.

Like that of phloem in vascular plants, the sieve element translocation system of kelp integrates plant growth by facilitating the transport of materials from areas where they are in excess, such as mature surface blades, to areas where they are in short supply, such as haptera and sporophylls growing in low light near the bottom. Experiments by Fox (2013) have also shown that when mature frond biomass is removed, transport from the remaining mature fronds facilitates the growth of new fronds (frond initials) near the base of the plant. The biomass–frond growth relationship is the basis of the positive relationship between plant biomass and productivity discussed in Chapter 6. This movement of materials has been called source-sink transport. Sieve elements are found in the Laminariales and some Fucales but are most developed in the large, rapidly growing giant kelp *Macrocystis* (Sykes 1908, Parker 1971). The primary longitudinal elements originate from the outer filaments of the primary medulla, develop sieve plates with pores, and together form longitudinal rows called sieve tubes (as sieve elements in figure 2.3C) (Schmitz and Srivastava 1974). Sieve elements are similar to the sieve tube members of vascular plant phloem but lack companion cells. In kelps, stretching produces a trumpet-like morphology that is widest at the sieve plate, and so these have been called trumpet hyphae. The primary longitudinal elements are connected through the cortex to the meristoderm via pits and plasmodesmata so that flow from cell to cell most likely occurs through the cytoplasm (a process called symplastic; Buggeln et al. 1985). Studies using radioactive and dye tracers have revealed flow rates in the primary elements of up to 70 cm hour$^{-1}$ (Parker 1965). The mechanisms causing flow are not understood well but appear to involve both an energy-assisted flow and mass flow caused by differences in osmotic pressure between solute-rich sources and solute-poor sinks (review in Buggeln et al. 1985). The most abundant materials transported are mannitol (around 70% of materials), amino acids (15%), and inorganic ions (15%; data and review in Manley 1983).

Iodine is especially abundant in plants relative to seawater and may function as an antioxidant and an antimicrobial agent (Küpper et al. 2008). The surface canopy of giant kelp is exposed to airborne particles and aerosols, and can rapidly take up and concentrate iodine as well as other radionuclides. Fronds take up radionuclides in the water column. These attributes make it a natural dosimeter that has been used to detect the spread of radioactive iodine, $^{131}$I, as seen most recently from the damaged Fukashima nuclear power plant (Manley and Lowe 2012). Sieve sap can also be used as an indicator of metals associated with coastal runoff (Fink and Manley 2011).

## SPORES

Sporophytes in good growing conditions can begin asexual reproduction via sporulation in less than a year (Neushul 1963, Foster 1975a, Buschmann et al. 2004). Plants can begin producing spores when they are quite small and, in good growing conditions, probably when they are only a few months old. The smallest spore-producing plant found by Neushul (1963) had four fronds that were 1.5–4 m tall. Plants with only 1–2 fronds may become reproductive in southern Chile (M. Graham, pers. comm.). As sporophytes become mature, their sporophylls become darkened with developing sori that can eventually cover nearly all of both sides of mature, reproductive blades (Neushul 1963; figure 2.4). Sori are dense aggregations of unilocular (i.e., single cell) sporangia that develop, along with elongate sterile hairs or paraphyses, from the meristoderm of the sporophylls (Henry and Cole 1982a; figure 2.3). There is one report of sori developing on what are normally vegetative blades (Leal et al. 2014). *Macrocystis* sori usually develop from the apex to the base of the sporophylls, and active spore release is indicated by the somewhat transparent whitish areas on sporophylls from which spores have been released. As in other Laminariales (review in Fritsch 1945, Neushul 1959, Henry and Cole 1982a), the nucleus in the developing sporangium produced by the meristoderm in *Macrocystis* divides meiotically. The resulting haploid nuclei divide mitotically 2–3 times and the sporangial contents develop into 16–32 biflagellate haploid spores. These are sometimes referred to as meiospores because, at least initially, they result from meiosis. Neushul (1959) saw that spores are not released individually from the sporangium, but are "ejected" as a group in a transparent packet. The packet ruptures and spores escape into the surrounding water. The haploid spores have 16 chromosomes (Walker 1952, Cole 1968) although Yabu and Sanbonsuga (1987) found haploid gametophytes with 32 chromosomes but suggested these were polyploids.

Although the biomass of fertile sporophylls is a relatively small and highly variable percentage of total plant biomass (from <1% to 10%, Neushul 1963), the small size of spores and the high density of sporangia on both sides of the sporophylls result in extremely high fecundity. For example, on a plant with 17 vegetative fronds, Neushul (1959) measured the area of fertile sori at $2 \times 10^6$ mm$^2$ and estimated a sporangium density of $10^4$ mm$^{-2}$ of sorus. Assuming that each sporangium contains 32 spores, this plant bore around $10^{11}$ zoospores. This is comparable to the $10^{12}$ spores produced by a single fruiting body of what is thought to be the most prolific fungus, the giant puff ball *Calvata gigantea* (Li 2011). It is also an underestimate relative to the life span of the

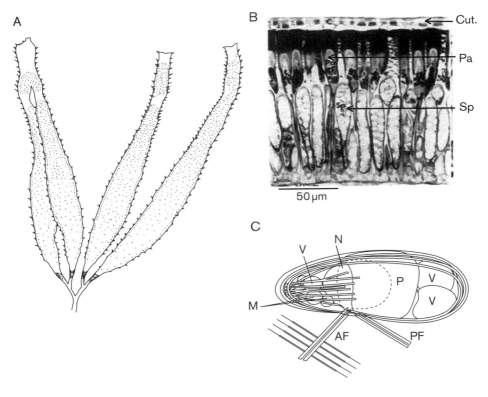

FIGURE 2.4

Sporophyll, sporangia, and spores. (A) Fertile (darker, stippled areas) *Macrocystis* sporophylls, about 35 cm long (from Neushul 1963). (B) Section through the surface of a typical, fertile Laminariales sporophyll (from Henry and Cole 1982a). (C) Typical Laminariales spore (stylized). Encircling bands are cytoskeletal microtubules. Adhesion vesicles around the periphery of the cell not shown, and flagella are not shown at full length. Spore is about 7 μm long (from Henry and Cole 1982a).

Cut. = cuticle; Pa = sterile paraphysis; Sp = sporangium. AF = anterior flagellum with hairs; V = vesicles with stored fat; M = mitochondrion; N = nucleus; P = chloroplast; PF = posterior flagellum.

SOURCE: (B) and (C) are reprinted with permission from the Psychological Society of America.

sporophyte because mature plants can produce spores year-round as old sporophylls continue to grow and produce new sori, and new sporophylls are produced (Neushul 1963). Sorus turnover can occur in less than 10 days (Graham 2002). Fecundity is, however, a function of vegetative biomass (Neushul 1963). It decreases as vegetative biomass is reduced by storms or grazers (Reed 1987, Graham 2002), or by artificially removing the canopy (Geange 2014), and its recovery can lag behind vegetative recovery (Graham 2002). This relationship, plus the effects of changes in temperature, nutrients, and water motion, can cause considerable temporal variation in sorus production and spore release (discussed in Chapter 4). This is particularly true in Chile where there are large geographic differences in spore production including annual sporophyte populations and some perennial *Macrocystis 'integrifolia'* populations that do not produce spores in autumn (Buschmann et al. 2004).

Planktonic spores, the primary dispersal phase of attached and drifting *Macrocystis*, are the beginning of the microscopic stages that link successive generations of sporophytes. These spores are tiny (around 5–7 μm in diameter; Henry and Cole 1982a) but well equipped for dispersal, settlement, and development into gametophytes (figure 2.4). As for most Laminariales, giant kelp spores lack a cell wall, are slightly pear-shaped, have two laterally inserted flagella, a single chloroplast, and storage and adhesion vesicles (Henry and Cole 1982a). The anterior, forward-directed flagellum (called a flimmer) provides propulsion and is relatively long with rows of stiff hairs, while the shorter, trailing, posterior flagellum is smooth (whiplash), an arrangement referred to as heterokont (Henry and Cole 1982a, Graham et al. 2009). Spores are photosynthetic, and if light is adequate, photosynthesis exceeds respiration even while swimming (Amsler and Neushul 1991). Storage vesicles in the posterior end of each spore contain lipids that make up 20–70% of spore carbon. Lipids are used to support metabolic activity in the dark and during the early development of gametophytes. The presence of low-density lipids probably reduces sinking rates and so increases time to disperse, and their location within the spore may cause the spore to tilt downward so that swimming tends to move the spore toward the bottom (Brzezinski et al. 1993, Reed et al. 1999). The combination of photosynthesis and metabolism of stored lipids enables spores to swim for over 72 hours (Reed et al. 1992). Spore swimming speeds have not been determined for giant kelp, but assuming they are similar to the maximum of 160 μm sec$^{-1}$ recorded for *Saccharina (=Laminaria) japonica* (Fukuhara et al. 2002), spores are very poor swimmers relative to typical water movement in the nearshore ocean. Their sinking rate of around 1.2 μm sec$^{-1}$ is also very slow and similar for live and dead spores. At this rate, and from a release height of 40 cm for the average sporophyll bundle, it would take 100 hours for them to sink to the bottom in still water (Gaylord et al. 2002) or around 10 hours if they swam straight down. They do not swim straight down, however, as they are neither phototactic nor geotactic (Henry and Cole 1982a). They are positively chemotactic to a variety of essential nutrients, but this most likely occurs at very small scales in the benthic boundary layer (Amsler and Neushul 1989). One can imagine these spores transported into calm water very near the bottom by water motion, then sinking very slowly with their anterior end down due to an overall slightly negative buoyancy with less dense fat in their rear, and finally swimming along a nutrient gradient to suitable microsites on the bottom. As Reed et al. (1992) pointed out, this behavior near settlement surfaces is the primary value of spore motility. The anterior flagellum appears to provide the initial adhesion to the substratum. When contact is made, the flagella are discarded and the spore attaches with materials from the adhesion vesicles (Henry and Cole 1982a).

Surprisingly, the planktonic phase may be further extended by planktonic gametophytes. It has been shown that if the spores stop swimming in the water column they can germinate into gametophytes that produce viable sporophytes when settled (Reed et al. 1992). These attributes of spores make longer distance dispersal via water movement possible, even though spore abundance in the water column and settlement density (measured by sporophyte recruitment) are usually greatest within a few meters of adults plants within stands (Anderson and North 1966, Reed et al. 1988, Graham 2003). Models of dispersal (Gaylord et al. 2002) and assessments of sporophyte and gametophyte recruitment at sites well away from possible spore sources (Reed et al. 1988, 2004, Gaylord et al. 2006a, 2006b) indicate planktonic spores and possibly gametophytes can be dispersed up to a few kilometers. Adult sporophytes can also contribute to dispersal when storms detach entire plants or plant parts containing sporophylls. Surface drifting "rafts" of detached giant kelp can survive for over 100 days (Hobday 2000a) and travel hundreds of kilometers while continuing to produce spores (Macaya et al. 2005, Hernández-Carmona et al. 2006). Exposure to high irradiance near the surface while dispersing, however, may reduce spore viability and development (Cie and Edwards 2008).

## GAMETOPHYTES

Our knowledge of gametophyte biology comes largely from laboratory studies (e.g., Cole 1968) and experiments that outplanted settled spores or gametophytes into the field on artificial substrata and assessed results as subsequent sporophyte recruitment (e.g., Deysher and Dean 1986a). Gametophytes are only a few microns or perhaps tens of microns in length, although under laboratory conditions that inhibit gametogenesis they can grow into visible masses of filaments (e.g., Sanbonsuga and Neushul 1978, Druehl et al. 2005). Because of their small size and similarity to other filamentous brown algae, there are as yet no techniques to study them directly or reliably in the field. Outplant experiments and recruitment of sporophytes on substrata placed in the field (e.g., Foster 1975b, Reed et al. 1997) and in clearings on the bottom (e.g., Reed and Foster 1984) clearly show that gametophytes occur naturally on hard substrata. Furthermore, as in some other kelps (Garbary et al. 1999), gametophytes can also occur as endophytes within other algae.

Settled spores begin to form a cell wall and develop into either male or female gametophytes with a 50:50 sex ratio (Roleda, Morris, et al. 2012a; figure 2.5). Their growth and reproduction depend on environmental conditions, particularly the requirement of exposure to blue light for gametogenesis (Lüning and Dring 1975, Lüning and Neushul 1978). As described in detail by Cole (1968) for populations in the Northeast Pacific, if conditions are optimal, spores developing into female gametophytes produce a "germ tube" through which most of the spore contents migrate. After migration a cross-wall forms at the distal end of the germ tube, isolating the single-celled gametophyte. The density of the cytoplasm increases and the cell enlarges to become pear-shaped and about $35 \times 10$ µm long as the gametophyte develops into an oogonium. Each oogonium extrudes an egg, about 18 µm in diameter that remains attached to the empty oogonium. In less favorable conditions the female gametophyte may divide to become multicellular. Although not yet determined for

Female gametophyte

1. Oogonium   2. Extruded   3. Zygote   4. Sporophyte   5. Multicellular females
                 egg

6. Male gametophyte          7. Antheridia

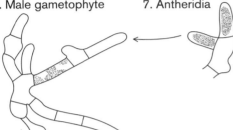

First cell
from spore

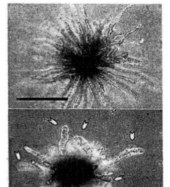

100 µm

Multicellular female gametophyte
(upper) with many developing
sporophytes (lower, arrows) after
fertilization.

8. Laminarian sperm

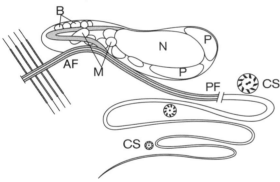

FIGURE 2.5

Sexual reproduction in *Macrocystis*. 1–4: Female gametophyte from the single-cell stage with developing oogonium to the four-cell sporophyte after fertilization. The diameter of the egg is 18 µm.

5: Multicellular female gametophyte (upper) with many developing sporophytes (lower, arrows)
after fertilization. Scale = 100 µm. 6: Multicellular male gametophyte (contents of only one cell are
illustrated). 7: Antheridia (shaded). 8: Diagram of typical Laminariales sperm, about 7 µm long after
release from antheridium.

AF = portion of anterior flagellum with hairs. B = "Type B" vacuoles (function unknown); M = mitochondria; N = nucleus; P = chloroplast; PF = posterior whiplash flagellum; CS = cross sections showing microtu-
bule arrangement.

SOURCE: Images 1, 2, 3, 4, 6, and 7 are from Papenfuss (1942), 5 is from Muñoz et al. (2004), and 8 is from
Henry and Cole (1982b) reprinted with permission from the Psychological Society of America.

giant kelp, filamentous gametophytes of some other kelps can survive at least 18 months in complete darkness (Dieck 1993). These gametophytes or portions of them can produce gametes when returned to optimal conditions in the laboratory (e.g., Sanbonsuga and Neushul 1978, Druehl et al. 2005). It has generally been assumed, based on work in the Northeast Pacific (e.g., Reed et al. 1988, 1997), that only one oogonium is produced per female, and females under optimal conditions are only one-celled. Muñoz et al. (2004), however, found that when female gametophytes from *Macrocystis 'pyrifera'* populations in southern Chile are grown under optimal conditions they produce extensive, branched, multicellular filaments with multiple oogonia and, after fertilization, multiple sporophytes (figure 2.5).

Male gametophytes follow a similar path from spore settlement to first gametophyte cell but then, similar to female gametophytes from southern Chile, divide to produce a multicellular, branched filament before producing antheridia (figure 2.5). The filamentous gametophytes can be distinguished by filament diameter, with males roughly half the size of females (about 5 µm vs. 10 µm; Lüning and Neushul 1978). This difference in size can be used to differentiate pre-reproductive males from females for hybridization experiments (e.g., Sanbonsuga and Neushul 1978). Antheridia can develop from cells within or at the tips of filaments, and each releases a single sperm. The cell that divides to form an antheridium can also become an antheridium, producing two sperm (Cole 1968). In optimal laboratory conditions, gametophytes of both sexes can become fertile in about 2 weeks (Cole 1968, Lüning and Neushul 1978). As for female gametophytes, filamentous male gametophytes under suitable conditions in the laboratory can be maintained in a nonreproductive, perennial state and portions removed and induced to become reproductive. One- to two-celled gametophytes maintained in a nonreproductive condition for at least 7 months can also be induced to become reproductive (Carney 2011). The degree to which such delayed development of gametophytes or small sporophytes (Kinlan et al. 2003) contributes to sporophyte recruitment in the field is discussed later (Chapter 4).

One of the most significant discoveries concerning reproduction in kelps is that when the egg is extruded from the oogonium it releases a pheromone that causes the sperm to be ejected from antheridia and chemotactically swim toward the egg (Lüning and Müller 1978, review in Müller et al. 1985; figure 2.6). The pheromone lamoxirene is effective at concentrations as low as $1 \times 10^{-11}$ M. Released sperm are biflagellate and pear-shaped, and similar in size and structure to spores (figure 2.5). They differ from spores in lacking adhesion vesicles but having a much longer, tapered anterior flagellum (Henry and Cole 1982b). Sperm possess chloroplasts, but their photosynthetic and swimming abilities, and longevity, are currently unknown. While the pheromone enhances the chances of fertilization and is active at extremely low concentrations, its dilution in the ocean is no doubt considerable, suggesting that male and female gametophytes need to be in very close proximity for fertilization to be successful. This may be facilitated by aggregation of settling spores associated with the hydrodynamic effects of variation in substratum rugosity at scales of centimeters to millimeters (Foster 1975a, Muth 2012). Müller (1981) suggested the pheromone was not effective beyond around 0.5 mm.

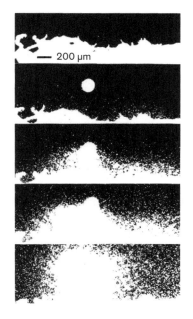

0

1

2

3

8

FIGURE 2.6

Sperm release induced by lamoxirene. The pheromone was absorbed on a $SiO_2$ sphere and the sequence
photographed at 1 min intervals. The sphere can be seen
in 1 and the increasing cloud of swimming sperm in 2,
3, and 8. The experiment was done with the kelp *Alaria
esculenta*, but results would be similar with *Macrocystis*.
SOURCE: From Müller (1988) reprinted with permission
from © Cambridge University Press.

In a series of innovative field experiments, Reed (1990a) outplanted natural substrata
seeded with *Macrocystis* spores at different densities and assessed fertilization success by
determining sporophyte recruitment. He found that a density of at least 1 spore $mm^{-2}$ was
necessary for successful recruitment, surmising that fertilization distance was critical.
Given the dilution effect on spore density with distance from source plants, it appears from
this requirement alone that the vast number of spores produced is necessary to maintain
populations. Giant kelp sporophytes can develop parthenogenetically from female gametophytes (Druehl et al. 2005). *Macrocystis* is also self-fertile and self-fertilization is likely,
especially if spore dispersal is limited (Gaylord et al. 2006a, 2006b). However, Raimondi
et al. (2004) found that sporophytes produced from self-fertilization suffered severe
reductions in fitness with much reduced sporophyll production and fecundity. They suggested this might be a cause of the periodic decline of populations on isolated reefs and
areas with slow currents where dispersal is limited and self-fertilization is therefore more
likely. The benefits of outbreeding have also been shown by crosses between plants from
different populations along the Chilean coast (Westermeier et al. 2011).

## BASIC ABIOTIC REQUIREMENTS

Interest in understanding the biology and ecology of giant kelp has led to considerable
research on the physical and chemical requirements and their amounts or levels needed
for growth and reproduction. For marine plants, the requirements are suitable levels of
temperature and salinity, amounts of light at particular wavelengths, and composition

TABLE 2.1   Abiotic requirements of *Macrocystis*.

| Stage | Salinity (‰) | Temperature (°C) | Nitrate (μmol L$^{-1}$) | Light (PAR 400–700 nm)[a] |
|---|---|---|---|---|
| Sporophytes | | | | |
|   Juvenile | >22[b] | 0–20[c] | >1[e] | 2–3 mol m$^{-2}$ day$^{-1}$ [f] |
|   Adult | | 8–23[d] | | |
| Spores | <43[g] | ND | ND | 90 μmol m$^{-2}$ sec$^{-1}$ [h] |
| Gametophytes | 25–35[i] | −1.5–26[j] | >1[l] | ~0.4 mol m$^{-2}$ day$^{-1}$ [m] |
| | | 10–17[k] | | |

a. PAR = photosynthetically active radiation. Surface PAR in southern California is ~20–60 mol m$^{-2}$ day$^{-1}$ (Dean 1985).

b. Lower salinity tolerance. May be a temperature–salinity interaction (Druehl 1978, Buschmann et al. 2004).

c. Lowest and highest surface temperatures based on total geographic range (Setchell 1932).

d. Range of surface temperatures for field populations in Chile (Buschmann et al. 2004, Vásquez et al. 2006).

e. Varies due to interaction between temperature and nutrients (Zimmerman and Kremer 1984).

f. Saturation of photosynthesis in field experiments. Data and review in Dean and Jacobson (1984, 1986).

g. Upper salinity tolerance for germination and germ tube formation (Bay and Greenstein 1993).

h. Saturation of photosynthesis in the laboratory (Amsler and Neushul 1991).

i. For gametogenesis (Sanbonsuga and Neushul 1978).

j. Tolerance limits for survival (Dieck 1993).

k. Optimal temperature range for female gametogenesis (Lüning and Neushul 1978).

l. Likely given data in Deysher and Dean (1986a).

m. For ~100% fertility in the laboratory at 12:12 photoperiod (Deysher and Dean 1984). Greater than 0.4 mol m$^{-2}$ day$^{-1}$ required for sporophyte production in the field (Deysher and Dean 1986a). Light requirements affected by temperature and nutrients. In addition to or as part of PAR, gametophytes require 260 μmol cm$^{-2}$ of blue light (360–500 nm) for gametogenesis (Lüning and Neushul 1978). This requirement is affected by the rate quanta are received (Deysher and Dean 1984).

and amounts of nutrients. The available data, summarized by life history stage, are presented in Table 2.1. Correlations between distribution and temperature suggest *Macrocystis* sporophytes do not survive well in very cold or very warm waters. Gametophytes in the laboratory can survive at higher and lower temperatures than those associated with sporophyte distribution, but optimal temperatures for reproduction are narrower. Correlations from natural distributions of sporophytes and lab experiments with gametophytes indicate that neither stage is tolerant of low salinities. Tolerance to high salinities has become of recent interest because of the potential effects of coastal desalination plants that discharge highly saline brine into nearshore habitats.

As might be expected given their different habitats and sizes, light requirements differ among stages of giant kelp. Gametophytes living on the bottom can grow and reproduce in very low light, about 1% of surface irradiance at 400–700 nm (as moles of photosynthetically active radiation in Table 2.1). This is in agreement with Lüning's (1981) conclusion that kelps generally require at least 1% of surface illumination. Gametophytes also require a particular amount of blue light to become fertile (Table 2.1). In

TABLE 2.2   Known nutrient requirements for *Macrocystis*.

| Elements[a] | Form normally used by plants[b] | Limiting in nature[c] |
|---|---|---|
| Carbon | $HCO_3^-$, $CO_3^=$ | No |
| Oxygen | $O_2$ | No |
| Nitrogen | $NO_3^-$, $NH_4^+$ | Sometimes |
| Phosphorus | $PO_4^=$ | Maybe |
| Manganese | – | Maybe[d] |
| Iron | Likely colloidal | Maybe |
| Cobalt | – | No[d] |
| Copper | – | Maybe[e] |
| Zinc | – | Maybe[e] |
| Molybdenum | – | No |
| Iodine | – | No |

a. Kwabara and North (1980).

b. Where forms are not given (–), it is assumed that the element is used as a free ion.

c. DeBoer (1981).

d. May limit growth in deep oceanic water (Kuwabara 1982).

e. Toxic as free ions in deep oceanic water (Kuwabara 1982).

contrast, juvenile and adult sporophytes require an order of magnitude more light to saturate photosynthesis although this can vary because the light required to saturate adult blades varies with their age and position in the water column (Wheeler 1980b). The amount of light required for the growth of embryonic sporophytes from a few cells to distromatic blades on the order of several millimeters tall is similar to that for gametophytes, but the embryonic sporophytes can grow under a wider temperature range and are not inhibited by high light intensities (Fain and Murray 1982).

Few macroalgae have been grown in defined culture media in axenic conditions, so little is known about their complete nutritional requirements. The assumption is that their inorganic requirements are similar to those of terrestrial plants (DeBoer 1981). North (1980) identified 38 elements in *Macrocystis* tissue, and Kuwabara and North (1980) found that at least nine elements were essential for the growth and reproduction of microscopic stages (11 including carbon and oxygen; Table 2.2). DeBoer (1981) suggested that of these, nitrogen, phosphorus, iron, and perhaps manganese and zinc may possibly limit macroalgal growth in nature. North (1980) suggested that copper may also be limiting for giant kelp. Much of the interest in nutrient limitation came from attempts to farm giant kelp using deep ocean water. In this un-natural circumstance, Kuwabara (1982) suggested that several nutrients could be limiting (Table 2.2). In nearshore waters where giant kelp grows naturally, however, only nitrate has been identified as limiting in some regions. Field correlations (e.g., Jackson 1977, Zimmerman and Kremer 1984) and lab and field experiments (e.g., Deysher and Dean 1986a, 1986b) in southern California have clearly shown that sporophyte recruitment declines and sporophytes begin to deteriorate if nitrate falls below 1 μmol $L^{-1}$ (Table 2.1). Giant kelp is particularly sensitive to low nitrate because its storage capacity for this nutrient is only a few weeks (Gerard 1982a, Zimmerman and Kremer 1986).

These abiotic requirements provide a general guide to interpreting variation in *Macrocystis* populations in nature. Taken together, the life history traits, morphology, internal structure, and abiotic requirements of giant kelp set its course in the complex sea of abiotic and ecological influences that constrain or enhance the success of individuals. These characteristics are manifested in the demography of the *Macrocystis* in different environments, its responses to abiotic factors and its role in ecologically complex kelp forests. We address this complexity in later chapters.

# 3

# THE ABIOTIC ENVIRONMENT

Nor can these variations excite surprise, when it is considered that this gigantic
weed is subject to every vicissitude of climate, of temperature, and exposure;
that it literally ranges from the Antarctic to the Arctic circle; through 120
degrees of latitude; that it lives and flourishes, whether floating or attached,
growing in bays, harbours, or the open sea when most distant from land; and,
lastly, that it equally adapts itself to the calmest or most tempestuous
conditions, to waters of uniform depths or those which rise and sink with the
tide, to dead water or to strong currents.

—Hooker (1847)


Environmental factors involved in seaweed ecology—light, nutrients, water
motion, and temperature—have similar stratified distributions. As a result, it is
difficult to determine which are the key factors. But it is easy to find a
relationship of any one with algal distribution.

—Jackson (1977)


## THE KELP ENVIRONMENT

The nonbiological environment defines the "fundamental niche" space occupied by
giant kelp. This includes the physical and chemical conditions that allow *Macrocystis*
settlement, growth, reproduction, primary production, and biomass accumulation.
These include the abiotic requirements for growth and reproduction as well as suitable
substratum characteristics and water motion regimes, and their temporal and spatial
variability. Of course, these factors rarely act separately, are often correlated with each
other, and interact with one or more life stages of kelp. All species, however, occur
within a climate envelope of multiple physical conditions that affect their growth,
reproduction, recruitment, and survival. For marine algae, these conditions may dif-
ferentially affect one or more life history stages that dictate whether or not a species will
flourish and form populations. These conditions apply across wide geographic scales,
among regional sites, and down depth gradients, and vary considerably over time. It is
hardly surprising, therefore, that we are still gaining significant new insights into how

*Macrocystis* responds to the abiotic environment and the long-term consequences of oceanic events such as El Niño/Southern Oscillation (ENSO), the Pacific Decadal Oscillation, and climate change.

The remarkable success of giant kelp ecologically and biogeographically is reflected in the extremely wide range of physical conditions in which it occurs. Delineating these conditions, however, has become a more expansive task over recent years following the genetic analyses of Coyer et al. (2001) and the taxonomic unification of Demes et al. (2009), who designated all species of *Macrocystis* as the single species *Macrocystis pyrifera*. Although the debate about the number of species goes back at least to Hooker (1847; see introductory quote for taxonomic classification), the recent consolidation into a single species means that earlier descriptions of the biogeographic and ecological limits of *M. pyrifera* and the conditions that bounded it (cf. Foster and Schiel 1985) need revision to accommodate what are now considered as ecomorphs or ecotypes.

With few exceptions, *Macrocystis* requires a hard substratum for settlement and attachment, water temperatures between about 4°C and 20°C, sea-bottom light intensities equivalent to 1% or greater than sea-surface irradiance, nitrate concentrations >1 μmol L$^{-1}$, oceanic salinities, and protection from extreme water motion. Obviously, almost all of its biogeographic range falls within these categories, but perhaps a more interesting question is why its range is limited in some places where these conditions seem to apply. For example, *Macrocystis* occupies much of the Pacific coast of California and Baja California, Mexico. To the south, it may be restricted by waters that are too warm or too low in nutrients (see sections below). To the north, *Macrocystis* 'pyrifera' ends in central California, primarily due to severe water motion (Foster and Schiel 1985, Graham 1997) and perhaps competition with the surface canopy kelp *Nereocystis luetkeana*, although another ecomorph (*M. 'integrifolia'*) occurs through British Columbia to Alaska. Giant kelp forests flourish and are particularly well developed on outer coasts between depths of around 5 m (which is usually outside the often intense swell conditions inshore) and 20 m, beyond which sea-bottom light is often too attenuated for effective recruitment and growth. Its presence along the coastlines of the extremely exposed subantarctic islands and outer coasts of Chile is usually restricted to areas behind headlands and in more sheltered bays outside of the full force of oceanic swells and storms (Villouta and Santelices 1984, Attwood et al. 1991).

*Macrocystis* is usually absent from estuaries and far inside of protected bays. This is most likely because of a shortage of rocky substrata, increased sedimentation, and reduced light. Reduced salinity can also restrict *Macrocystis* in bays and other areas with large freshwater inputs (Table 2.1). North (1969), for example, reported severe damage to adult giant kelp transplanted to Newport Bay, California (near Los Angeles) when salinity was lowered to 10 ppt during a storm.

In this chapter, we expand on these general characteristics and discuss their effects on *Macrocystis* and their variability.

## SUBSTRATUM AND SEDIMENTATION

*Macrocystis* spores are capable of settling and developing on just about any substratum, but, with few exceptions, giant kelp forests and the sessile communities associated with them require hard substratum to develop and persist. If these plants and animals do manage to attach and grow on sand, sediment, or shell debris, they are usually swept away in all but the calmest conditions. In this sense, therefore, there is an interaction between the physical factors of substratum type and water motion which affects the presence of kelp forests. Even in areas with massive rock for attachment, the hardness and friability of the rock may affect the survival of giant and other kelps during storms (Foster 1982). Some areas with limited rocky substrata, however, can support kelp forests. For example, between Oceanside and Del Mar, southern California (near San Diego), there are extensive but patchy areas of small boulders and cobble with a *Macrocystis* community (Schroeter et al. 1993). During storms, the drag on *Macrocystis* and other kelps is sufficient to dislodge plants, sometimes with their anchoring cobbles and small rocks still attached. These occasionally are moved within the kelp forest or transported to the beach or deeper water. There is also the case of kelp forests along 40 km of coast between Goleta and Gaviota north of Santa Barbara, California. Kelp forests in this region grew mostly on sand and were removed by the large El Niño storms of 1982–1983 (Foster and Schiel 1985, Schiel and Foster 1992). The kelp may have initially recruited to surfaces such as shells and worm tubes. Once established, the sand-filled holdfasts of very large plants probably acted as recruitment centers for subsequent generations (see Chapter 2, sporophytes). Once removed, there was no effective way for further recruitment. Numerous attempts to re-establish this bed proved unsuccessful (Foster and Schiel 1993) and only remnants attached to isolated rocky substrata along this coastline remain today. A few populations elsewhere also occur completely unattached (data and review in Gerard and Kirkman 1984), most likely sustained by the input of drifting fronds from other areas with attached plants. The vast majority of giant kelp forests worldwide, however, occur as attached plants on rocky substrata. The depth restriction of rocky reef in some areas, such as many of the fringing reefs of southern New Zealand where hard reef goes to sand at around 20–25 m depth, confines giant kelp to a relatively narrow band along the coastline (Schiel 1988, 1990, Schiel et al. 1995, Schiel and Hickford 2001).

Sediments also affect giant kelp forests and their associated species through scour and burial of rocky substrata (Weaver 1977, Reed et al. 2008) and recruitment inhibition (Devinny and Volse 1978, Carney et al. 2005). North (1971b) suggested that *Macrocystis* fronds are particularly susceptible to damage if covered with sediment, and Foster et al. (1983) concluded that changes in sediment cover could be responsible for some of the historical changes in the areal extent of kelp forests around San Onofre. Johnson (1980) recounted observations, made in the late 1800s on San Miguel Island (one of the Channel Islands of southern California), that kelp forests were destroyed by sand eroded

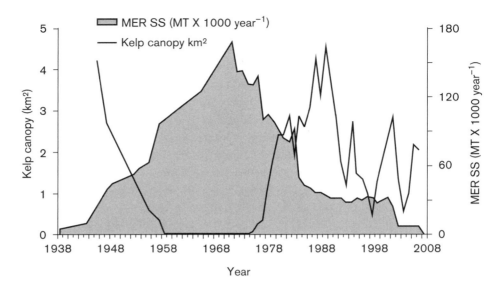

FIGURE 3.1

Palo Verdes, California, giant kelp canopy area and mass emission rates (MER) of suspended solids (SS) from ocean sewage discharges near this forest.

MT = metric ton.

SOURCE: Figure and data sourced from J. Gully, Los Angeles County Sanitation District, and Foster and Schiel (2010).

from the land. Reed et al. (2008) found that 80% of the rocky habitat at one of their sites was buried by sand during a storm, preventing the recovery of *Macrocystis* for at least 4 years. Kelp abundance, including that of giant kelp, is also reduced in areas affected by sediment plumes from landslides (Konar and Roberts 1996). Within kelp forests, Grigg (1975) listed burial as an important cause of mortality of *Muricea californica*, a gorgonian coral commonly found on reefs and in kelp forests south of Point Conception (central California). Burial can also kill young hydrocorals (*Stylaster californicus*) (Ostarello 1973) and other sessile animals (Weaver 1977), and may kill slow-moving invertebrates. Sediments can clog the filter-feeding apparatus of suspension-feeding invertebrates, and may be partly responsible for the generally higher abundances of filter feeders such as *Stylaster* on vertical surfaces where sedimentation is reduced (Ostarello 1973). A mosaic of sediment / sand and rock patches is common in kelp forests, and this pattern may change, particularly during storms. Small patches of shifting sediment are one disturbance that kills established organisms and creates space for recolonization. Such disturbances may therefore have significant effects on abundance and diversity of species within some kelp forests (Rosenthal et al. 1974, Foster 1975b, Grigg 1975).

A recent analysis of historical data concluded that the disappearance of large tracts of giant kelp near San Diego and Los Angeles in the 1950s was most likely due to sedimentation from postwar urban expansion and the vast amount of flocculent material discharged from sewage treatment plants into nearshore kelp forests. Foster and Schiel (2010) showed there was an inverse correlation between the amount of this material and the abundance of kelp (figure 3.1). It is known from laboratory experiments that even small amounts of sediment can greatly inhibit the attachment and growth of *Macrocystis* spores (Devinny and Volse 1978), a result similar to that of other large brown algae (e.g., Taylor and Schiel 2003, Deiman et al. 2012). Effects can occur both through inhibition of attachment to a stable substratum and through light inhibition if even a little sediment covers spores or developing microscopic algal stages. A finding by Dean et al. (1983), that sedimentation rates are negatively correlated with sporophyte recruitment, seems to be a general phenomenon. It is also highly probable that the small stages of many other algae and invertebrates in a kelp forest are negatively affected in this way.

## TEMPERATURE

The general correspondence between seawater temperature and kelp distribution geographically has long been known. Setchell (1893) wrote that "it is temperature that limits the distribution of the kelps and that too, the summer temperature, as heat not cold is inimical to their growth." In examining the geographic distribution of 17 benthic algal species, van den Hoek (1982) concluded that critical temperatures limited essential events in their life histories. For temperate species, he proposed six boundaries: a northern lethal boundary, beyond which a species cannot survive; a northern growth boundary, corresponding to the lowest summer temperature that permits sufficient growth; a northern reproductive boundary, the lowest summer temperature at which reproduction can occur; and their corresponding southern boundaries. His scheme, therefore, recognizes the critical interaction of temperature with life history stages and the demographic traits of growth, reproduction, and survival. As pointed out by Jackson (1977) and a review by Wheeler and Neushul (1981), however, temperature and other key abiotic factors are often highly correlated, which complicates any discussion of them separately. Of most importance for kelp forests and kelp productivity is the inverse relationship between temperature and nutrients (Jackson 1977) and its most important form for kelp growth, nitrate (Zimmerman and Kremer 1984, Zimmerman and Robertson 1985; see figure 3.2). These relationships are best known for the west coast of North America where strong seasonal upwelling occurs. Cross-correlation between environmental variables can also occur because of intermediate causal relationships. For example, low water temperatures and high nutrients can lead to phytoplankton blooms in surface waters, thereby attenuating light to benthic areas (Quast 1971a, Clendenning 1971b). On much larger spatial and temporal scales, ENSO events are associated with correlative changes in temperature, nutrients, severe water motion through storm activity, and alteration of the light environment because of the loss of canopy species, which together have caused the largest changes in giant kelp forests seen over the past century (e.g., Graham et al. 1997, Dayton et al. 1999, Parnell et al. 2010).

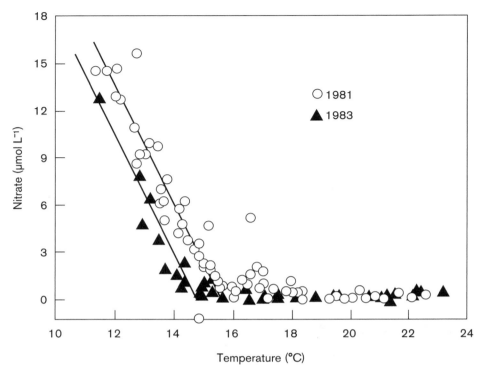

FIGURE 3.2

Scatterplot of the temperature–nitrate relationship during 1981 and 1983 at Catalina Island, southern California. Although there was no difference in slopes for the temperature–NO$_3$ relationship below 15°C between 1981 and 1983, the Y-intercept was significantly lower during 1983.
SOURCE: From Zimmerman and Robertson (1985).

The interaction between temperature and other physical factors has long been suggested (Druehl 1978) or demonstrated to be important for many kelps (e.g., Lüning and Neushul 1978, Dean et al. 1983). Some of these differences relate to latitudinal trends in species' occurrences and performance. For example, monthly mean surface temperatures of nearshore waters within the western north Pacific distribution of large stands of *Macrocystis pyrifera* vary from around 11–17° C in central California to around 15–26°C in Baja California, Mexico (Sverdrup et al. 1942, Edwards and Estes 2006). Edwards and Estes (2006) found, for example, that sea-surface temperature (SST) in some areas of central California exceeded 16°C for less than two weeks during summer 1997, but temperatures were greater than this for nearly a full year (April 1997 to March 1998) in coastal waters of southern California and Baja California. It is generally thought that adult giant kelp do not grow well above 20°C, but those in Baja California can experience temperatures beyond this for several weeks (North 1971a, Ladah et al. 1999, Edwards and Estes 2006) as can populations in northern Chile (Vásquez et al. 2006). The Baja populations are known to come and go, particularly around El Niño years, and their presence is probably an exception when cool water is upwelled (Dawson 1951, Edwards and Estes 2006). The deterioration of sea-surface canopies and sometimes entire plants, however, is undoubtedly due to the combined effects of thermal stress and severely depressed nutrients (see sections that follow). Zimmerman and Kremer (1986) found, for example, that *Macrocystis* frond elongation rates declined with increasing temperature (see figure 3.3A). They found no effect of temperature on growth between around 14.5°C and 18.5°C, and then a decline up to 21°C. However, they attributed this effect mostly to nutrient depletion at higher temperatures.

There may also be localized adaptations that allow warm-water populations to withstand higher temperatures, at least for short periods, although as discussed below (in the "Nutrients" section), since temperature and nitrate can be inversely correlated it is not always clear which factor is the cause of algal responses. North (1971a) found that sporophytes transplanted from Baja California, Mexico, to Newport Bay in southern California survived better during periods of warm water than did local plants. This may have been due to ecotypic differences in the ability to grow at low nitrate concentrations, as Kopczak et al. (1991) found in laboratory growth experiments using plants from regions with different temperature / nutrient regimes. This type of effect can also be manifested in latitudinal trends. For example, Oppliger et al. (2012) found a latitudinal difference in response to temperature of two "cryptic" variants of the kelp *Lessonia nigrescens* in Chile. The northern variant occurs from 16° S to 30° S latitude and the southern variant from 29° S to 40° S, and they appear to be reproductively isolated on a local scale (Tellier et al. 2009). The southern cold-water variant was unable to reproduce and survive at higher temperatures (ca. 20°C), whereas the northern variant could. Many other taxa show a latitudinal trend related to temperature (e.g., Peters and Breeman 1993). Similarly, the physiological performance of rafting *Macrocystis* declined from the south to the north along the Chilean coast, an effect largely ascribed to increased temperature (Rothäusler et al. 2011). This effect was compounded by the reduced length of blades, which underwent a loss of pigments possibly due to increased ultraviolet radiation, and increased cover of epibionts, particularly bryozoans, in warmer waters.

Besides the interaction of physical factors, another troublesome feature of assessing temperature effects is its measurement in natural conditions. The most common way of assessing temperature is now through satellite-derived data of the sea surface. However, this measurement of temperature may not truly reflect what an organism experiences in the field, especially when all its life stages are considered. In the case of *Macrocystis*, the requirements for the 1N stages are on the benthos and not in the water column or on the sea surface. The water column is often thermally stratified so that SST can be five or so degrees higher than on the bottom (Zimmerman and Kremer 1984). Furthermore, temperatures at the bottom of kelp forests can vary by 4–8°C in less than a day (Quast 1971a, Barilotti and Silverthorne 1972, Zimmerman and Kremer 1984). Rosenthal et al. (1974) found mean and maximum surface temperatures near their study area by San Diego, California, over a 5-year period to be 16.3°C and 24.6°C, respectively, whereas the mean

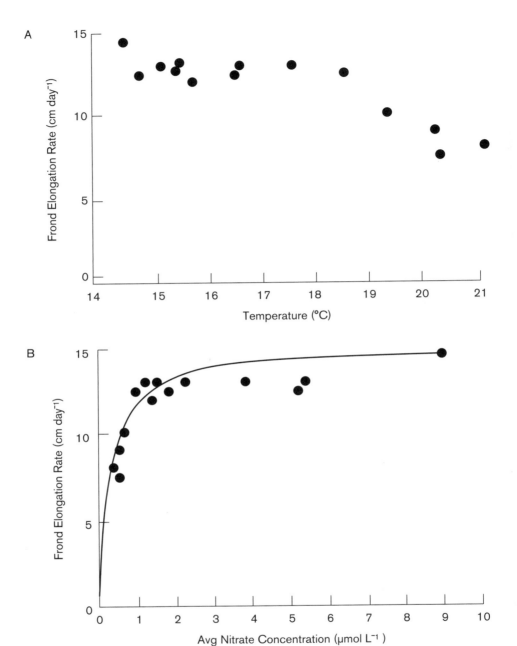

FIGURE 3.3

Frond elongation of *Macrocystis* in southern California as a function of temperature (A) and average ambient nitrate concentration (B).

SOURCE: From Zimmerman and Kremer (1986).

at 17 m was 13.0°C and the maximum 16°C. As expected, the thermocline is most pro-nounced in summer and often disappears over winter. Zimmerman and Kremer (1984) found, for example, that seasonal temperatures above the thermocline ranged from 15°C during winter to 20°C during summer, but below the thermocline they were more stable temporally, ranging between 12°C and 15°C seasonally. Furthermore, their thermistor profile data showed there was great variability, even over the course of a day, and espe-cially so in summer when there were incursions of cold, nitrate-rich water that were more frequent with increasing depth. These were linked to the semidiurnal internal tides, which can push these isotherms upward by as much as 20 m in 2–3 hours (Armstrong and LaFond 1966, Zimmerman and Kremer 1984, Zimmerman and Robertson 1985).

Temperature has also been suggested as being important in the geographic distribu-tion of kelp forest fishes and invertebrates (Quast 1971b, Gerrodette 1979). As for bull kelp (*Nereocystis luetkeana*) and gorgonian corals, major changes in the distribution of kelp forest species occur near Point Conception where the California Current moves off-shore, creating large changes in temperature within a short distance. Briggs (1974) also emphasized the importance of these changes near Point Conception in his review of marine biogeography. Similar breakpoints for many taxa are seen around 30° S latitude in Chile, the tip of South Africa, and Cook Strait in central New Zealand (Blanchette et al. 2009). Mechanistic understanding of these transition zones on species' distribu-tions inevitably involves a combination of effects due to temperature, nutrients, and changes of currents, including effects on larval dispersal.

## LIGHT

It is very difficult to define a critical depth—or depth range—below which photosynthesis will be controlled by light, but it is fairly certain that, wherever suitable substrate is available, the maximum depth at which benthic plants grow will be determined by the prevailing light climate.

—Dring (1981)

The light environment of kelp forests is complex across all spatial and temporal scales because of seasonal differences in light quality and angle of incidence, the influences of suspended sediments, phytoplankton blooms, shading by algal canopies, and a wide range of disturbances that can alter all of these. Both light quantity and light spectra are attenuated through the water column so that benthic organisms experience a far different light environment than those in the water column or on the sea surface. The methods and instrumentation used to measure light relative to the biology of kelp forest organisms took great strides in the 1970s (reviewed by Lüning 1981) and continue to this day with the availability of small and relatively inexpensive sensors capable of measur-ing quanta and different wavelengths, which are most relevant to algal requirements. Most modern measurements of irradiance in algal research are of photosynthetically active radiation (PAR) as photon flux density, which is particularly appropriate for plants because photosynthesis is a quantum process, in units of einsteins or moles per area per time, where $1 E = 1 mol = 6.02 \times 10^{23}$ photons.

A diverse array of light-absorbing pigments enables marine algae to use wavelengths between 400 and 700 nm (PAR) for photosynthesis (Haxo and Blinks 1950, Dring 1981), although other plant processes such as gametogenesis in kelps are sensitive to light outside this range (Lüning 1981). The amount of PAR required for growth varies among taxa and roughly relates to their pigmentation (Dring 1981). Most seaweeds in a kelp forest start life on the bottom, and light, as well as other factors affecting plant growth, must be suitable there. In the absence of other possible controlling factors, such as inadequate substratum or impacts of grazers, the lower depth limit of giant kelp and, therefore, giant kelp forests, is almost certainly determined by light. Lüning (1981) suggested that for most kelps, this limit will occur where irradiance is reduced to about 1% of that above the sea surface. Giant kelp generally grows deeper in clearer water, as seen in some central California kelp forests where depth distributions are broadly correlated with water clarity (Foster 1982, Spalding et al. 2003). On the negative side, there is at least anecdotal evidence that urbanization and coastal development have attenuated the depth distribution of some giant kelp forests, such as the outer, deeper margin of the Point Loma kelp forest (near San Diego) that has receded to shallower depths over several decades, probably due to increased turbidity (cf. Foster and Schiel 2010). On the positive side, even where inshore forests may by restricted in depth, deep kelp forests frequently occur in clear waters along relatively undeveloped coastlines or further out from the immediate coastal zone. For example, remotely operated vehicle (ROV) surveys by Spalding et al. (2003) found that giant kelp occurred as deep as 35 m in central California. The record depth for *Macrocystis* appears to be the 68 m found by Perissinotto and McQuaid (1992) at the Prince Edward Islands off South Africa. This was *Macrocystis 'laevis,'* which grew to around 25 m in length and never extended to the sea surface. They estimated that *M. 'laevis'* at 60 m depth received a photon flux of around 0.29–0.43 $\mu$mol m$^{-2}$ day$^{-1}$, which is just above the thresholds for gametophyte growth and sporophyte formation (Table 2.1).

An exciting discovery related to this is the relatively common occurrence of deep-water kelp populations in the tropical zone. Graham, Kinlan, et al. (2007) used a "synthetic oceanographic and ecophysiological model" to predict the location of tropical kelp forests, and postulated a deep-water refuge for kelps. Using this model, they predicted that the kelps *Eisenia galapagensis*, *Laminaria brasiliensis*, and *L. abyssalis* would occur as deep as 200 m and that they would be far more common across the tropical zone than had previously been supposed. They verified that *E. galapagensis* occurs at least to 60 m depth off the Galapagos shoreline. They speculated that genera with vicariant and disjoint distributions, such as *Laminaria*, *Eisenia*, and *Ecklonia*, may be connected by these deep-water refugia.

Although *Macrocystis* and other kelps generally require about 1% of sea-surface light to grow (Lüning 1981), this is a simplification because light requirements differ for the vari-

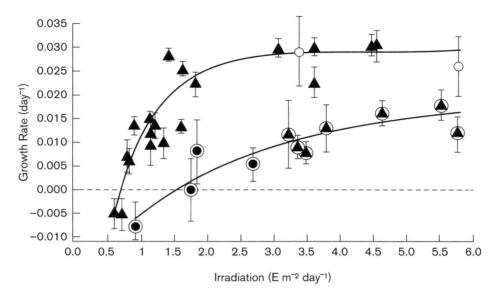

FIGURE 3.4

Production–irradiance curves for *M. pyrifera*. The average growth rates during 1979–1982 (i.e., mean instantaneous relative growth rate ±95% confidence interval) of juveniles transplanted to the San Mateo and San Onofre kelp forests in southern California versus mean daily irradiance levels during transplant period. Circled values are for plants that were nutrient limited in different years. Others were not nutrient limited. Lines are von Bertalanffy fits. Horizontal dotted line indicates light compensation above which plants are in net production and below which they are in net respiration. SOURCE: From Dean and Jacobsen (1986).

ous life stages of the same plant, and ambient light varies considerably. Nevertheless, the 1% level translates into around 1 mol $m^{-2}$ $day^{-1}$, which is a reasonable reference for *Macrocystis* requirements (Table 2.1). The values in the table should be considered approximate because requirements can vary with geographic location, temperature, nutrients (Lüning and Neushul 1978), water motion (Wheeler 1980a), and, for photosynthetic measurements on adult blades, age and position in the water column (Wheeler 1980b).

All plants have a characteristically shaped curve describing the relationship between irradiance levels and productivity, called "P-E" or "P-I" curves. These translate into carbon production or growth (Miller, R.J. et al. 2011, Tait and Schiel 2011). One form of this is a growth curve generated for *Macrocystis*. It can be readily seen that the relationship between irradiance and growth will depend on other factors, in this case nutrients (figure 3.4). Dean and Jacobsen (1986) found that when nutrients were sufficient, plants were in net respiration below 0.5–0.7 mol $m^{-2}day^{-1}$. Above this level, there is increasing growth with increasing irradiance to around 2–3 mol $m^{-2}$ $day^{-1}$, beyond which a saturating level of growth (called "P-max") is achieved. When nutrients were far lower, suppression of growth occurred across the irradiance regime.

*Macrocystis* fertility will decline if less blue light is available. Dean et al. (1983) estimated that *Macrocystis* gametophytes outplanted on artificial substrata in a southern California kelp forest must receive the light necessary to become fertile within about 40 days. Beyond this time, mortality sources such as sedimentation and grazing are apparently so high that few gametophytes can survive. This life history stage can live and grow for much longer periods in the laboratory (Sanbonsuga and Neushul 1980), but the extent to which gametophytes can be long-lived on natural substrata in the field is unknown. This has a bearing on the duration of potential "seed banks" in kelp forests (Edwards 1999, 2000, Schiel and Foster 2006; discussed in Chapter 4). Larger stages may survive longer at suboptimal light levels because they are partly above the bottom and not as affected by these factors on benthic substrata.

As discussed earlier, adult *Macrocystis* plants are generally insensitive to changes in subsurface light because they usually form a surface canopy and can translocate the products of photosynthesis toward the holdfast (see Chapter 2). Light transmission to the bottom is potentially restrictive of kelp populations because of water clarity, the amount dissolved and suspended material in the water, and shading by attached organisms. The amount of surface light intensity varies with latitude, season, and cloud and fog cover, but the range of intensities is much less than that created by water characteristics (Dean 1985). Day length is an important trigger for growth and reproduction in some kelps associated with *Macrocystis* (review in Lüning 1993). However, Harrer et al. (2013) found that for understory assemblages at four reefs in southern California, saturating irradiance occurs for 1.3–4.5 hour day$^{-1}$, depending on the time of year. The vast majority of the variation in net primary production was due to the biomass of the algae and only 2% annually was explained by bottom irradiance.

Light is attenuated logarithmically with depth and each wavelength has a particular extinction coefficient (Jerlov 1968). The extinction coefficient also varies with turbidity. In clear water, blue light is transmitted further than green light, while in more turbid coastal water, the reverse occurs (Jerlov 1968). Overall light transmission declines with increasing turbidity and, within the coastal water types designated by Jerlov (1968), Lüning (1981) estimated that the depth where irradiance is reduced to 1% of surface varies between 3 and 30 m, and Spalding et al. (2003) estimated this to be down to 40 m depth along the Big Sur coast of central California. Harrer et al. (2013) found that over the course of a year, the percentage of surface light reaching the bottom (at around 9 m depth) ranged from around 4% to 10%. Gerard (1984) measured PAR beneath and outside of canopies of giant kelp (figure 3.5). She found that at 1 m depth, just below a thick giant kelp canopy in southern California, an average of only 17% of above-surface light was available. This was reduced to around 2% at 11 m depth. She also found that open water areas had about four times the light of areas beneath a giant kelp canopy. The variance of light delivery was much greater in near-surface waters than in deeper water. She also found there was a logarithmic decline of light transmission with an increasing number of blades of *Macrocystis*, highlighting the great effect that shading by giant kelp canopies has on the light environment of communities below them. Some of the earliest studies in kelp bed ecology postulated the effects of plant shading on the bottom (e.g., Kitching et al. 1934). Numerous studies have shown that *Macrocystis* canopies can reduce irradiance by over 90% (Neushul 1971b, Dean et al. 1983, Reed and Foster 1984, Santelices and Ojeda 1984a), and dense surface canopies of giant kelp are often associated with a relatively sparse algal flora in the understory (Dawson et al. 1960, Neushul 1965, Foster 1982). Giant kelp canopies may also inhibit their own recruitment (e.g., Reed and Foster 1984). These effects are discussed in detail in Chapters 4 and 7.

Within a locality, understory algal cover (Foster 1982) and *Macrocystis* recruitment (Rosenthal et al. 1974, Reed and Foster 1984) can vary inversely with *Macrocystis* canopy cover. Pearse and Hines (1979), Reed and Foster (1984), and Dayton et al. (1984) demonstrated with experimental plant and canopy removals that giant kelp canopies can inhibit the recruitment and growth of the algae beneath them. Moreover, kelp recruitment on a local scale usually coincides with times when the surface canopy is reduced, including as a result of harvesting (Rosenthal et al. 1974, Kimura and Foster 1984). Santelices and Ojeda (1984b) also reported an increase in *M. pyrifera* recruitment when the surface canopy was experimentally removed. In contrast to some of the studies above, however, understory kelp biomass decreased with removal of the surface canopy. Differences among studies may be due, at least in part, to whether entire plants or only surface canopies were removed, and also the scale of the removal, as they all can affect shading from surrounding plants. Furthermore, understory kelps such as *Pterygophora californica*, *E. arborea*, and *Laminaria* spp. in the northern hemisphere, and *Ecklonia* and *Lessonia* species in the southern hemisphere can cause further light reductions, with consequent effects on the heavily shaded communities below them (discussed in Chapter 7).

Given the numerous factors that affect light quality and transmission, it is hardly surprising that light levels to the sea floor are frequently below those needed for growth of juvenile sporophytes (figure 3.6). Dean and Jacobsen (1984) showed that light levels 1 m above the bottom of a kelp forest in southern California varied greatly throughout the year, but in the summer months, irradiance fell below compensation irradiance, at less than 0.5 mol m$^{-2}$ day$^{-1}$ in areas away from kelp canopies because of phytoplankton blooms.

Water clarity or turbidity is influenced by terrestrial runoff, sediments resuspended by wave surge (Quast 1971a), plankton abundance (Quast 1971a, Clendenning 1971b), and probably dissolved and particulate matter produced by kelp forest organisms (Clendenning 1971b). We have observed the first three of these to produce near darkness on the bottom in kelp forests at mid-day. Moreover, changes in water masses with changing current conditions and tides can cause a rapid alteration in water clarity. For these reasons, short-term measurements of light on the bottom, although useful in comparing nearby areas at the same time, need to be used with caution in characterizing the light regime of a site. Light regimes, particularly if they are to be used for correlations with algal recruitment and growth, should be determined with in situ continuous recorders.

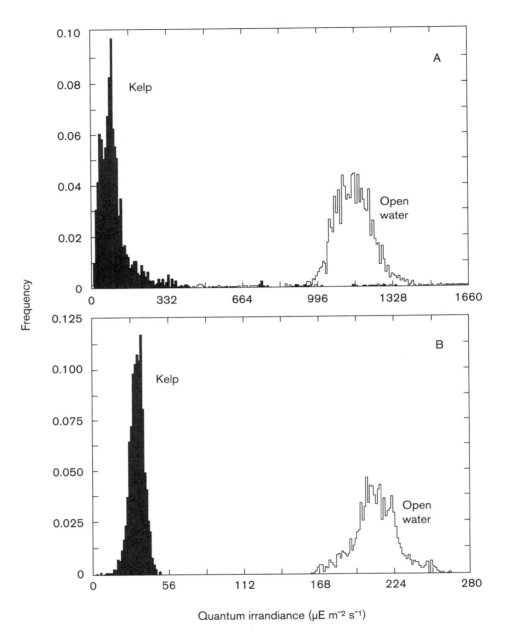

FIGURE 3.5
Frequency distributions of instantaneous quantum irradiance at (A) 1 m and (B) 8 m depths
in a *M. 'pyrifera'* forest in southern California and in adjacent open water.

NOTE: Recording interval = 1 sec, taken in late morning, swell height = 30 cm, swell period = 10–17 sec.
Current velocity = 3–6 cm sec⁻¹ at edge of canopy.

SOURCE: From Gerard (1984).

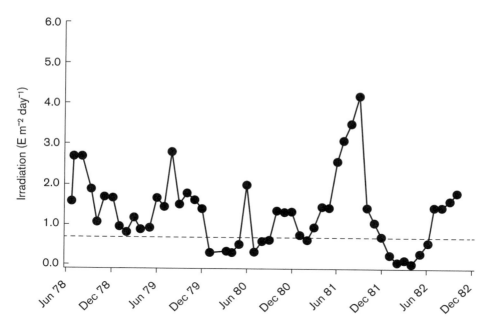

FIGURE 3.6

Monthly integrated levels of quantum irradiation for the San Onofre kelp forest in southern California at 13.2 m depth 1 m above the bottom. Dashed line indicates compensating irradiation level for juvenile *Macrocystis* growth.

SOURCE: From Dean and Jacobsen (1984).

Geographically, the light environment undoubtedly compromises the photosynthetic ability of giant kelp near its northern limit in the northern hemisphere, but this probably acts in concert with other factors such as wave exposure to affect the northern distribution of giant kelp in Alaska (Arnold and Manley 1985, Jackson 1987). Jackson's (1987) model showed that photosynthetic production is exceeded by respiration during winter in high latitudes, and he concluded that "seasonal variation is so extreme that the plant could not last the year at 53° N in 12 m of water, although it is able to survive the year in shallower water." How these photosynthetic parameters compare to those of other Alaskan kelps is unknown, although there is a rich flora and abundance of kelp species such as *Eualaria (=Alaria) fistulosa*, *Nereocystis*, and *Saccharina (=Laminaria) groenlandica* in these cold waters (Duggins et al. 2003), which may well outcompete *Macrocystis* as its own production capabilities decrease (e.g., Dayton 1975).

## NUTRIENTS

Seaweeds, including *Macrocystis*, obtain their nutrients from the surrounding waters because unlike most terrestrial plants they neither have true roots that can derive nutrients from the soil nor is there soil. Kelp holdfasts are attached to solid substrata and appear to serve no special nutrient uptake functions. Moreover, except for the occasional frond in the surface canopy, kelp tissues are not exposed to air and therefore all metabolic processes occur in water.

The known nutrient requirements of giant kelp are listed in Table 2.2 and, of these, nitrogen in its various forms has by far received the greatest attention as a possible limiting nutrient, particularly as it may affect *Macrocystis* in southern California (Jackson 1977, Wheeler and North 1981, Gerard 1982a, 1982b, 1982c, Zimmerman and Kremer 1984). Gerard (1982b) found that inorganic nitrogen concentrations in the surrounding water must be in the order of 1–2 µmol L$^{-1}$ (1 µM or 1 µmol = 1 µg atom) to support a typical giant kelp growth rate of 4% increase in wet weight per day. Inorganic nitrogen concentrations vary widely in nearshore waters, but are particularly great during upwelling or when there is terrestrial runoff (North et al. 1982). Nitrate levels inshore are often low (often less than 1 µM) during summer and early autumn along the coast of southern California and Baja California. This is especially so above the thermocline when the water is thermally stratified and during periods when warm water masses move into the region from the south (Jackson 1977, Wheeler and North 1981, North et al. 1982, Hernández-Carmona et al. 2001, Ponce-Díaz et al. 2003).

The most common ionic and reactive forms of dissolved inorganic nitrogen (DIN) are ammonium (NH$_4^+$), nitrite (NO$_2^-$), and nitrate (NO$_3^-$). Nitrate is most used for giant kelp growth and production due to its high concentration, although other forms such as urea can account for some uptake (Hepburn and Hurd 2005), as can ammonium from sediment efflux and excretion by reef consumers when nitrate delivery is low (Brzezinski et al. 2013). At Catalina Island in southern California, the daily variation in nitrate concentration is frequently as much or greater than the mean seasonal variation (Zimmerman and Kremer 1984). Reed et al. (2011), using 9 years of data, showed that monthly averages of DIN (nitrate plus nitrite) in central California were always above 3 µmol L$^{-1}$, that the springtime peak of DIN was a month later and over two times the concentration of that in southern California, and that DIN averaged less than 1 µmol L$^{-1}$ for the six consecutive months of June through November in southern California. The relationships between nitrate uptake and giant kelp growth parameters have been well described for sites in southern California (Zimmerman and Kremer 1984, Kopczak et al. 1991, Kopczak 1994). For example, Zimmerman and Kremer (1984) and Zimmerman and Robertson (1985) showed there was a hyperbolic, saturation relationship of frond elongation rate with increasing levels of nitrate (figure 3.3B). Growth was considerably reduced below 1 µM but reached a plateau beyond around 2 µM. A similar relationship was found for frond initiation rate and nitrate concentration, although the saturation level was less defined (Zimmerman and Robertson 1985).

Nutrients and their delivery act in tandem with several other abiotic factors. One of the best known is the inverse relationship between temperature and nitrate (figure 3.2). This relationship varies between sites and years, but generally shows a strong correlation between nitrogen and seawater temperature below around 15°C, although this can be highly variable (e.g., McPhee-Shaw et al. 2007, Sherlock et al. 2007, Lucas et al. 2011). At higher temperatures, nitrate and DIN generally fall below the 1 μM required for growth of giant kelp. McPhee-Shaw et al. (2007), for example, found that at temperatures less than around 11°C, the relationships for two regions of southern California converge. They found that these cold temperatures occurred only during upwelling of deep, outer-shelf water masses, and that site similarities at the coldest temperatures reflected the homogeneous nature of nutrient concentration in deeper waters. A similar relationship between nitrate and temperature is seen near the southern limit of the northern hemisphere distribution of giant kelp at Baja California, Mexico, although the inflection point for maintaining giant kelp growth was found to be slightly higher at 18.5°C during an El Niño year (Hernández-Carmona et al. 2001, Edwards and Estes 2006). Also reflecting the relationship between nutrients and temperature is the reduced growth rate–irradiance relationship seen during El Niño years when nutrients are depleted along the U. S. west coast (e.g., figure 3.4). This relationship may not be similar in other geographic areas, as Buschmann et al. (2013) reported that annual giant kelp populations in Chile declined at around 17°C when nitrate was above 3 μmol L$^{-1}$.

In southern California, giant kelp canopies commonly deteriorate during summer when inorganic nitrogen is low, although senescence can also be a factor (Rodriguez et al. 2013). Reduced nitrogen concentrations were probably also responsible for the massive loss of *Macrocystis* during the warm-water period of the late 1950s (North 1971a, Jackson 1977, North et al. 1982), and plants deteriorate at many locations during El Niño events due to a combination of high temperatures, storms, and severe nutrient reduction (e.g., Dayton and Tegner 1984, Edwards and Estes 2006, Parnell et al. 2010). Evidence that low inorganic nitrogen, and not temperature, is limiting under conditions of low nutrients and high temperatures comes from fertilization experiments. Dean and Deysher (1983) found that more sporophytes were produced on fertilized artificial substrata inoculated with giant kelp spores and placed within a kelp forest than on similarly treated but unfertilized controls. Hernández-Carmona et al. (2001) did a similar experiment that increased sporophyte recruitment as well the survivorship of juveniles transplanted to fertilized areas. Furthermore, others have found that there is not a typical growth response curve involving temperature alone. Zimmerman and Kremer (1986) found, for example, that frond elongation was constant between 14.5°C and 18.5°C, but above this range there was a linear decline in growth rate. It may be the case, however, that in summer other nutrients or temperature may be limiting once the nitrogen requirements of the plants are met. Laboratory studies by Manley and North (1984) suggested that phosphorous may be particularly important. North (1983) added nitrate and phosphate to seawater of ambient temperature (18–23°C) flowing into a large tank containing adult *Macrocystis*. These plants maintained healthy canopies and had high tissue concentrations of nitrogen, whereas nearby plants in the natural kelp forest, exposed to similar temperatures but not fertilized, suffered canopy losses and had low tissue concentrations of nitrogen.

Physiological processes within giant kelp plants can further obscure the relationship between nitrogen in the water column and plant growth. Gene expression for nitrogen acquisition varies along fronds. This may be a response to the distribution of nitrate in the water column but complicates an assessment of nitrate–growth relationships (Konotchick et al. 2013). Moreover, seaweeds can store nitrogen when the concentration in the surrounding water is high, sometimes called "luxury consumption," and then use these reserves for growth when the surrounding concentration drops (Chapman and Craigie 1977). However, there are considerable differences among species of kelps in their abilities to do this because of their differential capacities for storage and the duration they can go without new nitrogen input (e.g., Gerard 1982a, Korb and Gerard 2000). *Macrocystis* can accumulate nonstructural nitrogen compounds (Wheeler and North 1981, Gerard 1982a) including nitrate, and then use these reserves to maintain growth for at least two weeks, but probably not greatly longer, in low nitrogen environments (Gerard 1982a). Therefore, both the frequency of environmental sampling for inorganic nitrogen and the presence of tissue reserves can affect the interpretation of growth rates with respect to inorganic nitrogen.

The use of inorganic nitrogen in the water by giant kelp is also affected by water motion (Gerard 1982c, Wheeler 1982). Increased water flow over plants enhances uptake by increasing nutrient transport through the diffusion boundary layer (Neushul 1972, Gerard 1982c). Gerard (1982c) and Wheeler (1982) found that nitrogen uptake by *Macrocystis* increased with increasing current speed up to a maximum at around 2–4 cm sec$^{-1}$. Current velocities in kelp forests are often lower than this (Wheeler 1980a). However, Gerard (1982c) showed that water flow caused by wave surge can be equivalent to that of the current speeds above, and pointed out that because the plants are attached to the bottom and each blade is attached to a fixed point on the plant, very small waves can produce flag-like blade movement. This motion plus small currents and surge are sufficient to saturate nitrogen uptake even under very calm conditions. Furthermore, blade morphology is a highly variable trait and is responsive to different flow regimes, which can increase uptake of nutrients for photosynthesis and reduce drag in exposed conditions (Druehl and Kemp 1982, Koehl et al. 2008).

Geographically, nutrients, particularly nitrogen, undoubtedly play a key role in the southern limit of giant kelp in the northern hemisphere. The inverse relationship between temperature and nutrients is driven to a great extent by upwelling of cold, nutrient-rich water along the coastline (Jackson 1977, Konotchick et al. 2012). The southernmost giant kelp population along Baja California, Mexico, is somewhat ephemeral, occasionally dying off during periods of high temperature (around 23°C) and nutrient depletion (Edwards and Hernández-Carmona 2005). This is usually associated with El Niño conditions, which also produce large storm waves (Edwards 2004). Some concern has been expressed that with increased temperatures due to a changing climate, the southern limit of giant kelp may be decreased by several hundred kilometers (Ladah et al. 1999, Edwards and Estes 2006). However, it is noteworthy that *Macrocystis* has few populations in the southern part of Baja California, and most of these are greatly influenced by El Niños. Because of the particular circumstances of upwelling and high nutrients that allow these southern populations to exist in generally warm subtropical waters (Dawson 1951), it may be the case that its occurrence, rather than its disappearance, is the exception. The southern distributional limit of giant kelp may, in part, also be the result of its limited ability to store nutrients (e.g., Gerard 1982a, 1982b) when they are available. Might the northern limit (in the northern hemisphere) reflect the inability to store carbon when sufficient light is available? *Macrocystis* certainly does not have the abilities of some high latitude kelps to accumulate and store carbon reserves to fuel growth several months later. For example, the Arctic subtidal environment is highly constrained seasonally in light and nutrient availability, and kelps there have varying responses to these conditions (Kain 1979). For example, the high Arctic circumpolar kelp *Laminaria solidungula* lives for much of the year in dark conditions under sea ice. Plants achieve their greatest growth rates during iced-over conditions when there is little light but high levels of inorganic nitrogen ($NH_4^+ + NO_2^- + NO_3^+$) (Chapman and Lindley 1980, Dunton and Schell 1986). During the ice-free period, when growth might be expected, there is almost no measurable N in the water. However, the amount of light reaching plants during the open-water months affects thallus tissue density and carbon content (Dunton 1990). A reserve of carbohydrate, in the form of the polysaccharide laminaran, is built up as light becomes available, and this supports the carbon demand of growth in low-light periods. *Macrocystis*, however, can store carbon for only a matter of weeks (Rosell and Srivastava 1985) and so would almost certainly be limited by the low light and inorganic carbon availability in far northern waters. When light levels are high, stored carbohydrate may not be necessary in *Macrocystis* (Zimmerman and Kremer 1986), but this condition probably does not occur for long in far northern latitudes.

To close this section, we discuss studies that have dealt with the integrated nature of nutrient supply, uptake, and primary production with other factors. The delivery of nitrogen across inner shelf areas involves a series of complicated processes and includes alongshore currents, Ekman transport, and upwelling, internal waves, vertical stratification, and terrestrial runoff, and can also be affected inshore by upstream kelp forests (Jackson 1977, McPhee-Shaw et al. 2007). McPhee-Shaw et al. (2007) did a 2-year study on mechanisms of nutrient delivery to inner shelf areas of southern California. They found that spring and winter upwelling provided between 70% and 85% of the DIN to the inner shelf region, internal waves provided 8–14%, and terrestrial runoff from 4% to 35%. Others have speculated that the input of nitrogen, mostly as ammonia, from excretion by benthic animals and fishes makes a small contribution to the nitrogen used by algae (Bray et al. 1981, 1988). The analyses of Brzezinski et al. (2013) indicate that ammonium can be a significant source of DIN to meet the N demand of giant kelp during periods when nitrate delivery is inadequate. Some recycling occurs (i.e., water to kelp to animal to water) but some, especially by blacksmith (*Chromis punctipinnis*), is imported as the fish eat plankton at the edge of the forest and then excrete it in shelters within the forest at night (Bray et al. 1981). This contribution is probably minor, given the other much larger sources of N. However, it has been reported that frond-dwelling organisms may make important nitrogen contributions to giant kelp. In southern New Zealand, Hepburn and Hurd (2005) found that colonial invertebrates encrusting the blades of *Macrocystis* did not have a negative effect on growth through shading and by acting as a barrier to nutrient uptake, as was anticipated. Instead, during periods of low inorganic nitrogen levels in seawater, heavily colonized giant kelp fronds had greater growth than less colonized fronds. They speculated that hydroids, through ammonium excretion, could provide over 70% of nitrogen required by fronds when seawater nutrients were low.

Finally, a remarkably thorough study on physical pathways and use of nitrate by giant kelp was done by Fram et al. (2008), which will probably stand for some time as the most comprehensive study on this complicated topic. They quantified biological, physical, and chemical factors affecting nitrate uptake, measuring nitrate concentrations for 13 months and characterizing delivery with respect to thermal structure and current velocities inside and outside a kelp forest in southern California. They found that nitrate concentrations varied enormously around and within the kelp forest, and especially so during a period of stratification when nitrate varied by up to two orders of magnitude within a single day. Nitrogen acquisition in growing canopy fronds varied by a factor of six over a year even though there were two orders of magnitude differences in nitrate concentrations over the same period. Surprisingly, they showed that the kelp bed used up only around 5% of the nitrate supplied to the forest. They also found that the percentage of available nitrogen used by giant kelp is proportional to its residence time in the forest, and that this averaged only around 1.1 hours. This compares to the much longer flushing times of once weekly, with nitrate uptake of 4 hours, reported for the much larger Point Loma kelp forest where flow is dominated by cross-shore currents (Jackson and Winant 1983). Fram et al. also found that nitrate from upwelling periods supported 50% of net nitrogen acquisition by the kelp forest during their 13-month study. Assessing the influence of internal waves was difficult because their activity co-occurred with other processes. Nevertheless, they calculated that internal waves were strongest in late spring to early summer when seawater nitrate was low and kelp abundance was relatively high. Internal waves were found to account for up to 27% of nitrogen demand by kelp during summer and autumn. They also found that the nitrate from freshwater runoff was low because there was little rainfall during their study. Finally, they estimated that excretion of ammonia by *Membranipora*, a bryozoan inhabiting kelp blades, contributed up to 0.12 $\mu$m g$^{-1}$ month$^{-1}$ for fronds during warm stratified periods, a source of nitrogen initially elaborated on by Hepburn and Hurd (2005).

## WATER MOTION

To anyone visiting a wave-swept temperate zone coastline after a storm, the effects of water motion are quite obvious in the huge abundance of macroalgae tossed up on shore. Piles of *Macrocystis* can be several meters deep in such circumstances inshore from a giant kelp forest (figure 2.2). However, the effects of water motion are usually not so direct and obvious. In fact, they are pervasive and influence all aspects of the morphology, physiology, demography, and probably evolution of kelp. Water motion in all of its complexities and manifestations, therefore, constitutes one of the most fascinating factors in the biology and ecology of kelp forests.

We have argued that to understand the complex life histories of kelps and other habitat-forming seaweeds, they must be evaluated within the context of their highly dynamic nearshore environments, particularly with respect to physical disturbances (Schiel and Foster 2006). Stevens et al. (2001) framed the effects of water motion on kelp within two research themes. These were the ability to withstand wave forces that stretch fronds and affect survival in wave-swept environments, and understanding the boundary layers and flows around fronds that affect nutrient transport. In addition to these themes, however, are the interactions of water motion with numerous other factors. As discussed in the sections above, water motion affects the underwater light environment and therefore light acquisition by algae through light flecking and turbidity (Dring 1987, Dromgoole, 1988, Wing and Patterson 1993). It affects the transport of nutrients, plankton, and materials into and out of kelp forests (Wildish and Kristmanson 1997, Gaylord et al. 2007), sediment transport that can smother algae and impede recruitment (Devinny and Volse 1978, Taylor and Schiel 2003), and nutrient uptake by kelp fronds (Wheeler 1980a, Hurd et al. 1996, Denny and Roberson 2002). Numerous ecological processes are influenced by water motion including herbivory via sea urchin mortality (Ebeling et al. 1985), changes in sea urchin behavior (Cowen et al. 1982, Konar et al. 2014), alterations in available drift algae (Pearse and Hines 1979, Harrold and Reed 1985), propagule and larval dispersal (Reed et al. 1988, Gaylord and Gaines, 2000, Gaylord et al. 2002, 2006a, b, Graham 2003, Vásquez et al. 2006), algal reproduction (Serrao et al. 1996, Pearson et al. 1998), and settlement and recruitment dynamics (Vadas et al. 1990, Taylor and Schiel 2003, Taylor et al. 2010). Furthermore, climatic events such as ENSO and the Pacific Decadal Oscillation affect water motion and long-term structure of kelp forest communities (Dayton et al. 1992, Parnell et al. 2010). Across this wide spectrum, there has been significant progress over the past 20 years in understanding the effects of water motion across a range of physical scales. This reflects not only a growing interest in better integrating biological, ecological, and physical processes but also the development of new technologies and instruments that can measure physical processes from oceanographic to boundary layer scales.

Currents and surge produced by wind, tides, and waves produce drag, acceleration, and inertial forces acting on kelp forests (Gaylord et al. 2008, 2012). Currents are unidirectional flows, often changing on short time scales, while surge moves back and forth over the bottom, as well as up and down above the bottom (figure 3.7). Current speeds in kelp forests are highly variable but are generally in the range of near 0–15 cm sec$^{-1}$ (Wheeler 1980a, Bray 1981, Jackson 1983, Gaylord et al. 2007), although Neushul et al.

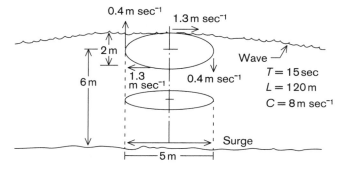

FIGURE 3.7

Water motion produced by swells in shallow water. Note that the
orbital motion of water particles flattens toward the bottom, eventually becoming entirely horizontal to produce surge. In the example
shown, a 5 m excursion of water was measured on the bottom at a
depth of 6 m, when a 2 m wave moved past at a velocity (C) of
8 m sec⁻¹, a wavelength (L) of 120 m, and a wave period (T) of 15 sec.
Orbital velocities at the surface are also shown.

SOURCE: From Foster and Schiel (1985).

(1967) measured maximum speeds of 40 cm sec⁻¹ near Anacapa Island off southern
California. The drag at higher speeds can pull *Macrocystis* over at angles up to 30° from
vertical, and the entire surface canopy may be submerged as a result (Neushul et al.
1967).

The kelp forest itself affects local currents. As a result of plant drag, current speeds
within kelp forests are often two to three times lower than the surrounding water (Jackson and Winant 1983), and if the forest is small, the incoming current will diverge
around it, producing a "bow wake" similar to that of a ship (Jackson 1983, Gaylord et al.
2007, Rosman et al. 2007). The reduction of surface waves by kelp plants is commonly
observed as "quiet water" inshore from kelp forests (cf. Darwin 1839), and artificial,
kelp-like tethered floats have been used in the past as breakwaters to reduce water movement in harbors (Isaacs 1976).

Surge speeds can be much higher than those of currents, particularly surges generated by long-period swells associated with winter or tropical storms. In central California, for example, 4 m high swells are typical in winter and produce water speeds of
over 1 m sec⁻¹ on the bottom at kelp forest depths (10 m). The force generated by water
moving this fast is equivalent to that produced by a wind speed of 126 mph (i.e., 56 m
sec⁻¹; Charters et al. 1969), which is quite sufficient to wrench benthic organisms from
the bottom and fracture the bottom in the process. Utter and Denny (1996) characterized the forces acting on giant kelp in the water column (figure 3.8). In their simplified
model, motion is considered to be in the vertical plane parallel to wave direction, and a

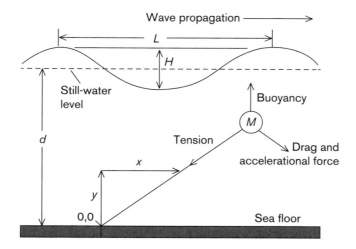

FIGURE 3.8

A schematic representation of the components of the numerical model of kelp mechanics.

NOTE: $d$ = water depth; $x$, $y$ = horizontal and vertical components of kelp position in the vertical plane; $M$ = mass of point element representing a kelp frond; $H$ = wave height; $L$ = wavelength. Not drawn to scale.

SOURCE: Reprinted from Utter and Denny (1996), with permission from Elsevier.

kelp frond is a point element tethered to the sea floor by a flexible and stretchable stipe. $M$ has a mass equal to that of the frond and is subjected to the same forces as the whole frond. The stipe has a tensile stiffness when stretched beyond its resting length and is attached at ($x = 0$, $y = 0$). Although empirically rates of kelp dislodgement were greater than those predicted by their model, it remains a useful depiction of forces that act on kelp.

## IMPINGEMENT AND INERTIAL FORCES

Gaylord et al. (2003, 2008) pointed out that the traditional view is that canopy-forming kelps move passively with the flows, which results in a decrease in water speed relative to fronds, thereby minimizing drag (e.g., Koehl 1984, 1986, 1999). However, when fronds are extended to their limit, there may be a jerk-like inertial force of deceleration acting on them for short periods (e.g., Denny et al. 1997, 1998, Gaylord and Denny 1997). They also clarified that "impingement" forces from breaking waves have far less of an effect on subtidal kelps than on those such as *Egregia menziesii* in the very shallow subtidal zone. Peak force transients associated with these can be raised by a factor of three relative to drag, and these can get disproportionately large with increasing sizes of seaweeds.

Storm-associated surge is perhaps the most important source of mortality for adult *Macrocystis*, at least in California (ZoBell 1971, Rosenthal et al. 1974, Gerard 1976, Foster 1982, Dayton et al. 1984, Reed and Foster 1984, Reed et al. 2008) and at sites studied in Argentina (Barrales and Lobban 1975). Plants are often torn loose in patches during storms and this can have considerably enhanced effects when these individuals become entangled with other plants, increasing the drag on their holdfasts and causing further detachments (Rosenthal et al. 1974). If the substratum is very hard and holdfasts are firmly attached, water motion may remove mainly the long fronds, leaving holdfasts and small fronds that may continue to grow (Foster 1982, Reed et al. 2008).

The susceptibility of *Macrocystis* fronds and plants to removal by water motion may translate into site and regional differences in plant densities and the extent of surface canopies. For example, in southern California, the east–west orientation of the shore-line, the protection afforded by offshore islands, and the distance from the northerly source of winter storms combine to make kelp forests relatively protected from frequent large swells (Cavanaugh et al. 2011). Surface canopies in this region typically vary in extent over a cycle of 3–4 years (North 1971a, Rosenthal et al. 1974, Cavanaugh et al. 2013), probably related to an increased susceptibility of older and larger plants with deteriorating holdfasts to removal by water motion. A similar cycle of growth, holdfast deterioration, and removal by storms occurs along the coast of Chubut in Argentina (Barrales and Lobban 1975). Canopies around Santa Barbara are even less variable, with occasional catastrophic losses from large swells or warm water periods (North 1971a). The patterns of large losses have been clearly demonstrated by the large winter swells associated with the El Niño events since the early 1980s (e.g., Grove et al. 2002). During the 1982–1983 El Niño, storm waves removed nearly 70% of the adult *Macrocystis* at some sites in the Point Loma kelp forest near San Diego (Dayton and Tegner 1984), and over 90% of the surface canopy along the Palos Verdes Peninsula, Los Angeles (Wilson and Togstad 1983), whereas some protected sites around Catalina Island (offshore of Los Angeles) still had a surface canopy of *Macrocystis* (Ebeling et al. 1985, Foster and Schiel 1993).

In contrast, most canopies in central California undergo a regular seasonal change, with growth in spring and summer leading to maximum development in early autumn, and then frond and plant loss during storms in late autumn and winter (Miller and Geibel 1973, Cowen et al. 1982, Foster 1982, Kimura and Foster 1984). In addition to such seasonal changes, there are year-to-year differences correlated with the severity of winter swells (Foster 1982). As in southern California, this was especially evident in the winter of 1982–1983 when swells over 7 m high with a 21-second period were recorded in central California (Seymour 1983). Almost all of the *Macrocystis* canopies were removed by large swells along the entire coast of California over this period (Reed et al. 2006—"An experimental investigation").

Reed et al. (2011) made significant inroads into understanding the interaction of water motion, nutrients, grazers, and production of giant kelp. In examining 17 sites in central and southern California over 9 years, they found that "significant wave heights" (i.e., the top one-third of wave heights, which removes the "noise" of smaller waves) were on average about 1 m higher through the year in central than in southern California (Reed et al. 2011). Sea urchin densities in southern California averaged 3–14 m$^{-2}$ compared to near zero in central California because of sea otter predation. Nutrients (nitrate and nitrite) averaged up to six times greater in central California. Nevertheless, net primary production was almost double in southern California because storms consistently reduced the standing crop of giant kelp in central California before the beginning of the season of maximum growth. Almost all *Macrocystis* canopies were removed in winter storms annually in central California, but only about 50% of the canopy was removed annually in the south. Furthermore, there was relatively little interannual variation in central California but great variation in the south because of occasional storms every few years that did remove much of the canopy. They concluded that cycles of disturbance and recovery were very important and easier to detect in giant kelp because of its fast growth, relatively short life, and its great proportion of biomass on the sea surface that makes it susceptible to wave disturbances. Cavanaugh et al. (2011) used satellite imagery of the canopies of kelp beds over the entire Santa Barbara Channel in southern California to examine their dynamics with respect to climatic data. They found a strong positive polynomial relationship between maximum wave height and the percentage loss of canopy (figure 3.9A), and that only the extreme wave events seemed to control regional kelp biomass. They also found recovery of canopy biomass in spring was negatively associated with SST (figure 3.9B) but not with wave height or other lower-frequency climate indices.

Of additional importance are effects on light caused by removal of overstory canopies during storms. Graham et al. (1997) showed that the seasonal removal of giant kelp canopies had flow-on effects in population dynamics. The numbers of recruit *Macrocystis* increased when canopies were removed, presumably because of increased irradiance, and were able to grow to adult sizes if adults were below a "threshold" density of around 10 plants per 100 m$^2$.

Other studies have concluded that if swells are too extreme, *Macrocystis* may not be able to persist; stands of all three ecomorphs of giant kelp are consistently described as occurring on "moderately exposed" to "protected" shores (e.g., Druehl, 1978, Villouta and Santelices 1984, Attwood et al. 1991, Vásquez et al. 2006). In the absence of biotic factors such as competition (Santelices and Ojeda 1984b), swells can determine the shoreward depth limit of kelp forests (North 1971a, Graham 1997) because, for a given set of swell conditions, surge speed increases as depth decreases.

On a biogeographic scale, increasing wave exposure may be a significant factor in the northern distributional limit of *Macrocystis* 'pyrifera' along the west coast of North America. It does not occur in large stands in the wave-driven coastline north of Año Nuevo Island (near Santa Cruz) in central California, and is largely replaced by the surface canopy-forming bull kelp *Nereocystis luetkeana*, which is extremely resistant to

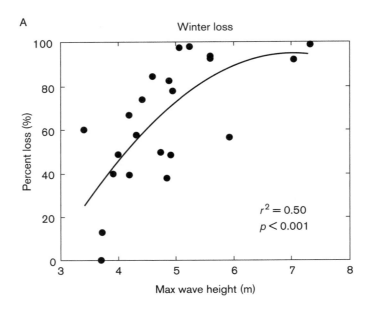

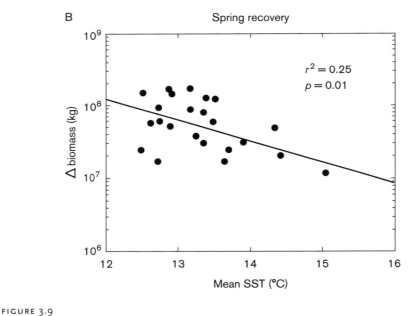

FIGURE 3.9

Regression analyses between (A) winter giant kelp canopy biomass losses and maximum significant wave height and (B) spring–summer kelp canopy biomass recovery and mean sea-surface temperature (SST) across the kelp forests of the Santa Barbara Channel in southern California. Winter losses were calculated as the percent change in kelp canopy biomass from the fall (September to November) maximum to the winter–spring (December to May) minimum. Recovery represents the $\log_{10}$ change in kelp canopy biomass from the winter–spring (December to May) minimum to the summer (June to August) maximum. Maximum wave height and mean SST for each year were calculated over the same periods.

SOURCE: From Cavanaugh et al. (2011).

breakage from water drag (Koehl and Wainwright 1977, Koehl et al. 2008). Surge is also thought to be the reason that giant kelp is restricted to calmer waters in northern and central Chile, replaced by the kelp *Lessonia trebeculata* on exposed shores (Villouta and Santelices 1984). Differences in the occurrences of surface and subsurface canopy kelps may also be related to differences in life history characteristics, growth rates, and susceptibility to grazing.

Other forms of *Macrocystis* are affected by water motion in various ways. *M. 'integrifolia'* occurs in more sheltered areas of the coast of California and British Columbia and rarely extends beyond around 6 m depth (Druehl 1981). In southern Chile, these distributional patterns may be different. Buschmann et al. (2006) examined populations of *Macrocystis* in southern Chile. They found that *M. 'pyrifera'* had higher densities and less variation in abundance in protected sites compared to exposed sites. Furthermore, populations in protected areas showed an annual cycle with most recruitment during late winter and fertile sporophytes in summer and autumn, but with loss of plants through to the following recruitment period. *M. 'integrifolia,'* however, had similar densities throughout the year in exposed and protected sites, with similar reproductive periodicity.

## BOUNDARY LAYER EFFECTS

Moving now from larger to smaller scales, Hurd and Pilditch (2011) pointed out that the relationship between blade morphology, water motion, and resource acquisition and allocation in kelp has been discussed for a century (Cotton 1912, Wheeler 1980a, Norton et al. 1981, 1982, Hurd 2000, Roberson and Coyer 2004, Koehl et al. 2008). In recent years, however, there have been significant advances in understanding the effects of water motion on giant kelp with respect to boundary layers around kelp blades and how water motion interacts with blade morphology and nutrient uptake. This topic was thoroughly reviewed by Hurd (2000), who discussed how the hydrodynamic environment around macroalgae influences their physiology and ability to acquire essential nutrients. Some of the earliest studies in this area were by Wheeler (1980a, 1982), who pointed out that the fixation of carbon during macroalgal photosynthesis is governed by a complex series of interactions between biological and environmental variables. These include the transport of $CO_2$ through the surrounding fluid, cell wall, membrane, and cytoplasm to the sites of carbon fixation. The boundary layer is the thin layer of fluid through which heat, momentum, and mass are moved from the fluid to the plant surface. The thickness of the boundary layer depends on water motion, and so by increasing water motion and decreasing boundary layer thickness, the transport distance is decreased. Wheeler (1980a) used experiments in water tunnels and found that by increasing water speeds over a *Macrocystis* blade surface from 0 to 4 cm sec$^{-1}$, the photosynthetic output increased by 300%. Furthermore, a critical velocity was around 1 cm sec$^{-1}$, beyond which the boundary layer adjacent to a blade was turbulent. He also concluded that morphological changes in kelp blades under different water motion regimes can be at least partially explained by hydrodynamic adaptations. Gerard (1982a) showed that boundary layer transport limited macronutrient uptake by *Macrocystis* when flow rates over blades were less than 3 or 4 cm sec$^{-1}$, and that nitrate uptake was saturated beyond around 2.5 cm sec$^{-1}$. However, she found that flow rates in the field always exceeded this value, even in a dense kelp forest during conditions of low currents and a calm sea state.

It has long been surmised that in some kelp species, blade morphology and behavior are adaptations for increased mixing at low water velocities (Wheeler 1980a, Hurd et al. 1996). In flow, waves of deformation pass along kelp blades and cause them to flap (Gerard 1982a, Hurd and Stevens 1997, Denny and Roberson 2002). This was tested in British Columbia, where Hurd et al. (1996) studied the effects of water motion on nitrogen uptake in distinct forms of *Macrocystis 'integrifolia'* from protected and exposed sites. They found for both blade morphologies that ammonium and nitrogen uptake increased with increasing water velocity, with a maximum reached at 4–6 cm sec$^{-1}$ and concluded that blade morphology does not enhance nutrient uptake in low flows. Further work on sheltered and exposed blade morphologies showed that "there were no obvious differences in how structural features such as marginal spines affected seawater flow around the different blade morphologies" (Hurd et al. 1997).

Hurd and Pilditch (2011) used novel techniques with electrodes to examine the characteristics of boundary layers around *Macrocystis* in southern New Zealand. They argued that although there is a compelling hypothesis that blade morphology in slow flow enhances nutrient flux to blade surfaces by increasing turbulence and overcoming limitations in mass transfer, there is little empirical evidence to support it. They used a micro-optode, with a tip diameter of 20 μm, to measure $O_2$ produced at the surface of actively photosynthesizing blades to indicate the thickness of the boundary layer. The diffusion boundary layer thickness varied from 0.67 to 0.80 mm and did not vary with mainstream velocities between 0.8 and 4.5 cm sec$^{-1}$. They further found that, as hypothesized, boundary layer thickness decreased with increasing velocity both for wave-exposed and wave-sheltered morphologies. They found that the boundary layer was thicker within corrugations than along apices of blades, which effectively traps fluid at the blade surface in corrugations and that corrugations are not turbulence-generating structures. They went on to discuss several potential roles of corrugations in wave-exposed plants, including increasing the surface-to-volume ratio for nutrient uptake, reduction of drag, and providing a quiescent region into which extracellular enzymes with important roles in carbon and nutrient acquisition may be safely deployed. For wave-sheltered plants, they concluded that thinner boundary layers were due to the lack of small-scale features that "trap" water at the surface. They concluded that when flows were as low as 2 cm sec$^{-1}$, the boundary layer can be so thin that it causes minimal resistance to the passage of nutrients to blade surfaces, perhaps accounting for slower growth of *Macrocystis* in wave-sheltered sites in southern New Zealand during autumn when nitrogen is limited. It should be kept in mind, however, that a large, thin blade morphology has a primary function of intercepting light (as in terrestrial leaves) and providing a large surface-to-volume ratio for nutrient and gas exchange. Furthermore, it was conjectured over a century ago that corrugations add strength to blades so that they do not tear easily (Skottsberg 1907).

## CONCLUDING COMMENTS

The inclusion of greater consideration of the abiotic environment, coupled with modern instrumentation, across spatial scales from submillimeter to oceanwide, and temporal scales up to decades, has heralded great advancement in our understanding of giant kelp settlement, growth, and dynamics. This is yielding a far more comprehensive view of kelp forests. We are gaining a much fuller appreciation of the fundamental processes and mechanisms affecting the kelp plants themselves across their full life histories and morphologies in different conditions. Rather than being precisely adapted to its environments, giant kelp is highly labile and responsive to environmental variation at its different life stages. We are gaining a much greater understanding about the underlying mechanisms affecting the variability of occurrence and performance of plants and populations in their continuously dynamic nearshore environments. Of all facets of kelp forest biology and ecology, this area of research, combined with great advances in understanding the ecology of early life history stages, has progressed the most in the 29 years since our last treatise on the ecology of giant kelp forests appeared (Foster and Schiel 1985). We anticipate that a greater union of climatic, oceanographic, and biological data, combined with a greater understanding of relevant spatial and temporal scales, will continue to progress and will offer a more holistic understanding of these important components of coastal ecosystems. They may also lead to better predictions of anthropogenic impacts and a changing climate (discussed in Chapters 10 and 13).

# 4

# DEMOGRAPHY, DISPERSAL, AND CONNECTIVITY OF POPULATIONS

Malthus (1798) saw the statistical properties of populations as common to man, animals, and plants. "Through the animal and vegetable kingdoms, nature has scattered the seeds of life abroad with the most profuse and liberal hand. She has been comparatively sparing in the room and nourishment necessary to rear them. The germs of existence contained in this spot of earth, with ample food, and ample room to expand in, would fill millions of worlds in the course of a few thousand years. Necessity, that imperious all pervading law of nature, restrains them within the prescribed bounds. The race of plants and the race of animals shrink under this great restrictive law."

—Malthus (1798) from Harper and White (1974)

The presence, absence, status, and connectivity of kelp forests are essentially a numbers game based on demographic characteristics of settlement, growth, mortality, and reproduction. Unsurprisingly, the demography of giant kelp is as variable as its morphology, responses to the abiotic environment, and interactions with other members of its community. Because giant kelp lives mostly in dynamic nearshore waters, occupying much of the water column, its susceptibility to partial or complete removal from severe water motion is a driving force in its ability to persist and replenish populations. Density-dependent and density-vague processes come into play as canopies and plants are thinned or removed and light gaps are provided to areas below. Furthermore, the kelp beds or patches themselves have a demography, often called "patch dynamics" within populations or "metapopulation dynamics" between populations, which are related to but somewhat separate from that of individuals within populations. There is a tendency to think of kelp forests as being continuously distributed along superficially similar areas of coastline, but this is far from the case. *Macrocystis* is largely bound to rocky substrata, which is patchily distributed. Patches, therefore, have sizes and spatial distributions that affect the probability of dispersal and connectivity of kelp beds spatially and temporally. These discrete patches can show great variability in going extinct and recolonizing over time. Also, within larger kelp patches, there is considerable spatial variability that results from a wide array of interactions. In any case, the basic prerequisite for any replenishment of populations and re-establishment of patches or areas in which plants have disappeared is the arrival of viable spores. Therefore, the production of spores, their behavior in the water column, dispersal abilities, settlement characteristics, and subsequent development through life history stages become critical. If you think of discrete kelp forest patches along a coastline, the concept of metapopulation dynamics applies because these must be connected at least occasionally through time both in terms of outbreeding and for re-establishment if patches go temporarily extinct, such as through major storm events. In this chapter, we discuss the demography and metapopulation ecology of giant kelp and the processes that underlie connectivity of patches, including spore production and dispersal.

## DEMOGRAPHY

The topic of demography and population biology of large brown seaweeds was reviewed by Schiel and Foster (2006). They contrasted *Macrocystis* and many other kelp species, as well as comparing kelps with the other major taxon of habitat-dominating seaweeds, the fucoids (order Fucales), finding considerable differences in reproductive output, growth, survival, and longevity (figure 4.1). One critical difference between kelps and fucoids is the "alternation of generations" of kelps, requiring spores to settle, grow into gametophytes, become fertile, and achieve fertilization to 2N sporophytes, processes that occur over several weeks. In contrast, fucoids begin with much bigger eggs, around 50–150 μm in diameter (Taylor et al. 2010) that are released directly from adult plants, have no obligate period of development to achieve a 2N stage but instead are fertilized directly at or soon after release, and so have orders of magnitude greater survival than kelp spores. Furthermore, fucoids can be enormously long-lived, with some species reaching several decades or more (Åberg 1992, Jenkins et al. 2004). Consequently, the population dynamics of these major orders of large, habitat-forming seaweeds are quite different, not only in their demographic properties but also in their susceptibilities to disturbances, population turnover, and abilities and ways of replenishing and sustaining populations. These contrasts between kelps and fucoids are not pursued further here but, nevertheless, they stand as good reminders that even in the same environments, there are many modes and mechanisms of population persistence for even superficially similar species with many shared characteristics of dominating areas and facilitating the presence of other species.

The first demographic study on giant kelp was done between 1967 and 1973 in a forest offshore of Del Mar in southern California. Rosenthal et al. (1974) recorded the positions of every *Macrocystis* plant within 1 m either side of a 100 m transect in 1967, adding a further three 50 m transects a year later. The original mapped plants declined steadily for 3 years when only 1 of the original 35 plants remained. Most mortality of adult *Macrocystis* over the 6-year study was from detachment during storms, entangle-

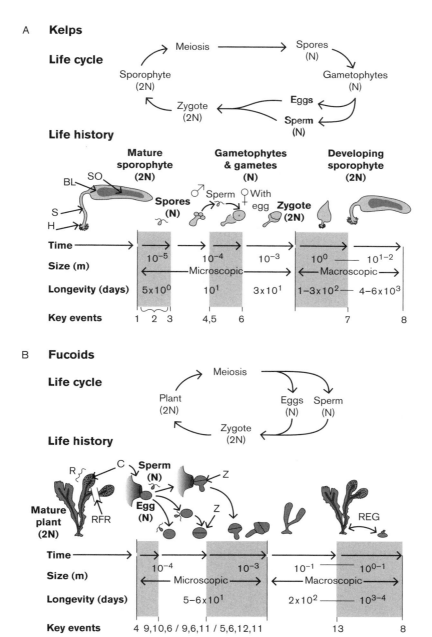

FIGURE 4.1

Kelp (A) and fucoid (B) life cycles and life histories with emphasis on stages, sizes (plant length), longevities, and events of particular relevance to population dynamics. The fucoid life history illustrates three possible sequences leading from gamete production to attached zygotes or embryos.

C = conceptacle (cavity in which gametangia develop); BL = blade; H = holdfast; R = receptacle (terminal portion of branch bearing conceptacles); REG = regeneration; RFR = reproductive frond release (in some fucoids); S = stipe; SO = sorus; Z = zygote.

1: spore production; 2: spore dispersal; 3: spore settlement; 4: gamete production; 5: sperm dispersal; 6: fertilization; 7: sporangia development; 8: sporophyte / adult plant mortality; 9: sperm and egg dispersal; 10: egg settlement; 11: zygote settlement; 12: zygote dispersal; 13: development of gametangia.

SOURCE: From Schiel and Foster (2006).

ment with drifting adults, and kelp harvesting. The severe grazing by sea urchins seen in other kelp forests at the time (Leighton 1971) did not occur at Del Mar. Rosenthal et al. calculated that the oldest plant was around 7 years old. Their survivorship curve showed, as would be expected, that most mortality occurred soon after recruitment (figure 4.2A). Of 387 recruits, less than 10 survived after 6 months. Overall, they found a paucity of recruitment, which they attributed to sedimentation from the coastline and poor light quality from resuspension of sediments (discussed in Chapter 3).

Paul Dayton and colleagues, especially Mia Tegner, made major advances in our knowledge of giant kelp forests over the past 40 years through a wide range of empirical studies. Dayton et al. (1984) did a comprehensive study of the demography and patch dynamics of *Macrocystis* and other kelp species for 3–10 years within several forests of southern California. Using standard life tables they compared survival of several kelp species from their juvenile stages onward (see figure 4.2A). They found in the Point Loma kelp forest off San Diego that 1.6% of *Macrocystis* recruits survived for a year and that only one of these plants, or 0.007% of the original cohort, lived to 86 months. The understory stipitate kelp *Pterygophora californica* had greater survival and longevity, but their analysis of this species started with larger established plants and so is not strictly comparable to *Macrocystis*, at least in the early stages. Nevertheless, they found that, once established, some 25% of *Pterygophora* plants survived for at least 11 years, but the longest-lived *Laminaria farlowii* they found were about 6 years old. Other studies have calculated that *P. californica* can live at least 12 years (Hymanson et al. 1990) and perhaps up to 20 years (Reed and Foster 1984). Similarly, the understory kelp *Eisenia arborea* had around 30% of its original numbers after 11 years. It seems clear from the study of Dayton et al. (1984) that *Macrocystis* has a much faster turnover of individuals within populations than some of the dominant understory species. This is consistent with the susceptibility of the much larger *Macrocystis* plants to storm damage, which increases as plants encroach into shallow water. For example, Dayton et al. (1984) also found that cohorts at 5 m depth had only about half the rate of survival of those at 15 m depth.

Dayton et al. (1984) also considered three concepts of patch stability within kelp forests: the persistence of patches through generations, the inertia or resistance of patches to invasions or disturbance, and their resilience or recovery following disturbance. Resistance to invasion of these patches by other species is due to the overstory of giant kelp and the undercanopy of smaller kelps, both of which inhibit recruitment, and also to limits of effective dispersal. No kelp invaded *Macrocystis* patches, although giant kelp invaded patches of smaller kelps, most likely due to both the massive reproductive output of giant kelp and the fact that they need less than one successful recruit per square meter to eventually dominate an area. As found in earlier studies (e.g., Pearse and Hines 1979), recruitment and survival in all cases were far greater when canopies had been removed. Winter storms were responsible for most mortality of giant kelp, providing light gaps of various sizes for subsequent recruitment and also accounting for the spatial mosaic of kelp seen within forests. Furthermore, patch persistence within all

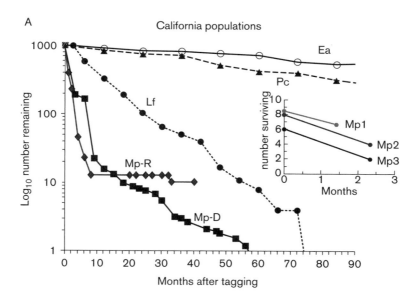

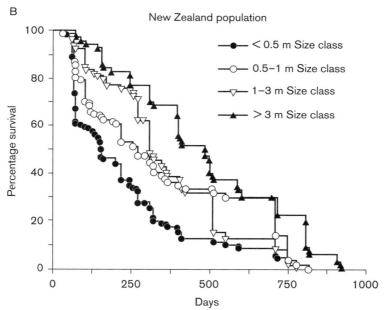

FIGURE 4.2

Survivorship curves of kelps. (A) From various kelp forests in southern California, kelps at around 15 m depth, adjusted to initial numbers of 1000. Both *Macrocystis* cohorts started as small recruits followed through time. *Laminaria farlowii* (Lf) also from initial recruits. *Pterygophora californica* (Pc) and *Eisenia arborea* (Ea) began as larger plants of unknown age at the start of the study. (B) Percentage survival of four size classes of *Macrocystis* in a harbor near Christchurch in southern New Zealand.

Inset: Mp1 = *Macrocystis*, settled spores to microscopic sporophytes from Deysher and Dean (1986a – reprinted with permission from Elsevier); Mp2 = *Macrocystis*, spores settled at high density, to visible sporophytes, from Reed (1990a); Mp3 = *Macrocystis*, spores settled at low density, to visible sporophytes, from Reed (1990a). Y-axis is log number surviving (orders of magnitude)

SOURCE: (A) Mp-R *Macrocystis* from Rosenthal et al. (1974); Mp-D and all other kelps from Dayton et al. (1984). (B) From Pirker et al. (2000).

three forests was similar, with unchanged patch borders throughout the study. A major conclusion of this study involved the fact that patches of *Macrocystis* can persist well beyond the generation time of this species, which in hindsight is probably obvious since forests have persisted in many locations in California since beds were first mapped in the early 1900s (see Chapter 1, History of Research).

There are few other field-based demographic studies on giant kelp outside of California. Westermeier and Möller (1990) found that an intertidal population of *Macrocystis* in southern Chile showed maximal recruitment in October to December (austral spring / summer) when most primary space was available after winter storms. The life expectancy of a recruit, calculated as "half-life," was only 45 days. They concluded that these populations of *Macrocystis* have an annual life cycle, which can extend to 2 years under favorable conditions. A 3-year study in southern New Zealand also showed considerable differences of the demography of *Macrocystis* compared to California populations. Although it occurs along portions of the open coast of the east coast of southern New Zealand, *Macrocystis* flourishes in the calm waters of large harbors. A study at two sites in Akaroa Harbour, near Christchurch, found that the sea-surface canopy was greatest during winter, when temperatures were low and nitrate levels high, and that canopy deterioration occurred during the summer months (Pirker et al. 2000, Pirker 2002). This is also the case in other large southern harbors of Lyttelton and Otago (Hay 1990, Brown et al. 1997). As found elsewhere, there was a negative correlation between temperature and nitrate levels, and nitrates were particularly low in summer. Average frond survival was 5.6–6.7 months and no fronds lived longer than 15 months. Survivorship curves varied considerably for the different size / age stages of plants, with maximum longevity of 2.5 years (figure 4.2B). One of the interesting findings of this study was that the forests died back considerably when a large amount of sediment was dumped into the water following heavy rainfall and a landslip. Sediment settled on *Macrocystis* on the sea surface and along the reefs, both causing the deterioration of canopy plants and impeding settlement and recruitment, so that the canopy fell to near zero the following year. These populations had high annual turnover, and individual plants rarely had more than several fronds at a time.

All of these studies show there is considerable size dependency in the demographics of *Macrocystis* across its life stages. Adults generally decline in numbers over 3–4 orders of magnitude from when they are visible recruits, with longevity depending on sites and conditions. During the initial phases of microsporophyte production, however, numbers decline by 4–6 orders of magnitude, and there is also massive mortality during the period from spore release to microsporophyte production. For example, around 10 billion spores of an annual kelp, *Alaria marginata*, produced around 400 visible sporophytes in a study in central California (McConnico and Foster 2005). The only other known estimate of survival from spore release to settlement for a laminarian was by Chapman (1984) for two *Laminaria* species. He found that each species released around 10 billion spores per square meter in kelp beds and that proportional survival from release to settlement was between $10^{-3}$ and $10^{-4}$. Estimates of settled spore survival are also small but orders of magnitude higher than free-floating spores. For *Macrocystis*, Deysher and Dean (1986a) found that spores settled at about $3.3 \times 10^8$ m$^{-2}$ produced $5 \times 10^6$ microsporophytes (figure 4.2A inset). Reed (1990a) seeded rocks with different densities of spores of *Macrocystis* and the understory kelp *P. californica*, placed them into a kelp forest, and counted the number of sporophytes that emerged. Although the results were highly variable, spore densities of $10^6$ and $10^8$ m$^{-2}$ produced $10^2$ and $10^4$ sporophytes per m$^2$, respectively.

Effective recruitment and replenishment of populations are affected, of course, by a multitude of factors influencing spore arrival, gametogenesis and growth, fertilization, and sporophyte recruitment and growth. It is interesting to note that from the early fine-scale experiments of Foster (1975b) onward, there is no evidence of any facilitation of recruitment in giant kelp. Although microfilms of bacteria and other microorganisms are potential facilitators, they are ubiquitous and develop very rapidly on new surfaces. *Macrocystis* can be among the first organisms to settle onto surfaces and grow if viable spores, space, nutrients, and light are adequate. From numerous studies it seems clear that natural disturbances of overstory plants are essential features of successional events in kelp forests.

Several mathematical models have been used to describe the growth and longevity of fronds or plants of *Macrocystis* (e.g., Anderson 1974, Jackson 1987, Nyman et al. 1990). Although interesting and instructive, no single model captures the innate variability of demographic processes of *Macrocystis* across the wide range of environmental conditions it inhabits. Burgman and Gerard (1990) developed a stochastic population model for *Macrocystis* that included five life history stages and took into account environmental and demographic stochasticity as well as density-dependent interactions on a monthly basis over 20 years. The five stages were gametes, microscopic sporophytes, blade-stage sporophytes up to 1 m long, subadult sporophytes up to 10 m long, and adults with canopy fronds over 10 m long. Their model recognized the transition from gametophytes to microscopic sporophytes through reproduction, and also the transitions of sporophytes through various growth stages. This model showed the population density of adult sporophytes fluctuating around 0.1 plants per m$^2$ on a seasonal cycle, which came close to that observed in the field-based studies of North (1971a), Rosenthal et al. (1974), and Dayton et al. (1984). It also identified that there were discrete, nonrandom "windows" of recruitment, comprising about 35% of the months that were simulated, which is similar to what was found in the empirical studies of Deysher and Dean (1986a, 1986b). Their model also found that there was a high probability of extinction of the population during El Niño events. Overall, this model incorporated many important features of populations and their variation and, in many ways, encapsulated the results of numerous studies showing the important density-dependent processes operating in *Macrocystis* forests.

Nisbet and Bence (1989) provided a model of "density-vague" population regulation of *Macrocystis*, particularly concerning 5–10 year fluctuations in numbers in a single bed. In density-vague regulation, density-dependent behavior is seen only in very low and very high densities, with populations fluctuating either randomly or in response to variation in the environment at intermediate densities. As in other studies, they concluded that self-shading can limit populations, that variable recruitment can greatly affect complex multiyear cycles, that factors not well understood at the time (labeled "stochasticity") such as light and nutrients could be crucial to population dynamics, and that *Macrocystis* populations often had the characteristics of density-vague regulation.

We now know most of the key biotic and abiotic factors operating in *Macrocystis* population regulation (see, e.g., Chapter 3) and the great variation in populations encountered in at least California over about 100 years. The overwhelming conclusion from these studies is the overarching effects of environmental variation on the dynamics of *Macrocystis* populations, acting in different ways across all its life history stages.

## METAPOPULATIONS

Variation in the demography and population dynamics of giant kelp greatly affects its "metapopulation" biology along a coastline. This occurs not only because of the presence, absence, or varying densities of kelp within forests but also through consequent processes of reproductive output, dispersal, and recruitment which provide connectivity within and among discrete patches (Levins 1969, Levin and Paine 1974, Hanski 1999) along a coast. Dan Reed and colleagues analyzed spore dispersal and *Macrocystis* patch structure to examine connectivity of populations over 34 years along 500 km of the California coast (Reed et al. 2006—"An experimental investigation"). This is the most thorough study yet done that analyzes the dynamics of discrete kelp patches over such large spatial and temporal scales. Most early studies have cited the turn-of-the-century maps by Crandall (1912, 1915), which was extended by North (1971a), to compare current with historical distributions of kelp patches in the Point Loma and San Diego area. Reed and co-workers, however, took advantage of detailed aerial surveys of *Macrocystis* beds from Point Conception south to the California–Mexico border done from 1958 onward by a kelp harvesting company (ISP Alginates, Inc.). From these surveys, the biomass of harvestable kelp (0–1 m depth below the sea surface) was estimated and then calibrated to the actual harvest. Data were ground-truthed in this way for 10 years and then standardized from January 1968 to October 2002 (418 months) across 500 km of coast between Point Arguello, California, to the United States–Mexico border. The aerial surveys had 2–20 km resolution, which was not fine enough to capture metapopulation dynamics of discrete patches, and so they augmented these with digital analysis of kelp canopies for the few years that data were available from aerial infrared images at 3–5 m resolution. By combining maps, they identified patches in which the *Macrocystis* canopy was either contiguous or separated by less than 500 m. In their analyses, patches were considered to be "occupied" if a surface canopy was present or "extinct" when no surface canopy was detected.

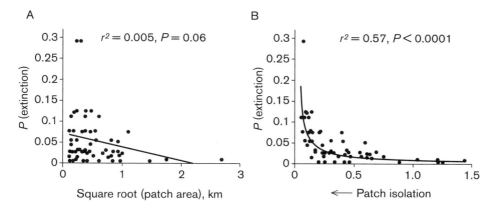

FIGURE 4.3

(A) Monthly probabilities for *Macrocystis* patch extinction versus patch size and (B) patch extinction versus patch isolation. Data are from 69 patches of giant kelp observed from 1968 to 2002 along the mainland coast of southern California. Patch size is defined as the square root of the area of the patch in square kilometers. Patch isolation is defined as the inverse distance-weighted average size of surrounding patches that are occupied in any given month, averaged over 418 months; low values of patch isolation indicate a high degree of isolation (indicated by arrow).

NOTES: Lines show fitted least-squares regressions:
(A) $P$ (extinction) = $-0.0316\sqrt{\text{(patch area)}} + 0.0708$.
(B) $\ln[P \text{ (extinction)}] = -0.8643 \ln(\text{isolation}) - 4.7217$.
$r^2$ and $P$ values for regressions are as shown.

SOURCE: From Reed et al. "A metapopulation perspective" (2006).

They found a surprising degree of patch extinction and recolonization over the 34-year period. In the extreme cases, close to 100% of patches were occupied at times by giant kelp but virtually none were occupied at other times, such as after the major 1982–1983 El Niño. Underwater surveys showed that subsurface *Macrocystis* plants inhabited many patches during this period, but they did not reach the canopy (and, hence, did not appear in the aerial surveys). Furthermore, there was not only nutrient stress affecting larger plants during this period but also recruitment failure because of the stressful conditions. "Extinction" and "recolonization" were measured as the monthly probability of a patch going from occupied to extinct, or vice versa. The monthly probability of a patch going extinct was between 0.005 and 0.292, and from being extinct to occupied from 0.023 to 0.200, showing, as in several previous studies, the highly dynamic nature of giant kelp populations. Patch extinction lasted from 6 months or so to as long as 13 years, although 80% of extinct patches were recolonized within 2 years.

Two of the most important findings of Reed et al. (2006—"An experimental investigation") relating to extinction involved the sizes of kelp patches and their distances apart. The probability of patch extinction was negatively correlated with patch size (figure 4.3A). All of the high probabilities of extinction were associated with small kelp

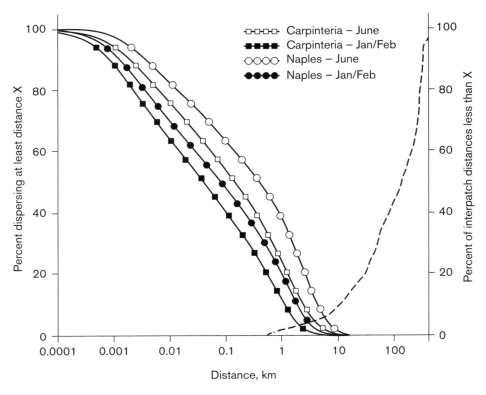

FIGURE 4.4

Dispersal among patches in southern California. Dispersal potential of giant kelp spores versus distance between discrete patches. Spore dispersal distributions (solid lines) were simulated on the basis of measured currents at Carpinteria Reef (squares) and Naples Reef (circles) from June 1 to 30, 2001 (open symbols) and January 15 to February 15, 2001 (closed symbols; see text for details). The left vertical axis indicates the probability of dispersing at least as far as the distance on the horizontal axis. The distribution of distances between patches (dashed line) was calculated from distances between all pairs of occupied patches in all months of the study period (418 months). The right vertical axis indicates the percentage of distances between patches that were less than the distance on the horizontal axis.

SOURCE: From Reed et al. "A metapopulation perspective" (2006).

patches, those less than around 0.25 km² (i.e., the square of 0.5 in figure 4.3A). Beyond an area of 1 km², the probability of extinction was near zero. Patch extinction was also highly related to patch isolation (figure 4.3B), far more so than to patch size. The mechanisms underlying recolonization of patches varied. These included recruitment from dormant microscopic stages, which probably last no more than several months in field conditions (Edwards 2000, Schiel and Foster 2006), spores derived from drifting adult plants, or spore dispersal from extant patches of adults. Because of the vast numbers of spores produced by established patches, it is probable that most recolonization of extinct populations occurs through spore dispersal. Such dispersal involves complex physical processes of transport (Denny and Gaylord 2010), the size of the spore source (Reed et al. 1997), and, in the case of discrete patches, the distance separating them. Using estimates of spore dispersal from a wide range of empirical studies and modeling, Reed et al. (2006—"An experimental investigation") found that the median dispersal distance of spores was about 40 m at one site and 400 m at another, with 10% of all spores dispersing 1 km at one reef and 4 km at the other (figure 4.4). Distances between neighboring patches, averaged over the 418-month period, ranged from 0.5 to 14 km, but 80% of patches were within 2 km of each other. In most cases, dispersal did not go beyond 20 m but in around 5% of cases dispersal was estimated to be more than 2 km (figure 4.4). The long tail on this distribution was greatly affected by currents and waves, two of the most influential processes in dispersal. They found that the level of connectivity among patches was only from 0.37% to 1.58%, depending on currents and inter-patch distances. Combining all of the calculations, they found that in summer conditions 42% of patches would exchange spores with at least one other patch along the coastline and 19% would exchange spores with no other patch.

This elaborate study highlights the highly dynamic nature of *Macrocystis* populations and the interaction of all factors over temporal scales from months to decades and spatial scales from meters to kilometers. The conclusion was that environmental variability in biotic and abiotic disturbances was the primary driver of extinction and recolonization. They found that the great majority of recruitment failures (what they termed "demographic stochasticity") was short term and that connectivity via spores was common. However, it seemed clear that many patches are near the limits of spore dispersal from other patches, which highlights perhaps their vulnerability to prolonged extinction should unfavorable conditions persist, such as through a changing climate.

## REPRODUCTIVE OUTPUT AND SOURCES OF PROPAGULES

The preponderance of evidence appears to be that spores produced by attached plants rather than from detached drifting plants are the major means of replenishment and recolonization of giant kelp forests. This conclusion is based largely on the observations that "drifters" tend not to persist for very long on inshore reefs where suitable habitat is available, spores released from drifters at the surface are unlikely to reach the bottom, and drifters tend to be somewhat ephemeral throughout the year. Furthermore, drifters found along reefs tend to show a geometric decline in numbers away from source reefs. For example, in southern California, Reed et al. (2004) found that beyond 1 km from source reefs, there were fewer than 0.5 drifting *Macrocystis* plants per 100 m² (figure 4.5), and over a 5-year period few of these lasted for even a few months. They concluded that local dispersal of spores from drifters contributed very little to recruitment of giant kelp among the reefs in their study. This is consistent with many other studies examining connectivity among populations. These studies do not preclude the possibility of

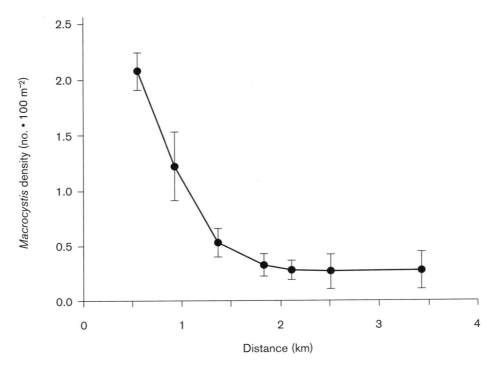

FIGURE 4.5
Falloff of numbers of *Macrocystis* drifters, with distance from nearest natural source. Data are averages (±1 standard error) of drifters per square meter of bottom on San Clemente Artificial Reef in southern California.
SOURCE: From Reed et al. (2004) reprinted with permission from Wiley & Sons Publishing.

dispersal of spores from drifting plants but they do indicate that this has a far lower probability event than dispersal of spores from fixed populations, especially on a local scale or where the array of discrete patches allow "stepping stones" for dispersal along a coastline. Given the abundance of drifters in some areas (Hobday 2000b, Macaya et al. 2005) and the enormous spore production of single plants, even low-probability long-range dispersal via drifters will have occurred, and this still seems the likely means of the spread of giant kelp from the northern to southern hemisphere and across the southern ocean (see Chapters 1 and 5).

Here, however, we consider the more frequent occurrence of dispersal within and among populations in ecological time frames. The "black box" of spore production, dispersal, and behavior of microscopic stages of kelp has been slowly prised open over the past 25 years. Most of what we know about the development and potential longevity of microscopic stages of kelp comes from laboratory studies on abiotic requirements and their effects (see Chapter 3). For recruitment to occur in nature, there must be an adequate supply, survival, and growth of these microscopic stages from the release of spores, development of gametophytes, fertilization, and production of sporophytes. Because of their small size and similarity to spores and filaments of other brown algae, giant kelp spores and gametophytes are extraordinarily hard to see or distinguish in natural conditions, but innovative experiments and techniques, such as underwater microscopes (Kennelly and Underwood 1984) and fluorescence (Graham 1999, Graham and Mitchell 1999), have allowed both visualization in some conditions and a growing understanding of how relevant processes operate on very small spatial scales. Unsurprisingly, it appears that the microscopic stages are influenced by as wide a range of abiotic and biotic factors as the larger, visible stages, the difference being that they are particularly challenging to work with.

As discussed in Chapter 3 on the abiotic environment, all processes of reproduction and growth are greatly influenced by environmental factors. Furthermore, ecological factors such as stand density and location affect sporophyte production and therefore reproductive output of giant kelp. *Macrocystis* can be reproductively active year-round, with individual plants having $3 \times 10^{11}$ zoospores at a time (see Chapter 2). The influence of stand location, plant density, and biomass of individual plants was tested in a multifactorial experiment in southern California. Reed (1987) selected two locations near Santa Barbara, California, and in each place established two 25 × 25 m plots with an average giant kelp density of about five plants per 10 m². One plot in each place was thinned to 0.3 plants per 10 m² and the other left as a control. He then trimmed 75% of fronds from five randomly selected plants within each plot to test how a reduction in plant biomass might affect reproduction. This experiment allowed tests of location, forest density, and plant biomass on reproductive output. As a proxy for reproduction, he used the dry weight of sporophylls produced in each treatment on selected plants. He first documented that there was a 99.5% correlation between sorus area and sporophyll biomass, showing that the amount of sporophyll mass was a good indicator of reproductive output (provided that sori were equally productive of spores across treatments). He found that the removal of vegetative fronds had a huge effect on reproduction by reducing sporophyll biomass by 80%. Reed surmised that the thinning of vegetative fronds decreased the organic carbon translocated to sporophylls, thereby reducing fecundity. Plant density had a significant effect on reducing reproductive output, but only at one location. He attributed the differences to the degree of shading of canopies, rather than to density per se. Fecundity is, however, a function of vegetative biomass (Neushul 1963). It decreases as vegetative biomass is reduced by storms or grazers (Reed 1987, Graham 2002) and its recovery can lag behind vegetative recovery (Neushul 1963, Graham 2002). These studies, therefore, showed that the localized conditions affecting stand density and frond density can have large effects on reproductive output.

It is known from numerous studies that the reproductive output of plants and forests is prodigious, especially when considering that *Macrocystis* plants are reproductively active year-round, but how this relates to the supply of spores and their variability was unknown. Graham (2003) addressed this problem using a detailed study to test how tightly coupled propagule output and supply were in the Point Loma kelp forest. He mapped giant kelp plants at different positions in the forest through most of a year, using a visual assessment of sporophyte fertility to gauge when sori were absent, present, and present and actively releasing spores; he categorized sporophyll "bundles" as small or large. At each monthly sampling time, he used a plankton pump with the nozzle a few centimeters above the substratum to collect released zoospores. In enumerating zoospores, he distinguished between those of *Macrocystis* and other kelps on the basis of species-specific absorption spectra of plastids within zoospores obtained from microphotometry (Graham 1999, Graham and Mitchell 1999). He found that zoospore supply varied from 250 to 54,000 $L^{-1}$, was greatest in March to April and declined to low numbers from July onward. He determined there was a poor correlation between "predicted" zoospore supply, based on his estimates of sporophyll production and condition, and actual supply. He concluded that current dampening within the kelp forest retained zoospores, but that forest edges tended to experience unidirectional currents that transported spores far from release sites, thereby decoupling local reproduction and zoospore supply. His results emphasized the importance of population size, location of release, and currents in the supply of zoospores.

## SPORE DISPERSAL AND RECRUITMENT WINDOWS

Given the vast reproductive output of giant kelp plants and forests, potentially numbering in the trillions, there is obviously a severe reduction in numbers through the processes encompassing release of spores to production of microscopic sporophytes. It could be assumed that in some way recruitment should be related to reproductive output, and yet across ecology there are numerous examples where this is not the case and where recruitment is more of a "lottery" (Sale 1978) or "sweepstakes" (Flowers et al. 2002). As shown in figure 4.6, landing on suitable substratum, avoiding other species, not being eaten, and chemically affected are only some of the filters that decouple spore production from recruitment. Importantly, it has long been noticed that recruitment often occurs in discrete pulses that seem to be dictated largely by the abiotic environment around kelp forests. These "windows" of recruitment have been clarified both by laboratory- and field-based experimental studies.

Beyond the basic requirement of blue light (400–500 nm) for gametogenesis to occur (see Chapter 2), the total amount of light affects fertility. Lüning and Neushul (1978) showed that 50% fertility of female gametophytes occurred at around 2.6 E $m^{-2}$. Similarly, Deysher and Dean (1984) found that 50% fertility in *Macrocystis* occurred at 5 µE $m^{-2}$ $s^{-1}$ and saturating irradiance at around 10 µE $m^{-2}$ $s^{-1}$. Furthermore, there is an interaction between light, temperature, and nutrients that affects successful recruitment of kelp. For example, laboratory experiments by Lüning (1980) showed that the total amount of light needed for gametogenesis in *Laminaria* increased exponentially as temperature increased. A series of innovative field-based experiments quantified this

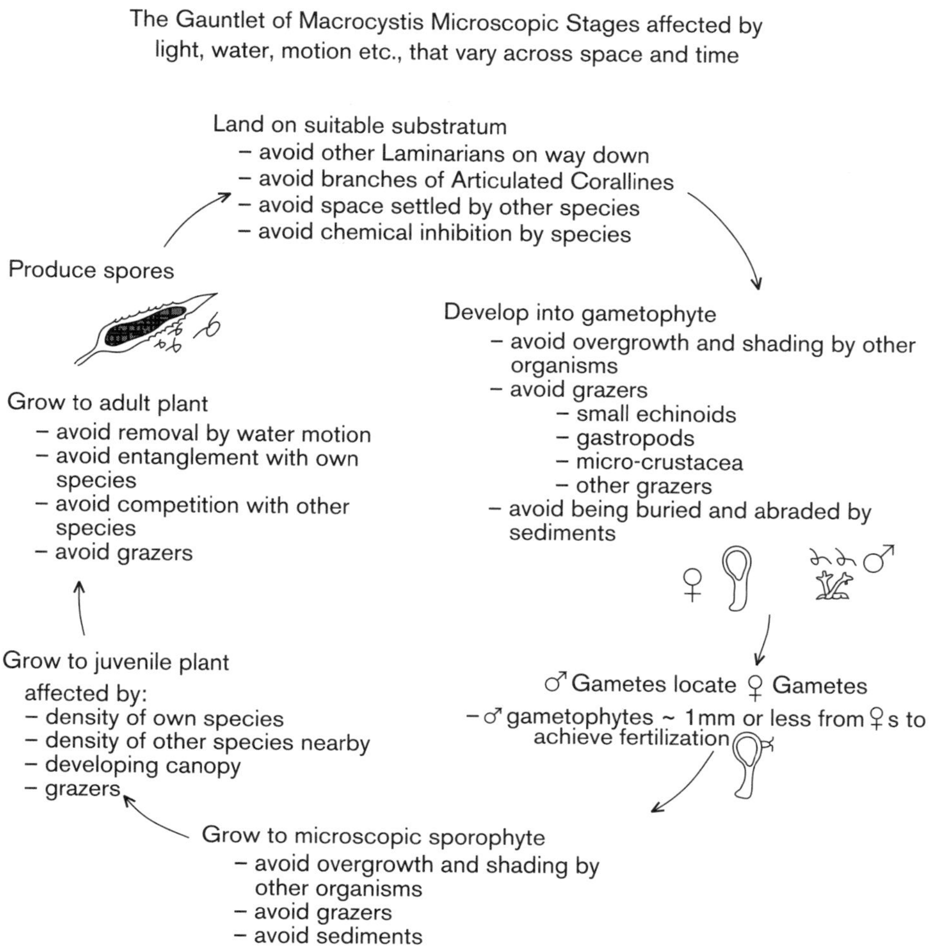

The Gauntlet of Macrocystis Microscopic Stages affected by
light, water, motion etc., that vary across space and time

Land on suitable substratum
– avoid other Laminarians on way down
– avoid branches of Articulated Corallines
– avoid space settled by other species
– avoid chemical inhibition by species

Produce spores

Grow to adult plant
– avoid removal by water motion
– avoid entanglement with own
  species
– avoid competition with other
  species
– avoid grazers

Develop into gametophyte
– avoid overgrowth and shading by other
  organisms
– avoid grazers
    – small echinoids
    – gastropods
    – micro-crustacea
    – other grazers
– avoid being buried and abraded by
  sediments

♀ ♀    ♂

Grow to juvenile plant
affected by:
– density of own species
– density of other species nearby
– developing canopy
– grazers

♂ Gametes locate ♀ Gametes
– ♂ gametophytes ~ 1mm or less from ♀s to
  achieve fertilization

Grow to microscopic sporophyte
– avoid overgrowth and shading by
  other organisms
– avoid grazers
– avoid sediments

FIGURE 4.6

The filters of microscopic stages of *Macrocystis* in the field.

SOURCE: From Foster and Schiel (1985).

for *Macrocystis*. Dean et al. (1983) settled *Macrocystis* spores onto small ropes in the laboratory and grew them through to the gametophyte stage before outplanting them to the field in southern California. They used in situ sensors and integrating recorders in the site and then counted the densities of sporophytes that eventually appeared over many different episodes of outplanting. They found there were threshold levels of light and temperature required for the production of different densities of sporophytes, effectively creating a window of recruitment (see figure 4.7). No sporophytes were produced above 17.6°C or below an irradiance level of 0.3 E m$^{-2}$ day$^{-1}$. Dense recruitment of >50 sporophytes per cm$^2$ occurred when the temperature was below around 16.5°C and irradiance above 0.4 E m$^{-2}$ day$^{-1}$. As temperatures approached 16°C, more light was required to

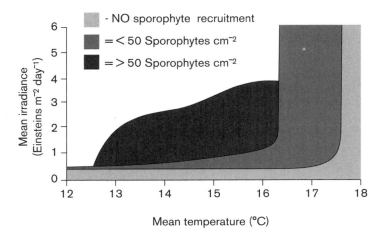

FIGURE 4.7

"Windows" of recruitment of *Macrocystis*, relating density of sporophyte recruits per square centimeter with respect to irradiance and temperature.
SOURCE: Data modified from field-based study of Deysher and Dean (1986a).

produce these high densities of sporophytes. Lower densities were produced between 16.5°C and 17.5°C regardless of light levels.

To refine these responses, further experiments by Deysher and Dean (1986b) were done in the laboratory in a crossed experimental design using 7 irradiance levels and 10 temperatures. These lab results were at variance with the earlier field study. There was no upper temperature threshold but instead nearly 100% sporophyte production from fertile female gametophytes between 17°C and 20°C at high light levels. They speculated that these differences from the field study were due to genetic variability among *Macrocystis* populations or to some factor correlated with temperature, such as nutrients. If nutrients were constant and plentiful, as in the laboratory experiments, fertility could stay high. However, temperature and nutrients are inversely correlated in natural conditions along coastlines (see Chapter 3) and so nutrients could be limiting for sporophyte production at high temperatures in field situations.

The experiments of Tom Dean and Larry Deysher greatly clarified the interactions of abiotic factors acting on microscopic stages of kelp and also provided a plausible mechanistic explanation for at least some of the great variation in recruitment that occurs in nature. Unfortunately, nature does not often come in convenient experimental units and therefore many other factors come into play in field situations. All of the processes affecting spore release, survival, germination, fertilization, and production of sporophytes are highly variable in nature. *Macrocystis* plants can become fertile in the first year (Neushul 1963) and release spores throughout the year, although with peaks occurring in early winter and in late spring—early summer (Anderson and North 1967, Reed et al. 1996). Reed et al. (1996) found that reproduction and the standing stock of spores of *Macrocystis* were positively correlated with the nitrogen content of plants and negatively correlated with seawater temperature. They showed that spore release was particularly poor during the 1991–1992 El Niño when temperatures were high and nutrients very low. Furthermore, at least part of the variability of sporophyte production and recruitment is related to the density of settlement of spores. Spore settlement must be $>1$ mm$^{-2}$ so that female and male gametes are close enough for fertilization (see Chapter 2).

## MECHANISMS OF DISPERSAL

Given the great sizes that *Macrocystis* can attain, their susceptibility to removal by storms, and their capability of staying afloat and drifting, it seems intuitive that drifting adult plants are a major factor in dispersal and connectivity of populations. This may be true over long time scales at great distances, and over evolutionary time scales, such as the spread of *Macrocystis* between hemispheres (as discussed previously), but on ecological time scales most of the evidence is to the contrary. Drifting plants, and probably the great majority of *Macrocystis* plants that come adrift, are either cast ashore (ZoBell 1971) or transported offshore. Many become entangled with attached plants and remained within established beds for up to several months (Rosenthal et al. 1974, Dayton et al. 1984). Reed et al. (2006—"An experimental investigation") concluded, however, that drifters do not provide a constant enough supply or an adequate residence time of reproductive sporophylls sufficient to account for recruitment patterns seen after disturbances (see earlier discussion).

Spore dispersal from attached plants, therefore, is the most likely source of recruits within and among populations. We have already discussed how spore production is influenced by the conditions of plants. Spores are tiny and their mobility has little if any effect over the distances they can travel in the water column, although it may help in microsite selection once they reach the bottom. The characteristics of the water column, however, dictate the distance spores travel and the likelihood they will make contact with the bottom.

Brian Gaylord and colleagues have provided much of what we know about dispersal in kelps, particularly *Macrocystis*. The numerous forces acting on spores and affecting dispersal across kelp forests were reviewed by Denny and Gaylord (2010) and Gaylord et al. (2012; see figure 4.8). These operate spatially over orders of magnitude from submillimeter boundary layer conditions to currents or even ocean-wide processes over many kilometers. Alongshore and tidal currents move within and across beds, but transport within the water column is influenced by stratification and internal waves. Because spores are tiny, neutrally buoyant, and have high drag coefficients in water (Vogel 1994), they sink slowly, if at all. Gaylord et al. (2002) found that *Macrocystis* spores sink at around 0.001 mm sec$^{-1}$ in seawater. Turbulent mixing is therefore required for them to reach the bottom (Denny and Shibata 1989, McNair et al. 1997). Turbulent eddies are

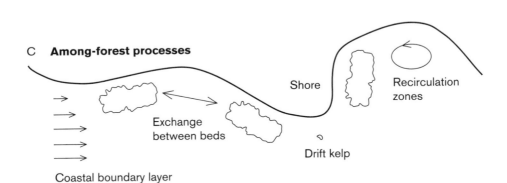

A **Individual kelp-level processes**

Current damping and wake effects

Surface gravity waves

Currents

Turbulent mixing

Bottom shear

Internal waves

*Macrocystis*

B **Forest-level processes**

Shore

Tidal currents

Kelp forest

Current attenuation within bed

Horizontal mixing

Current acceleration alongside bed

Propagating surface gravity and internal waves

C **Among-forest processes**

Shore

Recirculation zones

Exchange between beds

Drift kelp

Coastal boundary layer

FIGURE 4.8

Transport and mixing processes influencing kelp forests at various scales. (A) Fluid dynamic processes operating at the level of, or in the vicinity of, individual kelps. (B) Larger-scale phenomenon that influence entire forests. (C) Oceanographic factors that drive exchange of propagules and materials among kelp forests.

SOURCE: From Gaylord et al. (2012).

constrained near the seafloor and the spatial dimension of fluid motion is greater in the water column. Together, these hydrodynamic processes, interacting to some extent with the sinking speed of propagules and their release height above the substratum, determine the distance that spores travel, the proportion and numbers that are transported out of a kelp forest, and the probability they will settle somewhere else.

Spores behave as passive particles at scales greater than a few millimeters in moving seawater (Graham 2003). They do not have an obligate period of development, such as occurs in most marine propagules in the plankton, and can settle, attach, and begin to develop immediately after release (Santelices 1990). The swimming behavior of spores most likely aids in microsite selection as spores go through the benthic boundary layer near the bottom, and in settlement. They respond positively to nutrients, which can stimulate them to settle (Amsler and Neushul 1990) and can exhibit rheotaxis (response to water movement) and thigmotaxis (contact with surfaces) (Santelices 1990, Fletcher and Callow 1992). These behaviors probably help in concentrating spore settlement in appropriate microsites, leading to an increased success rate of gametophyte development and fertilization.

Frictional forces near the seafloor alter the flows produced by currents and waves, which results in a boundary layer of reduced velocity near the bottom (Gaylord et al. 2012; see figure 4.8). Turbulent eddies in the water column bring the neutrally buoyant spores into contact with the boundary layer and enable them to be in a position to settle (Gaylord et al. 2002, 2004, 2012, Denny and Gaylord 2010). The forces acting on spores, however, are complicated. Vertical mixing determines whether or not spores reach the bottom and how long they remain in the water column, and horizontal currents affect how far they are carried after release (Gaylord et al. 2002, 2007, Gaylord et al. 2006a, 2006b). Under most circumstances, it is probable that the vast majority of spores that reach the bottom do so within around 100 m of source plants (Gaylord et al. 2006b).

Dispersal distances are modulated by the size and reproductive output of forests, the duration of spore viability, and current speeds. In several experiments, for example, Gaylord et al. (2006b) tested dispersal around mature giant kelp plants that were outplanted on concrete blocks and acted as a source of spores, and checked for spore settlement on microscopic slides placed up to 120 m away in four directions. From these experiments, supported by modeling, they found that the modal dispersal distance from spores released from solitary plants was around 1 m and, therefore, the combination of dispersal kernels from many plants in a kelp forest causes high settlement within the forest. They concluded that this effect probably magnifies the strong density dependence seen in kelp forest dynamics (Reed 1990b, Reed et al. 1991). Clearly, turbulence and wave-driven processes dominated local spore transport, and currents dominated longer dispersal.

Gaylord et al. (2002) distinguished between "dispersal potential" and "colonization potential." The distinction is the difference between "getting there" and "being effective." Dispersal potential is indexed by the spatial scales over which propagules are carried, which is considerably different from the spatial scales over which successful establishment is possible. Spore clouds usually go through turbulent stirring and stretching after release and later undergo mixing through horizontal transport by orbital wave motion (Gaylord et al. 2012). Given a discrete source of propagules, either from a forest or from isolated individuals, there is a dilution effect and therefore a great decline in the density of spores with increasing distance. This greatly diminishes the chances they will settle in densities great enough to achieve the proximity required for fertilization to occur. However, multiple releases of spores over short time periods may increase the density of spores to the necessary densities (Reed et al. 1997). Gaylord et al. (2006b) showed that colonization via fertilization and recruitment can occur at least 1 km from the nearest source plants. This is a product of current speed, the size of forests and therefore spore supply, and the duration of viability of spores and gametophytes. However, dispersal and recruitment within kelp forests are much more probabilistic events.

## SEED BANKS

The shower of spores from giant kelp can provide a "bank of microscopic forms" (Chapman 1986) on the seafloor, but there is some debate about several aspects of its nature and longevity in a kelp forest, including whether it is truly analogous to a terrestrial seed bank. The term "seed bank" is well known and long established in terrestrial ecology. It starts as a "seed rain" as an inoculum for populations. "The store of seeds buried in the soil (the seed bank) is composed in part of seeds produced in the area and partly of seeds blown in from elsewhere"; seeds are in a dormant condition and they await adequate conditions before they germinate, which may take many years (Harper 1977). Analogies with marine seaweeds are obvious. There is no uncertainty about whether or not massive numbers of microscopic stages of marine plants can occur in marine habitats, but there are questions about their spatial variability, duration, and which microscopic stages are capable of delayed development and can still go on to produce viable 2N plants (see reviews by Hoffmann and Santelices 1991, Carney and Edwards 2006, Schiel and Foster 2006). If "storage" is simply defined as delayed or prolonged development, then some storage of microscopic stages undoubtedly occurs in large brown algae, but it is different from terrestrial seed banks. In annual seaweeds, prolonged development of microscopic stages is crucial to population dynamics because macroscopic plants may be absent for several months of the year (Dayton 1973, McConnico and Foster 2005, Schiel and Thompson 2012), while in perennial species, there is frequent large-scale loss of plants from grazing, severe water motion, and other abiotic stresses (Dayton et al. 1992, Underwood 1999). Unlike seeds, however, large brown algal propagules have no protective structures, although a possible exception is Saito's (1975) observation of *Undaria pinnatifida* gametophytes forming thick-walled "resting stages" at high temperatures. Furthermore, kelp propagules have no known innate dormancy, are far more transitory than most seeds, and the surrounding medium offers none of the protection afforded seeds in terrestrial soils. Rather than true dormancy, algal propagules simply appear able to delay or prolong development when conditions are unfavorable. The abiotic requirements for gametophyte development, gametogenesis, and sporophyte production in kelp were discussed in Chapters 2 and 3. Here we consider studies that explicitly tested delayed development within the context of assessing potential banks of microscopic stages.

Kelp gametophytes in culture can live for many months (possibly years; see Gametophytes in Chapter 2) and then complete gametogenesis if returned to optimal conditions, and the growth of microscopic sporophytes can also be delayed. However, the evidence of this occurring in nature is mostly conjectural. For example, disturbances from ENSO events often remove virtually all of the adult *Macrocystis* plants from sites. In sites at the extreme southern limit of *Macrocystis* distribution in Mexico, Ladah et al. (1999) found no adult plants over a period of at least 7 months from autumn (October 1997) onward during El Niño conditions and yet new recruits appeared in the following summer. Because the nearest adult populations were believed to be >100 km away, they surmised that recruitment was from resident micro-stages. However, Ladah and Zertuche-González (2004) found that during this El Niño, some *Macrocystis* adults survived in deeper water (25 m), although their surface fronds died down to 15 m depth, and may have been the source of subsequent recruitment.

Most of what is known about delayed development in giant kelp is from laboratory experiments, which have produced varying results. As discussed in Chapter 3, it has long been known that gametophytes can survive and grow for long periods in culture. However, there have been several tests to determine if these stages will have continued growth and development after being exposed to initially poor conditions of light and nutrients and then returned to adequate conditions. Both Kinlan et al. (2003) and Ladah and Zertuche-González (2007) found that growth of gametophyte stages could be delayed for 1–2 months when put into low-light and nutrient-poor conditions. However, when gametophytes were returned to normal light and adequate nutrients, they grew but did not complete gametogenesis and go on to produce sporophytes. Carney and Edwards (2010) put spores into low-light and low-nutrient conditions (10 μm photons m$^{-2}$ sec$^{-1}$, <1 μmol L$^{-1}$ of nitrate) and after 74 days transferred them to a series of treatments of adequate light and nutrients, which included nitrate, phosphate, trace elements, and vitamins. They found that growth occurred, although it was slow, when no nitrate was present, but that all nutrients were required for sporophyte production. Furthermore, when good conditions became available, sporophytes were produced within 14–24 days of receiving nutrients, compared to 54–96 days for nondelayed gametophytes. They concluded that one- to two-cell gametophytes of giant kelp can delay vegetative growth and reproduction for several months during low nutrient periods. They argued that the difference between their results and those of Kinlan et al. (2003) and Ladah and

Zertuche-González (2007) was that delayed gametophytes in their experiments had first been exposed to good conditions before delay was introduced with poor conditions, which might have affected their reproductive potential. Later experiments by Carney (2011) and Carney et al. (2013) corroborated earlier experiments in delayed development for the microscopic stages of several kelp species.

Laboratory studies provide some interesting insights into potential mechanisms of delayed growth and development of gametophytes and microscopic sporophytes, and offer clues about how these early life stages respond and survive in natural field conditions. For example, Deysher and Dean (1984, 1986b) found that these stages were relatively short-lived in the field. They found that densities of microscopic sporophytes on experimentally outplanted substrata seeded with *Macrocystis* spores peaked at around 40 days and then declined. The studies of Reed et al. (1997) further clarified many of the complexities relating to potential dormancy of microscopic life stages in natural conditions. *Macrocystis* can reproduce year-round and they found a high degree of reproductive synchrony during around half of their sampling periods over 2 years. These episodes of spore release varied from 1 week to 2 months. They placed rocks into a kelp forest and found no differences in recruitment on these and control rocks that had been in the kelp forest throughout the experiment. They concluded that *Macrocystis* had little capacity for dormancy and the vast majority of recruits come from recently settled spores. The work of Reed et al. (1997) does not preclude the possibility of some delayed development, but this is most likely unimportant when adult plants are nearby and actively releasing spores. Furthermore, their previous experiments on settled spores outplanted to the field showed that the vast majority of gametophytes died during their first week after settlement (Reed et al. 1988, 1994) and that only embryonic sporophytes, and not gametophytes, were found after 3 weeks. However, Carney et al. (2013) made the case for delayed development of gametophytes in the field. They cleared substrata within kelp forests by bleaching the bottom, thereby removing all algae, including their microstages. They found that this delayed kelp recruitment by around 2 months relative to controls. From these results, previous laboratory results on delayed development in kelp gametophytes, and population genetic analyses, they argued for "gametophyte banks" lasting at least 2–7 months. Although well argued, their conclusions were inferential and not based on direct observations of micro-stage development in the field. It is possible, however, that given particular conditions, any micro-stage may undergo delayed development and produce viable recruits.

Given the numerous intrinsic and extrinsic factors that affect early life stages, including grazing, scour, sediment, temperature stress, shading, and overgrowth (Vadas et al. 1992), the ability of micro-stages to persist for very long periods in adverse benthic conditions is probably not great. However, there remain several gaps in our understanding of how long these microscopic stages may persist in nature, the relative contributions of older versus younger stages to juvenile plant abundance, and, after disturbances of various degrees, how these contribute to population replenishment and persistence in such variable environments. Direct field-based observations and perhaps new genetic techniques for identifying species-specific microscopic stages may assist in this regard (Fox and Swanson 2007). In the meantime, it is not particularly fruitful to pursue inappropriate analogies to terrestrial seed banks but to simply consider these algal stages as what they are, microscopic plants.

## GENETIC STRUCTURE OF POPULATIONS

Because the vast majority of giant kelp spores fall within tens of meters of adult plants and there is a long tail to the dispersal distances of spores, populations are likely to be composed of a combination of plants originating from inbreeding, crossbreeding within populations, and crossbreeding among discrete kelp patches. Knowledge about the genetic structure of populations can therefore shed light on the relative probabilities of these events within kelp patches and help corroborate empirical and theoretical models on dispersal and population structure. The development of "DNA fingerprints" (Coyer et al. 1994, Macaya and Zuccarello 2010a) and microsatellite markers (Alberto et al. 2009) for *Macrocystis* is beginning to lead to a greater understanding of how reproduction biology and dispersal of kelp spores influence the genetic structure of populations over different scales.

Within giant kelp populations, Raimondi et al. (2004) evaluated the costs of self-fertilization. These costs result from inbreeding depression and can be seen in reduced survival, poor growth, and low fecundity. They postulated that there was a great potential for self-fertilization in giant kelp because the vast majority of spores fall near parent plants. To test inbreeding effects, they established laboratory populations with three mixtures of different ratios of sibling to nonsibling spores from 20 sporophytes. These yielded selfed zygote cultures of 100%, 50%, and 5%. They then used similar mixtures seeded onto natural rocks placed into a field site. They found there was a linear decrease in production of viable zygotes across the treatments, with the outcrossed mixture producing 40% more than the 100% mixture. However, there was no evidence of self-incompatibility. In the field trial, the densities of sporophytes were sampled three times over a year. They found that sporophyte abundance varied inversely with the level of self-fertilization. After 1 year, when plants were >10 m tall, densities were around 40% and 20% lower in the two selfing treatments (100% and 50%, respectively) compared to the outcrossed (5%) treatment. Furthermore, only 15% of the fully selfed plants became reproductive, compared to around 50% in the other two treatments, which also yielded 10 times greater sorus area than in the fully selfed plants. Gaylord et al. (2006b) estimated that selfing levels in the order of 10% may occur, based on dispersal distance and adult spacing. Raimondi et al. (2004) concluded that the costs of self-fertilization are extremely high and may be exacted when spore dispersal is limited, such as on isolated reefs in low current flow. They speculated that localized extinction of giant kelp is a potential result of inbreeding, but this has not yet been considered in the dynamics of kelp populations.

In a study along the coast of southern California, Alberto et al. (2010) tested the effects of geographic distance and habitat continuity on population genetic differentiation in giant kelp populations, using highly variable microsatellite markers for polymorphic loci developed by Alberto et al. (2009). They specifically tested the hypothesis of isolation by distance (IBD; Wright 1943), arguing that giant kelp populations were good models for this type of study because of the almost linear array of kelp patches at various distances apart along the coastline in the study area of the Santa Barbara Channel. They used GIS data of kelp surface canopies to estimate habitat continuity and geographic distances among pairs of sites. They then generated a matrix of geographic distances between all pairs of sites, and a matrix of kelp area per kilometer of coastline between all pairs of sites as an estimate of habitat continuity. They sampled 50 giant kelp plants from each of nine sites and genotyped them for 12 microsatellite loci. They found a high level of genetic diversity in the Santa Barbara Channel populations. The number of alleles per locus ranged from 7 to 50, and within a single population from 2 to 27. They found there was a steep increase in genetic differentiation in the eastern part of the study area where there was low kelp abundance at two sites. The best model of genetic distance and habitat variables showed that genetic distance increased as geographic distance increased (figure 4.9), and that genetic distance decreased as habitat continuity increased. Therefore, there were counteracting effects of distance and habitat continuity on genetic distance of the giant kelp populations. Using IBD simulation modeling, they found that the mean dispersal distance ranged from 1.8 to 2.9 km and from 0.6 to 0.7 km at high and low current velocities, respectively, which was consistent with the empirical results of Reed et al. (2004), and the physical transport model of Gaylord et al. (2006b), indicating that the scales of connectivity up to around 1 km were representative of most kelp forests. Their calculations of effective population size were only 0.2–9.6% of the empirically derived population size, which they argued agrees with those of other marine organisms with high rates of fecundity and juvenile mortality and high variance in reproductive success. They concluded that both geographic distance and habitat continuity are important factors operating independently to influence genetic distance and that the discrete kelp patches serve as stepping stones in the IBD model. They highlight that the consequences of habitat fragmentation on the metapopulation structure of giant kelp are likely to be severe.

Alberto et al. (2011) extended this work on the isolation by distance model by testing how much additional variability in genetic differentiation could be accounted for when transport time was added as a predictor. They examined the degree to which gene flow depended on seasonal ocean circulation, the asymmetrical nature of flows between sites, and the direction of connectivity based on an oceanographic model and genetic assignment tests. They noted that over demographically relevant scales, ocean currents tended to flow primarily in one direction, which affects gene flow between sites. They reanalyzed the data set of Alberto et al. (2009, 2010). They then used a Regional Oceanic Modeling System (ROMS) model for the Southern California Bight (Dong and

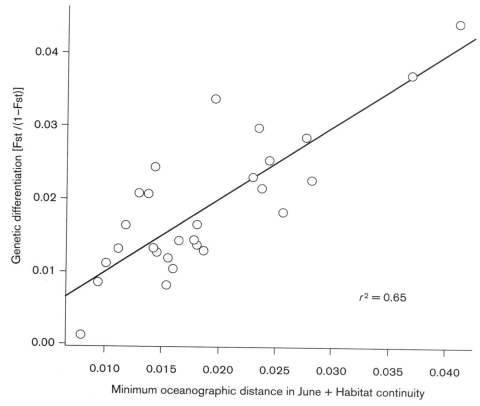

FIGURE 4.9

Two-factor regression model using minimum transport time and habitat continuity, as independent variables, to predict genetic differentiation in *Macrocystis* in the Santa Barbara Channel, southern California.

NOTES: Fst = a measure of genetic differentiation. Minimum oceanographic distance + Habitat continuity = a combination of minimum distance among pairs of sites and kelp forest area per kilometer of linear coastline.

SOURCE: From Alberto et al. (2011), reprinted with permission from Blackwell Publishing Ltd.

McWilliams 2007, Dong et al. 2009) to calculate transport times for a Lagrangian particle (i.e., one borne by water movement) to travel from a source to a destination site, using all pairs of 135 nearshore kelp patches.

They found that transport in the Santa Barbara Channel was primarily westward except during winter when around half of the flows were in the opposite direction. Combining their models and genetic analyses, they found that by incorporating oceanographic transport times between giant kelp patches, they nearly doubled the amount of genetic variation accounted for by geographic distance. One of their interesting conclusions was that they found little support for the notion that rafting sporophytes were a major source of dispersal. This was based on them finding the greatest explanatory power using summer transport times, especially during June when plants are highly reproductive but also when dislodgement of giant kelp is lowest. They also found that habitat continuity, one of the two predictors of Alberto et al. (2010), added only 6% more explanatory power to the best-fit model, although they also noted that transport times incorporate habitat features such as direction and duration of transport. Overall, they found strong support for ocean currents being a primary driver of population connectivity of giant kelp. This is not surprising, given the results of an increasing number of empirical and modeling studies on spore transport and recruitment dynamics, but this genetic verification of underlying structure and mechanisms is an important advance in understanding the population dynamics of giant kelp.

On a wider geographic scale, the relatedness of populations offers intriguing insights into the spread and colonization of giant kelp over long time periods. For example, Coyer et al. (2001) examined not only the evolution of *Macrocystis* (discussed in Chapter 1) but also the relatedness of populations in both hemispheres, using the noncoding rDNA internal transcribed spacer regions (ITS1 and ITS2). From their analyses, they inferred a northern-to-southern hemisphere dispersal of giant kelp because northern individuals were more diverse and had paraphyletic clades, but those in the south were less diverse and formed a monophyletic clade. They surmised that dispersal of giant kelp between hemispheres involved populations around Baja California, Mexico, and / or those around Catalina Island off southern California as the bridging or intermediate sources. They found little divergence in ITS sequences in southern hemisphere plants despite the vast distances between populations, probably due to kelp rafting in the strong flows of the West Wind Drift. They also found a strongly resolved Australia / Tasmania clade within the monophyletic southern hemisphere clade, indicating less gene flow between there and the rest of the southern hemisphere. Transport away from the coast in this region may be limited because of coastal currents and very low nitrate levels for sustaining kelp rafts, resulting in reduced gene flow between Tasmania and New Zealand across the Tasman Sea (Edgar 1987).

Knowledge about southern hemisphere populations and relatedness was expanded greatly by the extensive studies of Macaya and Zuccarello (2010a, 2010b) who examined the genetic structure of *Macrocystis* across the south Pacific region. They analyzed 770 samples using mitochondrial DNA and single-strand conformation polymorphism. Across the entire region they found seven haplotypes, five of which were along the 4800 km Chilean-Peru coastline (figure 4.10). One genetic break in haplotypes occurred at 42° S on Chiloe Island, Chile, which corresponded with a known biogeographic boundary. However, another genetic break at 33° S in central / northern Chile did not correspond to a known boundary for brown algae, although it was in line with some other marine taxa. However, this latitude is also the demarcation of ecomorphs, with *Macrocystis 'pyrifera'* occurring to the south and *M. 'integrifolia'* to the north, representing potential adaptations to warmer environmental conditions (Graham, Vásquez, et al. 2007, Demes et al. 2009).

Their results shed light on several potential processes. Long-distance dispersal via floating kelp is probably important over long time periods. They also noted that during the last glacial maximum (LGM, See Chapter 1) broad areas of southern Chile, including the fiords, were covered by ice sheets. In line with the conclusions of Fraser et al. (2009,

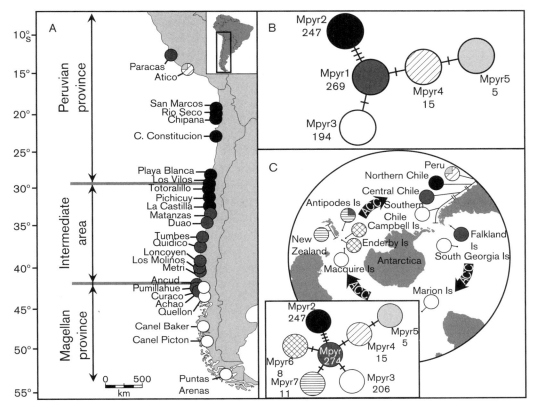

FIGURE 4.10

Distribution of giant kelp genetic diversity (haplotypes) along the Southeastern Pacific coast, subantarctic islands and New Zealand.

(A) Southeastern Pacific coast showing location of collecting sites. Different shadings represent different haplotypes designated as Mpyr1-5 in (B). Gray lines represent biogeographic breaks. Province names differ from those used in Chapter 5.

(B) Statistical parsimony network showing mutational pathways among haplotypes. Cross lines represent number of single-site substitutions. Numbers below haplotypes are a measure of haplotype frequency.

(C) Collecting sites and haplotypes in the subantarctic islands and New Zealand with additional haplotypes Mpyr 6-7. Inset is the parsimony network using the entire data set with cross lines and numbers below haplotypes as in (B).

ACC = Antarctic Circumpolar Current.

SOURCE: From Macaya and Zuccarello (2010b).

2010) for the large fucoid *Durvillaea antarctica*, they speculated that shared genotypes of giant kelp between southern Chile and other SE Pacific regions may reflect ice scour during the LGM. The distribution of one haplotype corresponds precisely to the extent of the Patagonian ice sheet and subsequent recolonization from transoceanic sources. They surmised that the Antarctic Circumpolar Current may have facilitated recolonization after the LGM through drifting kelp rafts.

With the advent of new genetic markers and new techniques, our knowledge of the genetic structure of kelp populations is still at an early stage. We will no doubt add to this emerging picture in the future, as new markers and techniques are developed and as we combine genetic, ecological, and oceanographic data of giant kelp forests across the vast regions they occupy.

---

This chapter on demography, dispersal, and connectivity concludes our discussion of the population biology of giant kelp. In going through the literature, it was evident that we have made enormous strides over the past 25 years in understanding kelp forest dynamics and the intimate relationship they have with the physical environment. The underlying themes of continual disturbances and replenishment come through strongly, which is unusual for a plant so large and dominant of its habitat. When considering its population biology across the spatial and temporal domains it occupies, there truly is no terrestrial analogue of *Macrocystis*.

# PART II

# THE GIANT KELP ECOSYSTEM

# 5

# GIANT KELP COMMUNITIES

To those whose experience with kelp beds has been confined to observations
only of the surface, all beds look very much alike, but divers having seen
Macrocystis in a number of localities state that no two beds have quite the
same aspect. The differences, indeed, are so pronounced from place to place
that one finds it difficult to diagram a generalized kelp bed.

—Dawson et al. (1960)

These species have evolved along unique paths, reminding us again that
extrapolation of results from one geographic area to another is unwarranted
even when physical conditions are similar.

—Moreno and Sutherland (1982)

Areas with giant kelp include a multitude of other species. Because giant kelp is usually dominant, however, is the most visually obvious species, and commonly has by far the greatest biomass, such areas are called giant kelp communities. They are also called giant kelp "forests" or kelp "beds." Between these terms we prefer "forests" to "beds" because it highlights the vertical structure in the water column and the surface canopy. The abundance of giant kelp varies considerably across areas, and adults may become temporarily absent for many reasons, such as the population being annual, or being removed by grazers, storms, or other episodic oceanographic events. We consider such temporal variation a characteristic of giant kelp communities that does not necessarily represent "phase shifts," "alternative stable states," or some other fundamental community change. "Community" can also imply a particular suite of species consistently associated with the dominant species, but this not the case even at local scales for *Macrocystis*, as the quote above from Dawson et al. (1960) indicates. Furthermore, given that *Macrocystis* occurs in both hemispheres across many different biogeographic provinces, there is clearly no single community of giant kelp across the regions it occupies. For example, considering only the South American coastline, giant kelp is distributed over two biogeographic provinces and eight ecoregions (figure 1.5) defined largely by the presence of distinct floras and faunas (Spalding et al. 2007).

While there are various arguments for and against the value of "community" as an ecological unit (e.g., Underwood 1986, Lawton 1999, Simberloff 2004), it seems to us that community studies are essential in providing a descriptive record of *Macrocystis* and its associated populations in one or more places at one or more times, and commonly including physical/chemical assessments, natural history observations, and experiments that provide insights into species' interactions. The accumulated information from such studies has been an enormous contribution to kelp forest science by providing the basis for model building and hypothesis testing. Moreover, giant kelp communities can vary at a scale of a few kilometers, so knowledge of local communities and the factors that affect them is essential for effective management. This has proved to be the case in many areas of the world where the status of giant kelp forests and their associated species have underpinned coastal management because of their natural, recreational, and societal value. Here we will use "community" in a purely descriptive or ecographic sense indicating the populations that occur in particular places and times, as defined by the investigator, but we knit the various studies together, along with broader-scale efforts, to examine generalities. Primary productivity and energy flow are also discussed, along with "who eats what" food webs; an examination of the wider causes and consequences of trophic interactions is reserved for other sections.

### A "COMPOSITE" COMMUNITY

The best-described *Macrocystis* forests are those that occur in the subtidal zone and are usually composed of *Macrocystis 'pyrifera.'* Although a very wide range of species can occur with giant kelp, we are unaware of any macroorganisms whose associations with giant kelp are obligate; they can be found in other kelp communities and on subtidal rocky reefs devoid of large brown algae (e.g., Pequegnat 1964). It is important to note, however, that most descriptions are based on what is visible to the naked eye, and we know little about a potentially rich microscopic community associated with giant kelp.

With those caveats in mind and to give a general idea of the structure of a kelp forest, figure 5.1 illustrates a "composite" California *Macrocystis* forest with emphasis on the large, visually obvious kelps that provide much of the structure of the community and for which we have the most descriptive information. The figure divides the community into habitats where particular organisms typically co-occur in California. It is a composite because of local variation in species' distributions and because the geographic ranges of some the organisms, such as sea otters (*Enhydra lutris*) and elk kelp (*Pelagophycus porra*), do not presently overlap. The figure indicates the potential complexity of the community, with multiple layers of vegetation (e.g., Dawson et al. 1960), over 50 species of fishes that commonly segregate into various habitats (e.g., Quast 1971b, Ebeling et al. 1980), and numerous invertebrates also found in particular habitats within the forest. North (1971a) listed 130 species of plants and almost 800 species of animals associated with giant kelp forests in southern California and northern Baja California, Mexico, although all

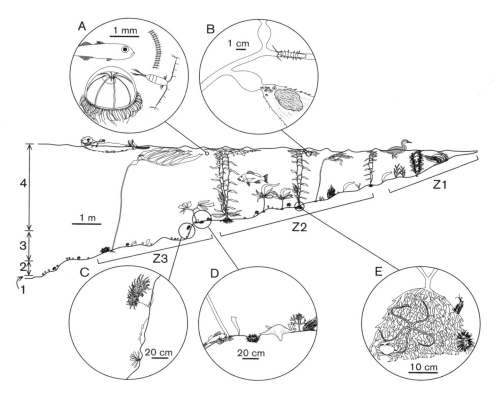

FIGURE 5.1

Cross section showing the inhabitants of a generalized *M. 'pyrifera'* forest in California. The numbers to the left designate vegetation layers: (1) small filamentous species and nongeniculate coralline algae; (2) bottom canopy plants including foliose red and brown algae, and geniculate corallines; (3) understory canopy kelps; (4) mid-water and surface canopy kelps such as *Macrocystis*, *Nereocystis*, *Pelagophycus*, and *Egregia*, and the fucoid *Stephanocystis*. Three broad zonal associations are common: Z1 inshore, Z2 within, and Z3 below the distribution of giant kelp. Various subhabitat associations are indicated within the circles: A = plankton in the water column (phytoplankton and zooplankton); B = animals associated with the surface of *Macrocystis* and other seaweeds including bryozoans, spirorbid polychaetes, and isopods (shown) as well as other crustaceans and molluscs; C = organisms characteristic of vertical surfaces (primarily sessile animals such as sponges, tunicates, bryozoans, sea anemones, and solitary corals); D = organisms characteristic of horizontal surfaces (sea stars, sea urchins, benthic fishes, algae); E = animals in giant kelp holdfasts (brittle stars, crustaceans, polychaetes [shown on the outside but which occupy spaces within the holdfast]).

NOTES: Various fishes occur in the water column, and birds (e.g., cormorants), seals, and sea otters forage in the water column and at the sea surface. Some of the organisms depicted here do not co-occur at any one site.

SOURCE: From Foster and Schiel (1985).

of them are not found together in any single kelp forest. For example, Pearse and Lowry (1974) listed 369 species of algae and invertebrates in central California compared to North's (1971a) total of 631 species of the same groups in all of southern and Baja California. The fauna in giant kelp holdfasts, the subject of the first kelp forest research using air supply diving in California (Andrews 1945), can include over 150 taxa (Ghelardi 1971). A few birds and mammals forage in the community, including cormorants, harbor seals, sea otters, and large elephant seals. California gray whales may cruise through forests feeding on crustaceans. Forests are embedded in a planktonic assemblage of mostly microscopic organisms, some of which are early stages in the life histories of larger members of the community. Sampling at one site in central California has also revealed a diverse, temporally variable bacterial assemblage on giant kelp blades. The phyla were similar to those in the surrounding water but distinct at lower taxonomic levels. Several of the blade bacteria were similar to those known to inhibit the settlement of larvae and growth of other bacteria (Michelou et al. 2013).

Subtidal giant kelp communities generally occur within a fairly narrow depth range. Even if suitable substrata are available, *M. 'pyrifera'* usually does not occur shallower than about 5 m and deeper than around 20 m. Shallower and deeper areas, as well as hard substrata under the canopy of giant kelp, are often inhabited by other kelps, many of which provide subcanopies or even surface canopies. We illustrate (figures 5.2 and 5.3) and discuss below several of these kelp forests by biogeographic provinces as defined by Briggs (1974) and further elaborated by Spalding et al. (2007). The forests discussed were selected on the basis of available data and to illustrate the range of variation within and among provinces. The emphasis is on large, abundant benthic organisms.

## VARIATION BY BIOGEOGRAPHIC PROVINCE

WARM TEMPERATE NORTHEAST PACIFIC · *Northern Pacific Baja California to Southern California*

Giant kelp forests are a ubiquitous feature of shallow rocky reefs along the 1000 km of coast and offshore islands from their southern limit at Punta San Hipólito, Bahía Asunción, Baja California (26.90° N, 113.98° W) to Point Conception, California (34.48° N, 120.56° W; figure 1.5). The southern limit is 300 km north of the southern end of the province at Bahía Magdalena, but *Macrocystis* occurs well north of the northern provincial limit. Kelp forests at their southern limit in the northern hemisphere are noteworthy because they occur in an increasingly subtropical environment where temperatures typically range from 14°C to 22°C (Hernández-Carmona et al. 2001). They generally are most prominent on the south–southeast sides of rocky points or shallow coastal shelves where upwelling decreases water temperatures and increases nutrients (Dawson 1951, Zaytsev et al. 2003). Low nutrients and severe storms associated with El Niño events depress upwelling and can increase storms, resulting in the complete loss of *Macrocystis* sporophytes (Edwards 2004). Moreover, the most southerly forests are

## A. Bahía Asunción

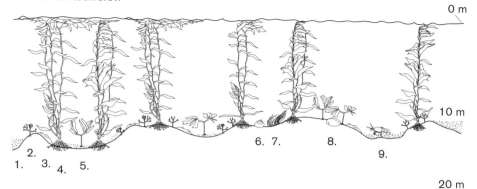

0 m

10 m

6. 7.    8.

9.

2.

1.   3.   4.   5.

20 m

1. Sand
2. Geniculate corallines (BC)
   *Dictyota flabellulata* (BC)
3. Sessile Animals
4. *Macrocystis 'pyrifera'* (SCK)

5. *Eisenia arborea* (UK)
6. *Megastraea undosa* (gastropod)
7. *Phyllospadix* (surf grass)
8. *Haliotis* spp. (abalone)
9. *Panulirus interruptus* (lobster)

## B. Point Loma

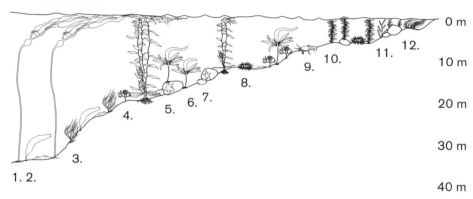

0 m

11.   12.

9.   10.

10 m

8.

20 m

5.   6.   7.

4.

30 m

3.

1. 2.

40 m

1. *Pelagophycus porra* (SCK)
2. *Laminaria farlowii* (UK)
3. *Lophogorgia chilensis* (gorgonid)
4. Geniculate Corallines (BC)
   *Rhodymenia* spp. (BC)
5. *Patiria miniata* (sea star)
6. *Pterygophora californica* (UK)

7. *Haliotis* spp. (abalone)
8. *Strongylocentrotus* spp. (sea urchins)
9. *Panulirus interruptus* (lobster)
10. *Egregia menziesii* (SCK)
11. *Stephanocystis osmundacea* (SCF fucoid)
12. *Phyllospadix* spp. (surf grass)

FIGURE 5.2

Distribution of conspicuous plants and animals found in five giant kelp forests (A–E) across the range of *Macrocystis* in the northern hemisphere.

BC = bottom canopy; SCF = surface canopy fucoid; SCK = surface canopy kelp; UF = understory fucoid; UK = understory kelp.

NOTES: The symbols in Panel A for sand (1), sessile animals (3), and *Macrocystis 'pyrifera'* (4) are generic to all figures unless otherwise noted. Organisms and horizontal axes are not to scale. Depth distribution is indicated on the right.

## C. Arroyo Quemado

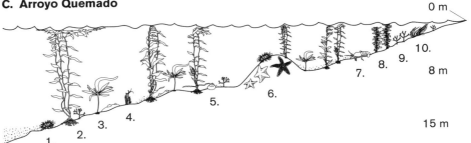

1. *Strongylocentrotus* spp. (sea urchins)
2. *Rhodymenia californica* (BC)
   *Chondracanthus corymbiferus* (BC)
3. *Pterygophora californica* (UK)
4. *Diopatra ornata* (polychaete)
5. *Megastraea undosa* (gastropod)
6. *Patiria miniata* (sea star)
   *Pisaster giganteus* (sea star)
7. *Panulirus interruptus* (lobster)
8. *Egregia menziesii* (SCK)
9. *Corallina chilensis* (geniculate coralline)
10. *Phyllospadix torreyi* (surf grass)

## D. Stillwater Cove

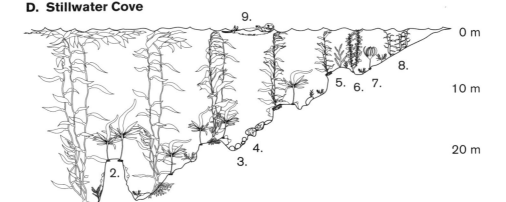

1. *Calliarthron* spp. (BC)
   *Plocamium cartilagineum* (BC)
2. *Pterygophora californica* (UK)
3. Boulders
4. *Patiria miniata* (sea star)
5. *Stephanocystis osmundacea* (SCF fucoid)
6. *Egregia menziesii* (SCK)
7. *Laminaria setchellii* (UK)
8. *Macrocystis 'integrifolia'* (SCK)
9. *Enhydra lutris* (sea otter)

FIGURE 5.2 *continued*

## E. British Columbia

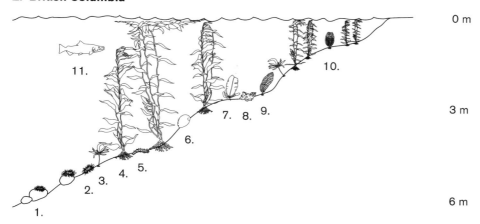

1. Boulders
2. *Strongylocentrotus* spp.
   (sea urchins)
3. *Pterygophora californica* (UK)
4. *Macrocystis 'integrifolia'* (SCK)
5. *Parastichopus californicus*
   (sea cucumber)

6. *Haliotis kamtschatkana* (abalone)
7. *Alaria marginata* (UK)
8. *Ulva* sp. (BC)
   Foliose red algae (BC)
9. *Saccharina groenlandica* (UK)
10. *Costaria costata* (UK)
11. Salmon

FIGURE 5.2 *continued*

very near or beyond the southern limit of several key grazers and predators that have been implicated in having large effects on community structure in southern California. These include red and purple sea urchins (*Strongylocentrotus franciscanus* and *S. purpuratus*; Rogers-Bennett 2013) and one of their predators, the labrid fish *Semicossyphus pulcher* (California sheephead; Eschmeyer et al. 1983). These forests are also at the southern limit of the historical range of sea otters (*Enhydra lutris*; Kenyon 1969). *Eisenia arborea* is the dominant understory kelp (figure 5.2A), and geniculate corallines (*Amphiroa* spp. and *Bossiella orbigniana*) and predominantly tropical / subtropical fleshy brown algae (e.g., *Dictyota* spp., *Padina* spp.) are very abundant on the bottom (Hernández-Carmona et al. 2001, 2011). The forests support fisheries for the abalone *Haliotis fulgens* and *H. corruguta*, and the spiny lobster *Panulirus interruptus* (Ponce-Díaz et al. 2003). The grazing mollusc *Megastraea undosa* and the spotted sand bass or cabrilla, *Paralabrax maculatofaciatus*, are abundant (G. Hernández-Carmona, pers. comm.). A forest near the southern limit at Islas San Benito can be seen in a poetic short film ("Mov. IB," by Alfredo Barroso; www.youtube.com/user/azuloceano).

The Point Loma kelp forest located off San Diego, California (32.69° N, 117.26° W) is one of the largest in the world, historically covering around 15 km² but now a maximum of around 6 km² (Foster and Schiel 2010). It is also one of the best-studied kelp forests, beginning with surveys by Turner et al. (1968) and continuing with numerous studies by

Dayton, Tegner and colleagues (e.g., Dayton et al. 1999). There is considerable variation within the forest both spatially and temporally, but general patterns are summarized along an idealized transect (figure 5.2B). The forest occurs on a broad, gently sloping mudstone–sandstone terrace with pockets of sand, cobbles, and boulders. Giant kelp is most abundant between depths of 6 and 25 m. The feather boa kelp *Egregia menziesii* and the fucoid *Stephanocystis* (=*Cystoseira*) *osmundacea*, which also form surface canopies, and surf grass (*Phyllospadix torreyi*) are abundant inshore of giant kelp. The kelps *Pelagophycus porra* and *Laminaria farlowii*, and the gorgonian *Lophogorgia chilensis*, are abundant offshore in deeper water. One of the commonest kelps of the coast of California, *Pterygophora californica*, forms dense patches beneath giant kelp. Bottom cover is dominated by geniculate coralline algae and fleshy red algae (*Rhodymenia* spp.). The sea anemone *Corynactis californica*, the solitary coral *Balanophyllia elegans*, the solitary tunicate *Styela montereyensis*, and numerous sponges are common on the benthos. The whelk *Kelletia kelletii*, bat stars (*Patiria* (=*Asterina*) *miniata*), and sea urchins (*Strongylocentrotus* spp. and *Lytechinus pictus* (=*anamesus*)) are common mobile invertebrates. Abundant fishes include blacksmith (*Chromis punctiipinnis*), señorita (*Oxyjulis californica*), California sheephead (*S. pulcher*), kelp bass (*P. clathratus*), and black-eyed gobies (*Coryphopterus nicholsii*).

The kelp forest off Del Mar, 25 km north of San Diego (not depicted in figure 5.2), well illustrates local variation in kelp forest community structure. In contrast to Point Loma, the giant kelp stand at Del Mar occurs on low-relief rock at depths of 14–16 m surrounded by a sand bottom (Rosenthal et al. 1974). Understory kelps (*P. californica* and *L. farlowii*) and foliose algae (e.g., the brown *Desmarestia ligulata* and the red *Rhodymenia pacifica*) are relatively sparse, and most of the bottom is covered by nongeniculate coralline algae. Rosenthal et al. documented that the water column and benthos harbor 98 species of benthic macroinvertebrates and 38 species of fish. Red and purple sea urchins were found on rock mounds and boulders but were not observed to graze actively on the surrounding vegetation. The forest at San Onofre a few kilometers north of Del Mar further illustrates local community variation as it occurs on boulders and cobbles surrounded by sand (Larson and DeMartini 1984) and has experienced large changes in populations associated with El Niños.

Similar to Del Mar, the benthos along the mainland coast at the northern end of this province near Santa Barbara is characterized by low-relief mudstone reefs interspersed with extensive sandy areas and occasional rocky outcrops. The area is relatively protected from large swells by Point Conception to the northwest and the Channel Islands to the south. The Arroyo Quemado kelp forest (34.47° N, 120.12° W), one of the Santa Barbara Coastal Long-Term Ecological Research sites (e.g., Reed et al. 2011), is representative of kelp forests on rocky reefs in the area (figure 5.2C). Descriptions from D. Reed (pers. comm.) indicate that the benthic organisms are similar to those described at Campus Point by Neushul et al. (1976), and the fish assemblage is similar to that at nearby sites such as Naples Reef described by Ebeling et al. (1980). *Macrocystis* occurs between 5 and 15 m depth. Extensive patches of sand occur above 5 m, interspersed with patches of the kelp *E. menziesii*, surf grass (*P. torreyi*), and geniculate coralline algae. The understory kelp *P. californica* and the fleshy red algae *Chondracanthus corymbiferus* and *R. californica* are common beneath giant kelp, as are sponges, tunicates, bryozoans, and the large tube-dwelling polychaete *Diopatra ornata*. Red and purple sea urchins, bat stars, the large sea star *Pisaster giganteus*, lobsters, and various molluscs are common. Kelp bass, various surfperches, rockfish, and California sheephead are the most abundant fishes. The forest terminates in a sand bottom at the seaward edge.

In contrast to the mainland, waters around offshore islands in this province are generally clearer and there is a greater amount of high-relief rock, conditions favorable to the growth of giant kelp and other rocky subtidal organisms. The majority of information on island kelp forests comes from the Channel Islands off southern California. Santa Catalina Island, near the southern end of this island group (33.41° N, 118.45° W), is bathed in warm southern waters. The kelp forests on the leeward side of the island experience little upwelling (Zimmerman and Kremer 1984) and are inhabited by many species with southern, warm-water affinities (Neushul et al. 1967). Somewhat oddly, the kelp forests are bounded at their shallow end by the stipitate kelp *E. arborea*, and at their deep end by this same species and two other understory kelps, *L. farlowii* and *Agarum fimbriatum*, which can form dense beds ranging to >30 m depth (Foster and Schiel 1993). These understory kelps are rare beneath the dense giant kelp canopy, where the understory is dominated by foliose brown and red algae, and coralline algae (Foster and Schiel 1993).

Giant kelp forests in the northern Channel Islands are more diverse than those of the mainland and if suitable substrata are available, giant kelp extends into deeper water (e.g., 35 m at Anacapa Island; Neushul et al. 1967). Ebeling et al. (1980) suggested that the continuity of well-developed rocky reefs, clearer water, and the high density of macroalgae and invertebrates (important sources of fish food) covering the bottom on the islands contribute to the increased density, biomass, and diversity of fishes that occupy these offshore forests. The physical presence of the Islands and the confluence of cold northern and warm southern currents produce complex circulation patterns that result in considerable spatial variation in temperature that correlates with floral and faunal differences among sites arrayed along the Channel Islands (Neushul et al. 1967, Murray et al. 1980, Hamilton et al. 2010).

COLD TEMPERATE NORTHEAST PACIFIC · *Central California to Southeast Alaska*

This large marine province extends from Point Conception, the boundary between southern and central California, through the Aleutian Islands in Alaska. *Macrocystis* in two ecomorphs (*Macrocystis 'pyrifera'* and *M. 'integrifolia'*) occurs along the ca. 4000 km of the coastline from Point Conception to Kodiak Island, Alaska (57.5° N, 153.3° W). Provincial boundaries are associated with the Alaska and California Currents, which are branches of the Subarctic Boundary Current. The Alaska Current moves relatively cold water (although warm for Alaska) northeast, while California Current moves water south before becoming a largely offshore current at Point Conception (figure 1.5), producing a relatively abrupt change in temperature and circulation that defines the southern provincial boundary. Many species that are common south of Point Conception are rare to the north, including California sheephead and spiny lobster, whereas others such as rockfishes (*Sebastes* spp.) are more common and diverse to the north. There is a translocated population of sea otters (*Enhydra lutris*) at San Nicholas Island, but otters are otherwise rare south of Point Conception, at least currently. They do occur and are often abundant along central California and portions of British Columbia and Alaska (Riedman and Estes 1988). Depressed nutrient levels, low enough to affect giant kelp growth and survival, are common south of Point Conception and in combination with large waves during El Niños cause widespread mortality (review and data in Edwards and Estes 2006). In contrast, low nutrients are rare north of Point Conception but waves are larger, causing large seasonal and year-to-year changes in subtidal *M. 'pyrifera'* abundance between Point Conception and Año Nuevo Island (37.11° N, 122.34° W; Foster 1982, Foster and Schiel 1985, Reed et al. 2011).

This difference in wave action is likely to be the major cause of the shift from *M. 'pyrifera'* to the bull kelp *Nereocystis luetkeana* (figure 1.1) at exposed sites within the range of *M. 'pyrifera'* (Foster 1982, Harrold et al. 1988, Foster and VanBlaricom 2001). Bull kelp forests become ubiquitous and *M. 'pyrifera'* forests are generally absent as wave disturbance increases from Año Nuevo Island north to southeast Alaska (Foster and Schiel 1985). Bull kelp sporophytes are well adapted to mechanical stress (Koehl and Wainwright 1977), are annual, and are generally absent in winter when more frequent and larger waves occur. They also appear to be less affected than *Macrocystis* by sea urchin grazing (Pearse and Hines 1979, Foster 1982) so the reduction in giant kelp north of Año Nuevo Island may also be related to increased grazing in the current absence of sea otters.

Stillwater Cove (36.57° N, 121.95° W) has a well-studied kelp forest that is semi-protected from waves in Carmel Bay, central California (figure 5.2D). It and a few others in the Monterey region were the first to be observed by diving (Andrews 1945). Other descriptions are provided by Foster (1982), Reed and Foster (1984), and Konar and Foster (1992), and from data obtained in surveys done since 1977 by the subtidal ecology class at Moss Landing Marine Laboratories (Pearse et al. 2013). The conglomerate and sandstone bottom of the cove is a mosaic of plateaus and pinnacles surrounded by relatively flat rock or fields of small boulders. A stand of *M. 'integrifolia'* occurs from the low intertidal zone to a depth of 1 m (figure 5.2D). This gives way to *Egregia menziesii, Laminaria farlowii, L. setchellii, Stephanocystis osmundacea*, and patches of surf grass to around 7 m depth. *M. 'pyrifera'* occurs at depths between 2 and 20 m, terminating in a sand bottom at the mouth of the cove. Dense stands of the kelp *Pterygophora californica* occur in the understory. *Pterygophora* grows very large at this site (>1 m tall) and is particularly abundant on the tops of plateaus where it can produce a nearly closed canopy. Geniculate coralline algae, *Calliarthron* spp., form dense mats of entangled branches up to 10 cm tall on the bottom surrounding the *Pterygophora* canopy and adjacent areas, and over extensive areas of nongeniculate (encrusting) corallines. *Cryptopleura farlowiana* and other fleshy red algae can be abundant in spring, often as epiphytes on *Calliarthron*. Bryozoans, sponges, solitary corals, and sea anemones are common on steeply sloping vertical walls.

A diverse suite of mobile animals occurs throughout Stillwater Cove. Turban snails (*Tegula* spp.) are extremely abundant, grazing on the benthos and on giant kelp plants throughout the water column and at the sea surface, as they do at other sites around Monterey Bay (Watanabe 1984a, 1984b). High densities of small purple sea urchins are found in the coralline mats (Kenner 1992). Large sea urchins and abalone are rare, probably because this site lacks suitable crack and crevice refuges from sea otter predation. Sea star diversity is high but only the bat star is abundant. Fish are abundant, especially juvenile rockfish, adult blue and kelp rockfishes, various surfperches, and greenlings (*Hexagrammos* spp.). Gobies and sculpins are common on the bottom. Brandt's cormorants (*Phalacrocorax pennicillatus*) dive for fishes in the forest, western (*Larus occidentalis*) and other gulls rest and forage at the surface, and great blue herons (*Ardea herodias*) and snowy egrets (*Egretta thula*) walk on the thick kelp canopies hunting for small fish. Harbor seals (*Phoca vitulina*) forage for fish within the forest (review in Foster and Schiel 1985).

This site also illustrates well how variation in topography and water clarity at the scale of hundreds of meters can affect forest distribution. The mouth of the cove terminates in sand at around 25 m depth, but the western edge is a rocky point (Cypress Point ) that slopes into the clearer waters of Carmel Bay to depths of over 30 m. This deeper area and similar areas to the south along the Big Sur coast were described by Spalding et al. (2003) from scuba and remotely operated vehicle (ROV) surveys. Giant kelp occurs to around 30 m, bordered at its deeper edge by the large, annual brown alga *Desmarestia ligulata*, and beds of the kelps *Eisenia arborea* and *Pleurophycus gardneri*, the latter to a depth of 45 m. Many of the same species of red algae seen beneath giant kelp in shallower water also occur in this deep region.

The kelp forest at Sand Hill Bluff, near Santa Cruz (36.96° N, 122.13° W), is one of the most northerly *M. 'pyrifera'* forests in the province and, similar to other forests in this wave-exposed region (e.g., Point Santa Cruz), has a low-relief mudstone substratum interspersed with patches of sand (Yellin et al. 1977, Pearse and Hines 1979, Breda and Foster 1985). The substratum is similar to that of the Arroyo Quemado kelp forest in southern California described above, and also terminates in a sandy bottom at around 16 m depth. The contrast between forests at Sand Hill Bluff and Stillwater Cove further illustrates the local variation that occurs among kelp forests within a region. Sand Hill Bluff is in the lee of a small point, slightly protected from northwest swells (Cowen et al. 1982, Foster 1982). At the time of sampling, the bottom at shallower depths (above 6 m) was dominated not by the geniculate corallines found at Stillwater Cove but by an assemblage of fleshy, thin-bladed red algae, especially *C. farlowiana*, *Polyneura latissima*, and *Phycodrys setchellii*, growing among dense mats of bryozoans, sponges, and tunicates. Giant kelp occurred between 6 and 14 m depths, with widely dispersed patches of the understory kelps *L. setchellii* and *P. californica*. Red algae also occurred under the giant kelp canopy but their abundance was reduced, presumably by light limitation. The brown alga *D. ligulata* was very abundant, particularly in spring and early summer after winter storms removed much of the giant kelp canopy. Sponges, tunicates, and pholad clams were common sessile animals. Over 50% of the bottom was unoccupied rock. Red sea urchins were common, clumped in small crevices and depressions. Turban snails were rare. The sea stars *Patiria miniata*, *Pycnopodia helianthoides*, and *Pisaster* spp. were common. Red algal abundance increased on rock in deeper water outside the canopy, as did that of sea urchins, tunicates, the anemone *Corynactis californica*, and the tube-dwelling polychaete *Diopatra ornata*.

Sand Hill Bluff was sampled qualitatively (Yellin et al. 1977) and quantitatively (Cowen et al. 1982, Foster 1982) in the late 1970s prior to sea otter (*E. lutris*) range expansion into the area, and qualitatively observed in 1983 and 1985 after otters arrived. These post-otter surveys (M. Foster, pers. obs.) noted a decline in red sea urchins that could have resulted from sea otter foraging, disease (Pearse et al. 1977), large swells associated with the 1982–1983 El Niño, or some combination of these and other factors. The structure of the forest, however, remained similar to that seen in previous surveys. Sea urchins at nearby Point Santa Cruz had larger effects, however, as documented by Pearse and Hines (1979). Prior to 1976, this forest ended at a depth of 6 m where it was replaced by high densities of sea urchins. The forest expanded to the rock–sand interface at 9 m depth following sea urchin mortality from disease, and sea urchins were further reduced throughout the forest when sea otters arrived at the site in 1977.

The large differences between Stillwater Cove and Sand Hill Bluff described above are most likely the result primarily of differences in substratum type and exposure (Foster 1982). Other forests similar to Stillwater Cove with hard rocky substrata and little sand influence have also been described in the region. Harrold et al. (1988), for example, described four sites along a wave exposure gradient and found that as exposure increased, bull kelp, understory kelps, and geniculate corallines increased. They argued this was probably the result of an indirect effect of an increase in light due to more thinning and greater seasonal loss of the giant kelp canopy. Foster and VanBlaricom (2001) sampled nine sites along the wave-exposed Big Sur coast, including a site near the one previously described by McLean (1962). They found that six of these sites were similar in composition to Stillwater Cove and to the exposed sites of Harrold et al. (1988). At three sites, however, sessile invertebrates rather than algae covered around 40% of the bottom. These sites were in areas with very high relief and often steeply sloping bottoms, a topography that at small scales within kelp forests is also dominated by sessile invertebrates. Based on similarities and differences among all subtidal kelp forests described from the Big Sur region to Año Nuevo Island, Foster and VanBlaricom (2001) suggested that they could be grouped into four general "kelp forest types" associated with differences in substratum and wave exposure: (1) wave-exposed forests on low-relief, soft mudstone characterized by high seasonal disturbance and low light with variable abundances of giant and bull kelp and an understory dominated by rapidly growing, thin-bladed foliose red algae; (2) wave-protected forests with a hard, moderate relief substratum, moderate-light conditions, persistent giant kelp canopies, and an understory of thick-bladed, relatively slow-growing foliose red algae; (3) wave-exposed forests on hard, moderate to high relief substratum with a seasonal reduction of kelp canopies and an understory dominated by coralline algae and various kelps; and (4) wave-exposed forests with hard, steeply sloping substrata dominated by seasonally variable kelp canopies and an understory of sessile invertebrates.

Occasional stands of *M. 'integrifolia'* also occur along the open coast in this region, forming "mini-forests" in shallow water. In central California, Oregon, and Washington, these are usually found from the low intertidal zone to a few meters' depth in the extreme lee of points, as at Stillwater Cove. They can also occupy tide pools. For example, "The Great Tide" pool at the southern end of Monterey Bay, featured in John Steinbeck's novel "Cannery Row," is a mini-kelp forest that people can wade through at low tide. It contains a high diversity of invertebrates typically found in large tide pools in the area (M. Foster, pers. obs.). These types of shallow "forests" have not been quantitatively sampled, but our observations at low tide indicate they contain few organisms other than giant kelp. The bottom cover is usually dominated by the thickly intertwined, strap-like holdfasts of *M. 'integrifolia,'* and the very dense canopy not only reduces light but also brushes over and then lies on the bottom as the tide recedes.

These very shallow stands contrast with many of those in semi-protected bays and tidal channels of Washington, British Columbia, and Alaska where *M. 'integrifolia'* occurs from the intertidal zone to depths of at least 12 m. Studies of these forests have generally focused either on harvesting giant kelp or on the effects of sea urchins and sea otters on the distribution of large brown algae, with few data on other associated organisms. Druehl (1978) found in British Columbia that giant kelp occurred at sites characterized by hard substrata, high salinity, and moderate water motion (waves or tidal currents). Druehl and Breen (1986) provided some information on associated algae and invertebrates at a site in Barkley Sound on the west coast of Vancouver Island, British Columbia. Giant kelp occurs from the low intertidal zone to 4 m depth where the substratum changes from rock to boulders and sand (figure 5.2E). The understory is inhabited by some species found in central California and others found only further north. Understory kelps include *P. californica, Costaria costata, Saccharina (=Laminaria) groenlandica* (a far northern species), and *Alaria marginata*, with the brown alga *D. ligulata*, the green *Ulva* spp., and various red algae on the bottom. Abundant large invertebrates include red (*Strongylocentrotus franciscanus*) and green (*S. droebachiensis* (a far northern species) sea urchins, the abalone *Haliotis kamtschatkana* (another far northern species), and the sea cucumber *Parastichopus californicus*. This forest was similar to those described by Shaffer (2000) in the Strait of Juan de Fuca in Washington. Shaffer (2000) noted that these forests can be important habitats for resident rockfish and also for migrating salmon, which feed on various small fishes along the forest edge.

High densities of sea urchins occur along the lower edge of some of these forests and can have a large, persistent influence on the lower limits of giant kelp (Druehl 1978), especially along the west coast of Vancouver Island, similar to conditions off Santa Cruz Island, California, before disease severely reduced sea urchin populations (Mattison et al. 1977). When the sea urchins were removed either experimentally (Pace 1981) or by sea otters (Watson and Estes 2011), the forest expanded to the limit of hard substrata (usually less than 10–12 m). After sea urchin removal, the expansion often begins with annual brown algae, followed by *Macrocystis* and perennial understory kelps. Annuals may increase again, depending on temporal variation in the abundance of perennial species at particular sites (Watson and Estes 2011). Based on historical distributional records, van Tamelen and Woodby (2001) suggested that at least some *N. luetkeana* forests in southeast Alaska became *Macrocystis* forests after sea otters reduced the abundance of sea urchins, as occurred on the Big Sur coast of California after the return of sea otters (McLean 1962).

WARM TEMPERATE SOUTHEASTERN PACIFIC · *Northern Peru to the West Coast of Southern Chile*

This very large coastal province extends from remarkably near the equator in the Gulf of Guayaquil at the Peru–Ecuador border (4.44° S, 81.33° W) to Chiloé Island, Chile (42.50° S, 74.19° W; Briggs 1974, Spalding et al. 2007; figure 1.5). The documented northern limit of *Macrocystis* is in the region of Piura—Islas Lobos de Tierra, Peru, at 6° S latitude, around 200 km south of the Gulf of Guayaquil (Neushul 1971a, Hoffmann and Santelices 1997). Giant kelp continues southward of this region along the west coast of South America across the southern provincial limit to the tip of the continent, a distance of around 6000 km, well exceeding the total distributional length of the species in the northeastern Pacific. The province is swept by the Peru–Chile (Humboldt) Current that originates from the cold, eastward-flowing Antarctic Circumpolar Current (West Wind Drift) where it intersects South America near Chiloé Island. The Peru–Chile Current gradually warms as it flows north, and veers away from the coast near the Gulf of Guayaquil. This and associated nearshore currents produce complex circulation patterns along the coast (reviews in Santelices 1991, Thiel et al. 2007) that affect upwelling. Cool, upwelled waters from Concepción north into Peru moderate the general warming trend and facilitate the presence of giant and other kelp populations (review in Vásquez et al. 2006). El Niño conditions alter the currents and can depress upwelling, with large negative effects on *Macrocystis* and changes in associated species. These oceanographic features and their effects on giant kelp forests are like those along the coast of the southern part of the northeastern Pacific, particularly Baja California.

The '*integrifolia*' and '*pyrifera*' ecomorphs of *Macrocystis* occur in the province, with *Macrocystis* '*pyrifera*' dominant in the south northward to Valparaíso, and *M. 'integrifolia*' dominant from Valparaíso north to the provincial boundary. The two ecomorphs overlap around Valparaíso and again at the northern end of the province in Peru (Neushul 1971a, Vásquez 2008). The exposed shallow subtidal zone north of Valparaíso (north

## A. South Central Chile

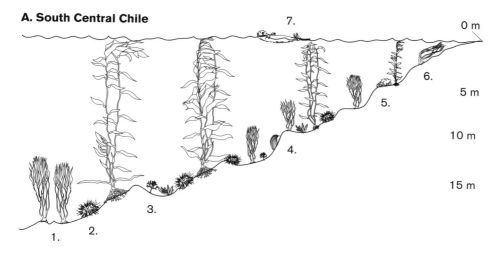

1. *Lessonia trabeculata* (UK)
2. *Tetrapygus niger* (sea urchin)
3. *Dictyota kunthii* (BC)
   *Asparagopsis armata* (BC)
4. *Concholepas concholepas*
   (Chilean abalone)
5. *Pyura chilensis* (tunicate)
6. *Lessonia nigrescens*
   (shallow water kelp)
7. *Lontra felina* (sea cat)

## B. Beagle Channel

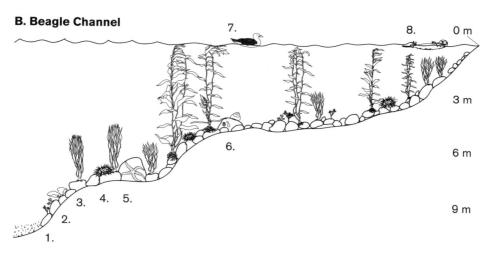

1. Boulders
2. *Epymenia falklandica* (BC)
   *Gigartina skottsbergii* (BC)
3. *Lessonia flavicans* (UK)
4. *Loxechinus albus*
   + 3 other species (sea urchins)
5. *Cosmasterias lurida* (sea star)
6. Gastropods *Tegula* sp.
   *Nacella mytilina Fissurella* spp.
7. *Larus dominicus* (gull)
8. *Lontra felina* (sea cat)

FIGURE 5.3

Distribution of conspicuous plants and animals found in five giant kelp forests (A–E) within the range of *Macrocystis* in the southern hemisphere.

BC = bottom canopy; SCF = surface canopy fucoid; SCK = surface canopy kelp; UF = understory fucoid; UK = understory kelp.

NOTES: The symbols in Figure 5.2 Panel A for sand (1), sessile animals (3), and *Macrocystis 'pyrifera'* (4) are generic to all figures unless otherwise noted. Organisms and horizontal axes are not to scale. Depth distribution is indicated on the right.

## C. Tasmania

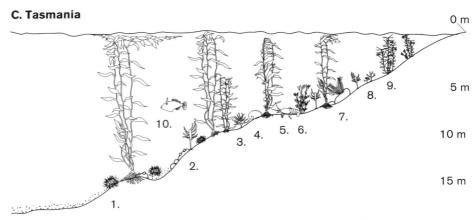

1. Sea urchins
   *Centrostephanus rodgersii*
   *Heliocidaris erythrogramma*
2. *Ecklonia radiata* (UK)
3. *Acrocarpia paniculata* (UF)
4. *Haliotis rubra* (abalone)
5. *Jasus edwardsii* (lobster)
6. *Cystophora* spp. (UF)
7. *Cenolia trichoptera* (crinoid)
8. Bottom canopy
   Fleshy red algae
   Nongeniculate corallines
9. *Phyllospora comosa* (SCF)
10. *Meuschenia freycineti* (leather jacket)

## D. New Zealand

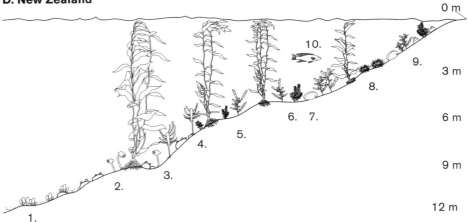

1. *Perna canaliculus* (mussel)
2. *Pyura pachydermatina* (tunicate)
3. *Ecklonia radiata* (UK)
4. Geniculate corallines (BC)
5. *Landsburgia quercifolia* (SCF, UF)
6. *Carpophyllum*
   *maschalocarpum* (SCF, UF)
7. *Haliotis iris* (abalone)
8. *Evechinus chloroticus* (sea urchin)
9. Other gastropods
   *Trochus viridis*
   *Cookia sulcata*
10. *Odax pullus* (butterfish)

FIGURE 5.3 *continued*

## E. Marion Island

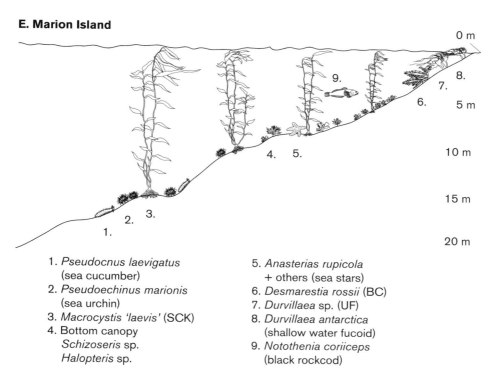

1. *Pseudocnus laevigatus*
   (sea cucumber)
2. *Pseudoechinus marionis*
   (sea urchin)
3. *Macrocystis 'laevis'* (SCK)
4. Bottom canopy
   *Schizoseris* sp.
   *Halopteris* sp.

5. *Anasterias rupicola*
   + others (sea stars)
6. *Desmarestia rossii* (BC)
7. *Durvillaea* sp. (UF)
8. *Durvillaea antarctica*
   (shallow water fucoid)
9. *Notothenia coriiceps*
   (black rockcod)

FIGURE 5.3 *continued*

central and northern Chile) is dominated by the kelp *Lessonia trabeculata*, the abundance of which (as biomass) is over 1000 times greater than that of *M. 'integrifolia'* (Vásquez 2008). Giant kelp is restricted to wave-protected sites and occurs from the intertidal zone to depths of around 15 m (Villouta and Santelices 1984, Santelices 1991, Vásquez and Buschmann 1997). A "typical" *M. 'pyrifera'* forest in south-central Chile is diagrammed (figure 5.3A), based on descriptions by Santelices (1991), Pérez-Matus et al. (2007), Vásquez et al. (2006), and Villegas et al. (2007). The large stipitate kelps *L. nigrescens* and *L. trabeculata* commonly co-occur with giant kelp, with *L. nigrescens* forming a band above the upper limit of the kelp forest. *L. trabeculata* occurs in the understory beneath giant kelp and can be abundant in deeper water outside the kelp forest. Cavities in the cortex of the blades of this species cause the thalli to float, and plants reach up to 2.5 m tall, appearing tree-like (Hoffmann and Santelices 1997). Bryozoans, the tunicate *Pyura chilensis*, the brown alga *Glossophora kunthii*, and frondose red algae (e.g., *Asparagopsis armata*) can be abundant beneath the kelps, as can a bottom cover of nongeniculate coralline algae, which are especially abundant in areas dominated by the black sea urchin *Tetrapygus niger*. The herbivorous gastropod *Tegula tridentata* can be abundant. Invertebrate carnivores include various sea stars and the large muricid gastropod *Concholepas concholepas* (Chilean "abalone" or loco). Many of the common fishes are from tropically derived families, including the herbivore *Aplodactylus punctatus*, and the benthic-feeding carnivores *Semicossyphus maculatus* (Chilean sheephead), *Cheilodactylus variegatus* and *Girella laevifrons*. The southern sea lion *Otaria flavescens* (Alonso et al. 2000) and the sea cat *Lontra* (=*Lutra*) *felina* (Larivière 1998) are found throughout the province and southward, and forage in a variety of habitats including kelp forests. Unlike its northern relative *Enhydra lutris*, *L. felina* rarely eats sea urchins. There are no comprehensive descriptions of *Macrocystis* forests in Peru, but observations by Fernández et al. (1999) suggest their composition is similar to that of northern Chile.

M. *'pyrifera'* populations at the southern end of the province are perennial in wave-exposed locations but annual in protected waters where they are affected by high temperatures and low nutrients in spring and summer (Buschmann et al. 2006). More species with southern, cold-water affinities are associated with giant kelp in this region (e.g., the sea urchins *Loxechinus albus* and *Arbacia dufresnei*, and the red alga *Sarcothalia crispata*). Grazing by the turban snail *T. atra* can negatively affect recruitment of young sporophytes in the annual giant kelp populations, but it appears to have little effect on the overall abundance of adult giant kelp or other seaweeds (Moreno and Sutherland 1982, Henríquez et al. 2011). The annual community is unusual not only because it is annual but also because it lacks understory kelps; the bottom is dominated by the slipper limpet *Crepipatella fecunda* that can facilitate giant kelp recruitment (Henríquez et al. 2011).

MAGELLANIC PROVINCE · *West Coast of Southern Chile–Southeast Argentina*

The intersection of the Antarctic Circumpolar Current with southern South America produces the Cape Horn Current that flows south along the coast from around Chiloé Island, among and through numerous islands and channels, and around Cape Horn. Cold water on the east side of Cape Horn flows north along the coast of Argentina as the Falkland Current, turning east and leaving the coast in the vicinity of Rio de Plata (at 35° S). These coastal currents bound the province (figure 1.5), which includes the Islas Diego Ramírez (56.49° S, 68.72° W). Giant kelp occurs around these islands (Contreras et al. 1983) and may represent the southernmost populations in the Southern Hemisphere. *Macrocystis 'pyrifera'* is the only ecomorph, and it occurs throughout the province, including the Falkland Islands. Neushul (1971a) did not observe plants north of Rio Colorado (40° S) although drift plants have been observed to Rio de Plata (Kühnemann 1970).

The structure and diversity of kelp forests in this province greatly impressed Darwin (1839). Dayton (1985b) described numerous forests from Isla de Los Estados off the tip of Cape Horn to Chiloé Island. Some of the best-studied forests are around Puerto Toro (55.08° S, 67.06° W) in Beagle Channel (e.g., Moreno and Jara 1984, Santelices and Ojeda 1984a, Vásquez et al. 1984, Castilla 1985; figure 5.3B). The general vegetation structure is similar to northern Chile, but there is a difference in species. *Macrocystis* occurs in forests covering several square kilometers at Islas Los Estados, but in Beagle Channel they occur in 50–100 m wide coastal belts to 15 m depth where rock meets soft substratum. The upper limit is bounded by dense stands of the kelp *Lessonia flavicans* (=*L. vadosa*; Guiry and Guiry 2012). This species occurs less abundantly under giant kelp and can form dense stands in deeper water outside the giant kelp canopy, a bimodal distribution similar to that of the kelp *Eisenia arborea* at Catalina Island in southern California described previously. Fleshy red algae such as *Epymenia falklandica* and *Gigartina skottsbergii* and patches of algal turf, all underlain by crustose algae, cover the bottom (Santelices 1991). *Loxechinus albus* is the most abundant (based on biomass) of the four sea urchin species in the region, and *Cosmasterias lurida* is the most common sea star. This sea star is a benthic predator that rarely eats sea urchins (Castilla 1985). The urchins that feed primarily on drift algae can occur in dense patches and locally reduce macroalgal abundance (Dayton 1985b, Newcombe et al. 2012), but appear to have no large-scale effects (Castilla 1985, review in Vásquez and Buschmann 1997).

Darwin (1839) marveled at the diverse invertebrates associated with giant kelp in the region. Quantitative studies have identified from 43 to 114 taxa associated with giant kelp holdfasts (Ojeda and Santelices 1984, Rios et al. 2007), and 63 taxa from holdfasts and fronds combined (Adami and Gordillo 1999). Although difficult to compare with other studies due to differences in sampling design, the sizes of organisms collected, and taxonomic resolution, this richness is certainly lower than that reported for holdfasts (>150 species) or fronds alone (114 species) in southern California (Ghelardi 1971, Coyer 1984). Species richness of algae, fishes, and most other major groups associated with giant kelp is also reduced relative to southern California, perhaps reflecting geographic isolation (Santelices 1991) and extinctions related to Pleistocene glaciation (review in Camus 2001). The fish fauna is composed mainly of the southern cold-water Antarctic cods (nototheniids) *Notothenia magellanica* and *Patagonotothen* spp., which are demersal fishes that feed on amphipods, polychaetes, and other benthic invertebrates associated with decomposing drift algae. Herbivorous fishes are absent from these descriptions. Moreno and Jara (1984) noted that the habitat and niche characteristics of the 18 fish species that occur in the region are similar to those in the Aleutian Islands, Alaska, even though their taxonomic composition is quite different. Various birds and mammals forage in and around the forests. The gull *Larus dominicus* is noteworthy because its foraging on the very abundant bivalve *Gaimardia trapesina* living on giant kelp fronds may represent the largest removal of energy by any high-level predator in the community (Hockey 1988). Predation may also prevent the fronds from sinking, as happened in southern California when another small bivalve, the scallop *Leptopecten latiauratus*, settled by the thousands on giant kelp fronds (McPeak et al. 1988).

Giant kelp forests along the southern coast of Argentina occur in gradually warming waters from south to north. Kühnemann (1970) described extensive stands at Puerto Deseado (47.7° S) in the intertidal zone and also subtidally, where there is understory kelp (*Lessonia* sp.) and a diverse assemblage of fleshy and coralline red algae. Sea urchins are rare. Seven forests were described along the coast of Chubut between 46° S and 42° S by Barrales and Lobban (1975). Forests occurred from the low intertidal zone to 15 m depth, with their lower depth limit defined by available sandstone substratum and their composition associated with variation in wave exposure. The understory in wave-sheltered forests is dominated by the green alga *Codium vermilara*, in moderately exposed forests by *Lessonia* sp. (also forming a band at the upper margin of the forests), various other brown (e.g., *Zonaria* sp.) and red (e.g., *Callophyllis* sp., *Corallina officinalis*) algae, and in the one exposed forest by foliose red algae *Hymenena falklandica* and *Epymenia* sp., and "plates" of nongeniculate corallines. The mussel *Aulacomya ater*, brittle stars *Ophioplocus* (=*Ophioceramis*) sp., and turban snails occurred in protected microhabitats in all exposures. Sea urchins were present but were neither large nor abundant, and only six species of fish were noted as common. The sea lion *Otaria byronia* was also seen within forests. The most important grazer appeared to be the isopod *Phycolimnoria* sp. that bores into and weakens holdfasts, contributing to the observed loss of plants in a 3–4 year cycle. At the time of their surveys, Barrales and Lobban (1975) noted that this section of the South American coast was largely uninhabited by humans, with little coastal development and few fisheries, making it an ideal area for studying giant kelp forests little disturbed by human activities. We are, however, unaware of any studies in the region since 1975.

OTHER TEMPERATE SOUTHERN HEMISPHERE PROVINCES

*Macrocystis* occurs in five other temperate provinces in the southern hemisphere: the isolated island groups Tristan da Cunha–Gough Islands groups (37.24° S, 12.44° W), Amsterdam–Saint Paul Islands (38.36° S, 77.58° W), the Agulhas at the southern end of South Africa, the southeast Australian shelf, and southern New Zealand (Setchell 1932, Womersley 1954; recent distributional records in Hay 1990, Perissinotto and McQuaid 1992). There are no descriptions of giant kelp communities for the island provinces or Agulhas. *Macrocystis* 'angustifolia' is the only ecomorph present in the Agulhas region, and it occurs only in an upwelling area (Bolton and Levitt 1987; figure 1.5) at protected sites inshore from forests of the dominant canopy-forming kelp in the province, *Ecklonia maxima* (Isaac 1937b in Womersley 1954). Upwelling occurs up the west coast of southern Africa into Namibia (part of the Benguela Province in Spalding et al. 2007) but, while temperatures appear suitable (Bolton and Levitt 1987), giant kelp is absent. Womersley (1954) suggested this is because currents do not favor the transport of drift plants, and faunal differences discussed by Briggs (1974) also suggest a lack of larval transport. Increased turbidity may also be a factor (R. J. Anderson, pers. comm.) as well as a very linear and wave-exposed coastline with little rocky habitat (Field and Griffiths 1991).

The southeast Australia province includes the cooler waters of Tasmania, Bass Strait, and the southeastern Australian mainland from near Robe, South Australia, to the east and north around the tip of the continent to Bermagui, New South Wales (Briggs 1974). Giant kelp occurs in the southern region of the province from its western end eastward to Walkerville, Victoria, along the mainland, inshore from fringing reefs at depths of 2–10 m and offshore at depths to 35 m (Womersley 1954, Shepherd and Sanderson 2013), and along western Tasmania shores and reefs in Bass Strait. Although *M. 'angustifolia'* is the most abundant ecomorph in the province, *M. 'pyrifera'* is also present and the only ecomorph in Tasmania. Cribb (1954) provided the first descriptions of these forests, noting that giant kelp was sparse along the mainland, but stands on the east and south coasts of Tasmania occurred as extensive fringes along protected shores as well as offshore. The large fucoid *Phyllospora comosa* and nongeniculate coralline algae usually formed a dense band in shallow water of the fringing forests in Tasmania and Bass Strait, and the deeper ends of forests were often determined by the presence of a sandy bottom (figure 5.3C). Surveys and experiments by Barrett et al. (2009) and Ling et al. (2010) described species associated with giant kelp in several Tasmanian forests. The stipitate kelp of Australia and New Zealand, *E. radiata*, a diverse assemblage of large fucoids (e.g., *Cystophora* spp., *Sargassum* spp., *Acrocarpia paniculata*), and various fleshy and encrusting red algae commonly occur beneath giant kelp, particularly at depths shallower than around 10 m. The invasive Japanese kelp *Undaria pinnatifida* now also occurs seasonally at some sites (Valentine and Johnson 2003). Invertebrates include bryozoans, sponges and sea stars, the crinoid *Cenolia trichoptera*, the sea urchins *Centrostephanus rodgersii* and *Heliocidaris erythrogramma*, and commercially fished lobster (*Jasus edwardsii*) and abalone (*Haliotis rubra*). Large fishes include several wrasses and the leather jacket *Meuschenia freycineti*. The weedy sea dragon *Phyllopteryx taeniolatus* occurs among algal fronds. Australian giant kelp forests have experienced severe declines over the last 30 years associated with El Niños, exceptionally large storms, and the increasing influence of the warm, nutrient-poor, southward flowing East Australian Current (figure 1.5). Data from a site in southern Tasmania indicate that the inverse relationship between temperature and nitrate found in California (Chapter 3) also obtains in Tasmania, with similar effects on giant kelp growth (Shepherd and Sanderson 2013). In addition, increasing temperatures and the shoreward movement of the warm water current have facilitated larval transport from northern waters of the diadematid sea urchin *C. rodgersii* to Tasmania, where it has deforested large areas (Johnson et al. 2011, Wernberg et al. 2011). These declines in Australian giant kelp forests led to their listing as an endangered ecological community in 2012 under the Australian Environmental Protection and Biodiversity Conservation Act (Shepherd and Sanderson 2013).

The Southern New Zealand Province includes the southernmost part of North Island from around 37° S, the South Island, Stewart Island, the Snares Islands, and the Chatham Islands 700 km to the east (Spalding et al. 2007; considered as a Region by Briggs 1974). *M. 'pyrifera'* is the only ecomorph in the province. It occurs north to Castle Point (40° S, 176.23° E) on the southeast coast of the North Island, but only to Jackson Head (43.98° S, 168.59° E) on the west coast of the South Island. The southern

FIGURE 5.4
Giant kelp forest (top) and dense aggregation of the abalone *H. iris* (bottom), at the
Chatham Islands, New Zealand. Note the
lack of understory algae in the forest. Abalone are abundant in shallow water (2–4 m)
where they feed on macroalgal drift.
SOURCE: Photos by M. S. Foster.

distribution is probably related to the influence of the Southland Current that brings cooler water from the region of Jackson Head along southern New Zealand and north-ward along the east coast (figure 1.5). Smaller-scale distribution is correlated with local variation in temperature and the availability of rock at suitable depths (Hay 1990). On the mainland, giant kelp occurs on the open coast in patches with suitable substrata, at the mouths of fiords, and within bays. It occurs in protected areas at the Chatham Islands (Hay 1990, Schiel et al. 1995). As in southeast Australia, one of the most dis-tinguishing features of these forests is the diverse and abundant flora of large fucoids. Schiel and Hickford (2001) used depth-stratified sampling to describe the giant kelp forest at Godley Head on the protected side of Banks Peninsula near Christchurch. *Macrocystis* occurs from 3 to 9 m depth, mixed with a sparse understory of the fucoids *Carpophyllum maschalocarpum* and *Landsburgia quercifolia* at 3–6 m, and stands of the kelp *E. radiata* at 6–9 m (figure 5.3D). Geniculate corallines are abundant at all depths. Encrusting sponges and tunicates, and the stalked tunicate *Pyura pachydermatina* were abundant from 9 to 12 m, and the mussel *Perna canaliculus* forms extensive patches at the deep end of this range where the forest terminates in a sand bottom. Abalone (*H. iris*), top shells (*Trochus viridis*), and the turbinid *Cookia sulcata* are abundant in shallow water, along with the ubiquitous echinometrid sea urchin of the New Zealand coast, *Evechinus chloroticus*. Common fishes include a suite of wrasses (*Notolabrus* spp.), leatherjackets (*Meuschenia* (=*Parika*) *scaber*), and the herbivorous butterfish *Odax pullus*.

In contrast to this mainland forest, similar surveys of two sites at the Chatham Islands (Schiel et al. 1995) found *Macrocystis* from depths of around 12 to >16 m. The fucoid *C. maschalocarpum* dominates in shallow water, and the abundant stipitate kelp of the New Zealand mainland, *E. radiata*, is absent. Illustrative of the great variation found among giant kelp forests in New Zealand is the fact that the bottom beneath giant kelp at the deep end of at least one forest, when surveyed, was a "barrens" without sea urchins, almost entirely covered by nongeniculate coralline algae and occasional sponges and tunicates (figure 5.4). Although heavily fished at the Chathams, abalone are very abundant in shallow water (figure 5.4). Sea urchin abundance was highly variable among sites and these grazers produced only small patches of deforested areas when their density was high. Wrasses and the very abundant blue cod, *Parapercis colias*, were the most common of the nine species of large reef fishes observed.

## SOUTHERN OCEAN

The Southern Ocean is a "Realm" as defined by Spalding et al. (2007), composed of four provinces, three of which are inhabited by giant kelp: the Subantarctic Islands which include those south of Australia and South Africa, Subantarctic New Zealand, and the Scotia Sea south of South America. The land in this Realm comprises mostly isolated islands or island groups, often referred to collectively as the Subantarctic Islands (which is somewhat confusing, given that this name is also used to designate a Province within the Realm). The Subantarctic is defined as the region between roughly 46° S and 66° S. The southern boundary defines the demarcation between the southern temperate zone and the Antarctic Circle, the northernmost latitude in the southern hemisphere where the sun does not go below the horizon in a 24-hour period. The Antarctic Circumpolar Current (also called the West Wind Drift) flows largely unimpeded through the Realm with average surface water temperatures of 4–9°C (NASA JPL 2013). Gale-force winds and large swells are the norm. The differences in the marine biota that distinguish the provinces are largely ascribed to differences in the age of islands, their geographic isolation, influence of ice ages, and current patterns that affect colonization (e.g., Briggs 1974). *Macrocystis* was noted on the lee sides and in protected coves and inlets at most of these islands during the early explorations by Hooker and others (see Chapter 1, History of Research, and references in Perissinotto and McQuaid 1992 for recent records). However, the species associated with giant kelp have been described in only a few locations and these descriptions vary, depending on the interests of the investigators. This is not surprising given the problems of access and weather at these remote locations.

*Macrocystis* occurs with quite a variety of different species across the Realm, which suggests further investigation among these provinces could greatly improve our understanding of the variety of community structures associated with giant kelp forests.

The most studied giant kelp forests across these regions are in the Subantarctic Province at Marion Island (figure 5.3E), one of the two Prince Edward Islands located at 46.9° S, 37.75° E (Hay 1986, Attwood et al. 1991, Beckley and Branch 1992, Perissinotto and McQuaid 1992). The smooth blades of giant kelp in this region were once thought to characterize a separate species, *Macrocystis 'laevis'* (see Chapter 1, Taxonomy). Even within the context of highly variable *Macrocystis* forests, these forests are unusual as they occur in two habitats. The more typical forests occur nearshore at depths of 10–20 m. However, very deep forests occur to depths of 68 m in the exceptionally clear water over rocky areas of the open shelf between Marion and Prince Edward Islands, the deepest known stands of giant kelp (Perissinotto and McQuaid 1992). The biota associated with these deep-water populations have not been described. These deep giant kelp stands do not reach the sea surface and this, plus great depth, probably contributes to their persistence in very high water motion.

The nearshore forests at Marion Island are 50–100 m wide, commonly bordered in shallow water by the large, kelp-like fucoids *Durvillaea antarctica* and perhaps *D. willana*, another large brown alga, *Desmarestia rossii*, and abundant nongeniculate coralline algae, all of which are taxa common in shallow water throughout the Subantarctic Islands. Beckley and Branch (1992) quantitatively described community composition and biomass at three sites. They found a high biomass of foliose red (e.g., *Schizoseris* sp.) and brown (*Halopteris* sp.) algae, and the green *Codium* sp. These declined with depth within the forest, while the biomass of the 200 species of macrobenthic animals found increased with depth, which they noted correlated with a decrease in water motion. As in some sites in southern Chile, the small bivalve *Gaimardia trapesina* was abundant on blades within the giant kelp canopy. Only one sea urchin, *Pseudoechinus marionis*, was seen. It occurred at densities of up to 50 m$^{-2}$ but, like many of the other animals, appeared to feed on detritus within and adjacent to the forests, with little evidence of direct herbivory (Attwood et al. 1991). Only three fish species were found, their low diversity being attributed to the isolation and relative youth (<300,000 years old) of the island (Blankly 1982). The nototheniid *Notothenia coriiceps* was most abundant and, like the other two species, feeds on a variety of small benthic polychaetes, crustaceans, and occasionally limpets. The plunder fish *Harpagifer georgianus* is commonly eaten by imperial cormorants, *Phalocrocorax atriceps*. Although there are some interesting differences such as a lack of barnacles and large bivalves, Beckley and Branch (1992) concluded that the fauna was generally similar to that of the Kerguelen Islands in the same province.

Kelp forests at the Auckland Islands (50.55° S, 166.05° E) in the Subantartic New Zealand Province have been described by Schiel and colleagues (Kingsford et al. 1989, Schiel 1990). Subtidal plant zonation was similar to other areas in the cold Southern

Ocean with dense stands of the fucoid *D. antarctica* in shallow water. In contrast to other areas, however, dense stands of another large fucoid, *Xiphophora chondrophylla*, occurred below *D. antarctica*, reflecting the diverse and abundant flora of large fucoids characteristic of Australia and New Zealand. Also abundant in the shallows of exposed outer reefs was the brown alga *Halopteris funicularis*, which forms dense feathery beds on sloping reef walls. The endemic kelp *Lessonia brevifolia* overlapped the lower limit of *X. chondrophylla* and the upper limit of giant kelp, which occurred from 6 to 20 m depth. The only abundant herbivore was the small (around 7 cm) white abalone, *Haliotis virginea*. Only eight species of large fishes were observed, the most common being the benthic, carnivorous nototheniids *Paranotothenia angustata* and *P. microcepidota*. The family composition and low diversity of the fish fauna are reminiscent of southern Chile, but the species had high affinities with the New Zealand mainland.

## WITHIN COMMUNITIES: HABITATS AND PATCHES

As in other communities distributed along environmental gradients, there are often zones characterized by visually obvious species or group of species. Giant kelp forests are often terminated by sand at their deep margin. However, when they occur on moderately wave-exposed rocky shores that extend from the littoral into deep water, they are commonly "zoned," bordered on their deep and shallow ends by other marine plants, usually other kelps and fucoids (figure 5.1). These patterns can be a direct result of changes in the physical environment (e.g., changes in substratum type, light penetration, and wave action with depth; Chapter 3) but may also be influenced by biological interactions (Chapters 7–9). Within a zone, organisms often occur in "patches" (e.g., Dayton et al. 1984), and particular species may occur in different parts of the forest as a result of resource partitioning (e.g., Foster et al. 2013). These smaller-scale, "within-zone" patterns are discussed below. They have long fascinated kelp forest ecologists, in part because their causes are most amenable to understanding with field experiments.

### BOULDERS, CREVICES, KELP HOLDFASTS

Arguably the most important contributor to local species diversity and abundance within giant kelp forests is the diversity of habitats within a forest that results from variation in rock type, topography, and structural complexity of large benthic organisms, relationships similar to those on coral reefs (review in Alexander et al. 2009). For example, boulder fields provide considerable structural complexity within the kelp forest at Stillwater Cover in central California (figure 5.2D). The tops of boulders have a high cover of the geniculate coralline *Bossiella californica* subsp. *schmittii* (Konar and Foster 1992). The undersides and spaces between boulders accumulate decomposing drift algae, provide some protection from predators, and are habitat for brittle stars, chitons, juvenile abalone, and clingfish that are relatively uncommon in other habitats within the forest (M. Foster, pers. obs.).

Rocks with cracks and crevices are often occupied by abalone and sea urchins, especially in the presence of predators (e.g., Lowry and Pearse 1973). Sea urchin abundances (Guenther et al. 2012) as well as overall community richness (Torres-Moye et al. 2013) have been positively correlated with substratum rugosity at sites in southern California and Baja California Sur, Mexico. Sessile species such as cup corals, bryozoans, tunicates, and sponges can be aggregated on topographic features (Rosenthal et al. 1974, Foster et al. 2013). These sorts of associations have also been described in Tasmania, where Alexander et al. (2009) found that rock type (e.g., boulders vs. bedrock) and refuge diversity (e.g., crevices, holes) have significant effects on the species richness of mobile invertebrates like crinoids, sea urchins, and abalone. The influence of topography is particularly evident on rock walls, which can be common in some forests and beds, and are often dominated by sessile invertebrates (figure 5. 1), while flat areas immediately above the walls are usually dominated by seaweeds (Goldberg and Foster 2002, Miller and Etter 2008). Experiments in central California indicate that this striking pattern in the distribution of the geniculate coralline *Calliarthron* spp. results from a combination of poor algal recruitment and slow growth (low light) on vertical surfaces (Goldberg and Foster 2002).

Considerable biogenic complexity is provided by algal holdfasts and blades, particularly those of *Macrocystis*. Skottsberg (1907) was the first to highlight the diverse and abundant fauna living in giant kelp holdfasts, and Andrews (1945) catalogued the inhabitants of holdfasts in Monterey Bay, California, as did subsequent studies by Ghelardi (1971) in southern California, Ojeda and Santelices (1984) in southern Chile, and Edgar (1987) in Tasmania and at Macquarie Island. The entangled haptera provide a fairly rigid framework with spaces of around 1 cm or less. Holdfasts can be quite large. For example, the largest holdfast sampled by Ghelardi (1971) had an average basal diameter of 83 cm and was 68 cm tall, with a volume of 120,143 cm$^3$. Large holdfasts have some habitat partitioning with older, dead, brittle haptera in the center surrounded by a ring of living haptera. Ghelardi (1971) found that gammarid amphipods, polychaetes, and isopods were the most abundant groups, and most animals occurred in both live and dead areas of holdfast, but some were more abundant in one or the other. Species identities and abundances over a range of holdfast sizes, in conjunction with observations from artificial holdfasts, indicated that the common taxa are found in all sizes and ages of holdfasts, that early colonists are not facilitative, and that later colonists do not displace the early colonists.

Holdfasts appear simply to accumulate species and individuals as they grow. Holdfasts sampled in Chile by Ojeda and Santelices (1984) were smaller than those in California (largest = 15,000 cm$^3$) and contained only 53 taxa, with sea urchins, polychaetes, and crabs being most abundant. Although less diverse, assemblage dynamics were similar to those found by Ghelardi (1971). Taxa accumulated rapidly and began to level off at holdfast volumes around 3000 cm$^3$. Faunal composition and colonization appeared to be little affected by seasonality, and invertebrate species composition was similar to nearby areas not occupied by holdfasts. This and rapid colonization suggest that source populations are local.

Feeding relationships among the holdfast fauna are poorly known, including whether some of the inhabitants move to the surface of the holdfast or elsewhere to feed. Many of the major taxonomic groups are detrital feeders, so the primary source of food is likely to be detritus that accumulates on and in the holdfast. This faunal composition is similar to that in other relatively hard, complex algal structures where isotopic analyses have shown that the food webs are largely detrital-based, with detritus in the form of sedimented particulate organic matter (Grall et al. 2006, Schaal et al. 2012). It is likely that the base of the food web in giant kelp holdfasts is similar but this remains to be investigated. Sea urchins (Tegner et al. "Sea urchin cavitation" 1995) and isopods (Barrales and Lobban 1975), however, may feed on haptera, leading to increased dislodgement of giant kelp during storms.

Benthic habitats are also created by other algae in giant kelp forests, especially those with relatively stiff, densely packed branches such as geniculate corallines. Dearn (1987) used a suction dredge to sample the fauna in 10–15 cm tall mats of *Calliarthron* spp. in the giant kelp forest at Stillwater Cove (figure 5.2D), and found 178 invertebrate taxa from 12 phyla, with a mean density of nearly 300 individuals per 625 cm$^{-2}$. The faunal assemblage was very similar to that found in giant kelp holdfasts, and a similar fauna developed in artificial (plastic) mats. As in holdfasts, the fauna in these mats appear to result from mat structure that provides microhabitats and accumulates detrital and planktonic food. Canopy removal experiments in Tasmania by Edgar et al. (2004) have also shown that these and other large canopy-forming algae can affect the abundance of species by providing a refuge from predators and sources of food.

## HABITAT ASSOCIATIONS OF FISHES

Understory algae and *Macrocystis* fronds in the water column and on the sea surface provide habitat for numerous invertebrates and fishes. Coyer's (1984) very thorough seasonal study of mobile invertebrates on giant kelp fronds at Catalina Island in southern California found 114 species, with copepods and amphipods being most abundant. Fronds also provide habitat for sessile species whose distribution on fronds varies with their life history characteristics but is also influenced by location. The common bryozoan *Membranipora membranacea* was found by Bernstein and Jung (1979) to be most abundant on fronds at the outer edges of a forest at Point Loma, California, because larvae were depleted by settlement and fish predation as currents swept them through the forest. Bryozoan abundance is also affected by water flow because it affects food availability and feeding (Arkema et al. 2009). Ebeling et al. (1980) reviewed studies of fish–habitat associations in California, and used cine-transects to sample the habitat and distribution of 51 fish species at various locations and depths within mainland and island kelp forests near Santa Barbara in southern California. The fishes segregated into four general habitats plus a group of "commuters" that co-occurred throughout the water column (figure 5.5). A similar study in a kelp forest with a cobble bottom near San Diego found that giant kelp could at least partially provide the "Kelp Rock" habitat

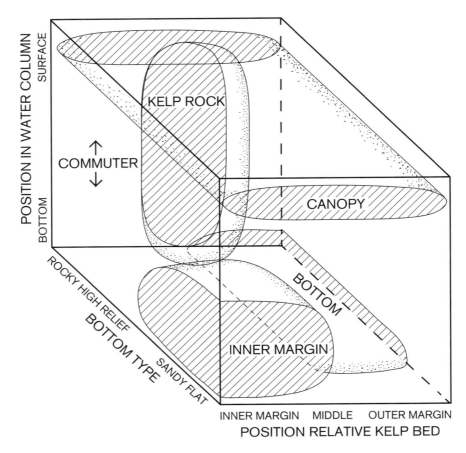

FIGURE 5.5

Daytime habitats occupied by particular groups of fishes in kelp forests near Santa Barbara, California. Examples within each habitat include the following. Kelp Rock: blacksmith (*Chromis punctipinnis*). Canopy: kelp surfperch (*Brachyistius frenatus*). Inner Margin: dwarf surfperch (*Micrometrus minimus*). Bottom: lingcod (*Ophiodon elongatus*). Commuters are fishes such as kelp bass (*Paralabrax clathratus*) that move between habitats.

SOURCE: Redrawn from Ebeling et al. (1980).

described by Ebeling et al. (1980) when bottom relief was low (Larson and DeMartini 1984). Reed et al. (2006—"Quantitative assessment") found that rock cover could have a larger effect than rock type on fish assemblages. Fishes in southern Chile also tend to be associated with particular types of bottom topography and habitat (Moreno and Jara 1984). The general relationship between algal and fish assemblage structure, however, is less clear. Holbrook et al. (1990) surveyed 26 reefs in southern California and concluded that algal structure (i.e., giant kelp density plus the abundance of foliose algae) had little effect on fish species richness and composition. They argued that this was a consequence of the generalized ecological requirements of many of the fishes which,

FIGURE 5.6

Sketch of a portion of the Point Loma kelp forest showing patches of algae with characteristic canopy guilds, including the floating canopy of giant kelp *Macrocystis 'pyrifera,'* the stipitate canopy of the understory kelps *Pterygophora californica* and *Eisenia arborea*, the prostrate canopy of the kelp *Laminaria farlowii* and the fucoid *Stephanocystis (=Cystoseira) osmundacea)*, and patches of turf and nongeniculate (crustose) coralline algae.

SOURCE: From Dayton et al. (1984).

in turn, reflected the highly dynamic nature of algal composition on the reefs. In contrast, Pérez-Matus et al. (2007) found higher fish species diversity at sites with both *Macrocystis 'integrifolia'* and *Lessonia trabeculata* in northern Chile. Fish assemblage structure also varied with depth, a likely reflection of changes in microhabitats.

## PATCHES

Benthic species commonly have aggregated distributions within kelp forest habitats (e.g., Rosenthal et al. 1974), and dense aggregations are referred to as "patches." Patches composed of dense stands of various seaweeds and sea urchins at scales from around 1 m² and larger are particularly common. Dayton et al. (1984) described these at Point Loma, the most visually obvious being discrete stands of *Macrocystis*, long-lived understory kelps, and areas with high densities of sea urchins and nongeniculate coralline algae (figure 5.6). Patches often occur in particular topographic settings. An example is the understory kelp *Pterygophora californica* that occurs in dense stands on elevated plateaus in Stillwater Cove. Its shading in this habitat reduces the abundance of giant kelp, but giant kelp is still abundant as it recruits in other patches where *P. californica* is not as dense (Reed and Foster 1984). The removal of algal canopies in eastern Tasmania, including that of giant kelp at one site, caused very little change in associated algae, a result attributed to the high cover and mosaic nature of bottom cover "turf" composed of green (*Caulerpa* spp.) and foliose red algae, and geniculate corallines (Edgar et al. 2004) that were apparently little affected by changes in light. Spatial variation also occurs within patches, an essential feature driving patch dynamics as physical and biological disturbances remove organisms or clear the substratum, initiating succession.

## TEMPORAL VARIATION

Previous discussion has focused on the spatial variation that characterizes kelp forests at patch-to-geographic scales. Based on the extant historical information (mostly from California) beginning in the early 1900s, long-term, persistent changes in kelp forest size have occurred at sites in British Columbia and central California as a result of sea urchin mortalities from predation by expanding sea otter populations, and disease (see Community discussions above; review for California in Laur et al. 1988). Fish (especially large ones) and abalone populations have also declined due to fishing (Foster and Schiel 2010; see Chapters 11 and 12), and some large kelp forests in southern California have declined due to deterioration in water and benthic quality associated with coastal development and sewage discharge (Schiel and Foster 1992, Foster and Schiel 2010). The considerable literature on these more persistent changes, often at only a few sites and driven by anthropogenic disturbance, has tended to diminish an appreciation of the naturally dynamic nature of kelp forest communities and the coastal environments where they occur. These dynamics are discussed below.

Before discussing temporal variation, however, it is necessary to comment on the methods used to assess it in conjunction with spatial variation. The extensive canopy of most giant kelp forests on the sea surface allows a measure of *Macrocystis* abundance as canopy cover, without getting wet. Canopy areas were first measured in California and Australia from boats that moved around canopy boundaries while taking positions using triangulation from measured landmarks, sextants, and other methods, and then calculating the area within (Crandall 1915, Cribb 1954). In both countries, boats were soon replaced by aerial photographic surveys from planes, using landmarks to scale the photos. This method can survey large areas of coastline relatively rapidly and take account of open patches within canopies, which is difficult to do from a boat.

Photosynthesizing algae reflect strongly in the infrared portion of the light spectrum, while seawater absorbs strongly at these wavelengths, so infrared photos give the best resolution. For consistency, current methods take photos only on clear days with a swell less than 1.5 m, tides of +0.3 m or less, and a sun angle of greater than 30° above nadir (e.g., MBC 2013). Similar results can be obtained using multispectral sensors. Deysher (1993) reviewed the pros and cons of various methods, including satellite images, for giant kelp canopy assessments in southern California, and Anderson et al. (2007) used and evaluated various methods for other kelps that reach the sea surface in southwestern South Africa. Suitable cameras could also be used on unmanned aerial vehicles (UAVs). Repeated surveys have provided a rich record of canopy dynamics in southern California and Tasmania that can be correlated with various environmental phenomena (e.g., figure 10.1). These aerial methods, however, may be replaced by images from satellites (see figure 11.11). Landsat 5 and successors with a multispectral Thematic Mapper (TM) have resolutions of 30 m or less and can scan the entire earth every 16 days. To date, the most thorough use of this method was by Cavanaugh et al. (2011) who analyzed 25 years of Landsat images with high resolution in the near-infrared (760–900 nm), combined with area-per-biomass relationships obtained in the field, to examine local to regional variation in the Santa Barbara, California, area. This method requires considerable image processing but has the potential to track changes in giant kelp canopy area and biomass worldwide.

The advent of scuba in the 1950s (see Chapter 1, History of Research) allowed in situ assessment of plant and frond densities and plant sizes, measurements that have been essential for demographic studies (Chapter 4). They are usually done in conjunction with assessment of the cover and density of other organisms occurring with giant kelp. Giant kelp frond and plant densities have also been determined using sonar (e.g., Grove et al. 2002). Studies focused on demography or spatial and temporal dynamics of the community, therefore, have generally used in situ methods and occasionally sonar, whereas those concerned with giant kelp abundance, for example as it relates to harvesting potential or responses to climatic variables, have increasingly used remote sensing methods.

PATCHES WITHIN FORESTS

The dynamics of patches within a kelp forest have been the focus of over 40 years of study by Dayton and colleagues, primarily at Point Loma (e.g., Dayton and Tegner 1984, Dayton et al. 1984, 1992). Repeated sampling along fixed transects during and after several changes in ocean climate have shown a dynamic patch structure driven by complex interactions related to the differential effects of various disturbances such as large waves, low nutrients, and sea urchin grazing, the life history features of particular species, and interactions among species. Further variation occurred during successional events after disturbances, depending on the degree of disturbance and the ability of potential colonists to recruit and survive, an ability affected by subsequent ocean climate and interactions among colonists. *Macrocystis* had especially strong effects on understory algal patches through its ability to affect light on the bottom. Patch survivorship varied among species, with patches of some understory kelps persisting for over 10 years (see Chapter 4). These studies indicate that, at least at Point Loma, large-scale oceanographic events can have great effects on the entire forest but recovery is driven by the aggregate of diverse responses of patches.

A very common feature of kelp forests are patches created by sea urchin grazing (see discussion of forests within provinces above). Foster and Schiel (1988) summarized the size of these patches based on information from 224 surveys done outside the range of sea otters in California. Patches within these sites ranged in size from <1 m² around crevices and depressions where urchins aggregate to thousands of square meters, where abundant *Strongylocentrotus purpuratus* and *S. franciscanus* actively forage. The great majority (86%) of patches, however, were between 1 and 10 m². The larger patches are formed when drift algae are reduced, inducing a change in sea urchin behavior from passive to active foraging (e.g., Ebeling et al. 1985, Harrold and Reed 1985) although large recruitment events of sea urchins may also be important (Watanabe and Harrold 1991). In California, conditions favoring kelp recruitment (Harrold and Reed 1985) or storms that remove the exposed sea urchins (Ebeling et al. 1985, Watanabe and Harrold 1991) commonly result in urchin-patch longevity of only a few years (Foster and Schiel 1988, Graham 2004). Such patches are dominated by sea urchins and nongeniculate coralline algae, and usually "deforested" (with few or no large kelps), but they are not necessarily "barren" of other algae and invertebrates. Graham (2004), for example, used data from repeated surveys of many sites around the Channel Islands off southern California, and found that of a total of 275 common species, only 25 were absent from areas characterized as deforested.

These within-forest dynamics also occur at smaller scales, as Foster (1975b) found using concrete blocks placed within a kelp forest. The subsequent biological composition on this initially unoccupied substratum changed with successional age and the time of clearing. Most algae (including *Macrocystis*) and many sessile invertebrates could be initial colonizers, depending on when the blocks were placed into the forest, and changes in dominance over time were largely the result of differences in growth rates and apparent competitive success. The presence of topographic edges at the scale of centimeters enhanced giant kelp recruitment, illustrating even smaller-scale patch structure resulting from turbulent flows around microsites.

Complete clearings within forests are not unusual, and they occur at various scales during large wave events that scour the bottom, uncover rock buried by sand, overturn boulders, and expose new surfaces by fracturing and moving rock (Ebeling et al. 1985, Dayton et al. 1989). As argued by Foster and Schiel (1993), numerous small areas among and within patches are usually available for recruitment, and recruitment success per area need not be high to produce adult *Macrocystis* densities of around 1 plant per 10 m² that characterizes a well-developed forest in southern California (see Chapter 4, Demography). Given these features, combined with a massive reproductive output, survivorship of microscopic stages for at least a few months, and the increase in light on the bottom when surface canopies are thinned or removed, the ability of *Macrocystis* populations to recover rapidly and persist should be expected, and a lack of recovery would be surprising.

EL NIÑO EFFECTS

A growing number of studies have documented local and regional spatial changes at seasonal and interannual time scales, particularly in California and Baja California.

The changes that have been most studied were initiated by high temperature and low-nutrient conditions, large waves associated with El Niños, and large wave events from storms (e.g., waves with a height around 6 m and period greater than 17 sec; Ebeling et al. 1985, Tegner and Dayton 1987). El Niños are often followed by La Niña periods with cold, nutrient-rich water that affects recovery of northern hemisphere populations (reviews in Tegner and Dayton 1987, Edwards 2004). The two most consistent effects of these disturbances are the severe depletion or elimination of giant kelp, and its generally rapid recovery. This was first noted in southern California after the 1957–1959 El Niño, but the effects were confounded by those from sewage discharge and other anthropogenic influences (review in Foster and Schiel 2010). By the time of the next large El Niño in 1982–1983, the effects of a large El Niño alone could be better assessed. The kelp forest at San Onofre, north of San Diego, illustrates many of the effects of high temperature and low-nutrient conditions. Fixed plots within this forest have been sampled since 1976 (to at least 2013) as part of monitoring for the effects of the once-through cooling system at the San Onofre nuclear power plant (e.g., SCE 2001). Most large organisms sampled, including giant kelp, severely declined after the 1982–1983 El Niño, with giant and understory kelp recovery occurring within 2–4 years but larger sea urchins (*Strongylocentrotus* spp.) taking 10 years (figure 5.7). The bat star, *Patiria miniata*, a ubiquitous inhabitant of California kelp forests that can be killed by disease at high water temperatures (Tegner and Dayton 1987), had not recovered after 20 years. In contrast, another echinoderm, the white urchin *Lytechinus pictus* (=*anamesus*), recruited in high numbers during the El Niño but went to near zero within 9 years. Subsequent El Niños generated great declines in *Macrocystis*, which always recovered within a few years, whereas the understory kelp *Pterygophora* underwent a general decline after repeated events (figure 5.7). The 1982–1983 El Niño essentially eliminated *Macrocystis* and understory algae (with the exception of nongeniculate corallines) within a giant kelp forest at Catalina Island off southern California but had little effect on other kelps and associated species in shallow (<3 m) and deep (>25 m) water above and below the giant kelp forest (Foster and Schiel 1993). Various ephemeral algae and the invasive fucoid *Sargassum muticum* colonized the area, but the formerly abundant species recolonized and the *Macrocystis* forest recovered in around 4 years. *Macrocystis* populations at their southern limit in the northeast Pacific are strongly affected by El Niños but generally recover in 1–2 years. In this area, the understory kelp *Eisenia arborea* can inhibit *Macrocystis* recovery and, at some sites with little suitable substratum and dense stands of *Eisenia*, this inhibition may last for 20 years (review in Edwards and Hernández-Carmona 2005).

In central California, the 1982–1983 El Niño caused relatively minor changes in temperature and nutrients but generated exceptionally large storm waves (review in Foster and Schiel 1985). We observed that many entire *Macrocystis* plants were removed in Stillwater Cove, and the few small fronds on plants that remained were eliminated by turban snails whose density greatly increased on the remaining plants when much of their habitat, giant kelp fronds, was removed. Strong wave surges stripped the blades

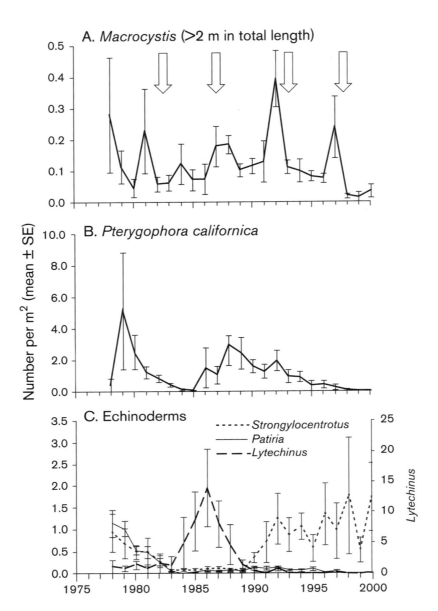

FIGURE 5.7

Temporal variation of common species in the kelp forests at San Onofre, near San Diego, California (A–C), and at Caleta Constitución, Antofagasta, Chile (D–F). San Onofre data are means (±1 SD) of counts in six stations and Caleta Constitución data (Panels D, E) are from counts along four transects perpendicular to shore; Panel F is from 20 quadrats in the kelp forest. Arrows and boxes indicate strong El Niño events.

NOTE: See text for species and ecomorphs.

SOURCE: Data from SCE (2001) and Vásquez et al. (2006).

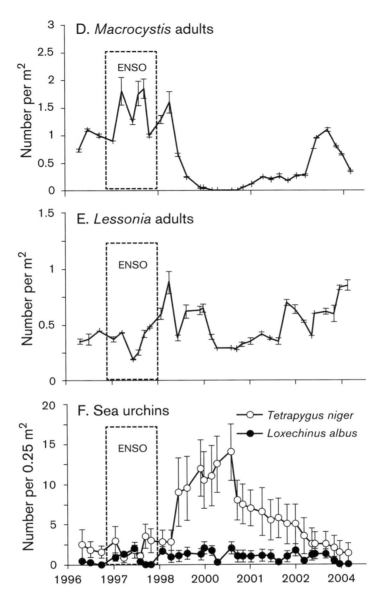

FIGURE 5.7 *continued*

from the understory kelp *Pterygophora californica*, dislodging some plants, and the bottom was scoured of most sessile animals. Coralline algal mats survived but were thinned. Blooms of the annual foliose brown alga *Desmarestia ligulata* developed during spring and summer following the massive storm disturbance, with some recruitment of *Macrocystis* and *Pterygophora. Desmarestia* then declined, and further recruitment and growth of kelps resulted in the return of pre-disturbance patterns of abundance and distribution by the summer of 1984, 1.5 years after the disturbance. With the exceptions of the near elimination of adult *Macrocystis*, changes in turban snail grazing, and *Desmarestia* blooms, these temporal changes were similar to those of typical seasonal changes at this site that result from disturbance by winter storms (Foster 1982, Reed and Foster 1984). At many wave-exposed sites in central California south of Monterey, the surface canopy changed from *Macrocystis* to *Nereocystis luetkeana* in the summer–autumn of 1983, but had changed back to *Macrocystis* by 1985 (M. Foster, pers. obs.). Foster and VanBlaricom (2001) found abundant bladeless stipes of the understory kelps *Laminaria setchellii* and *P. californica* beneath giant kelp canopies during surveys of this region in 1985, suggesting these species also increased after the El Niño but were killed by lack of light when giant kelp returned. Large waves alone can also cause changes in giant kelp forests in southern California, as shown by Ebeling et al. (1985) at Naples Reef near Santa Barbara. A large wave event in 1980 removed most giant kelp from the reef, and the consequent reduction in drift food caused sea urchins *(Strongylocentrotus* spp.) to switch from passive drift feeding to active foraging, removing most other algae except nongeniculate corallines. This caused a decline in surfperches that feed on invertebrates in the understory. A second storm in 1983 removed most of the exposed sea urchins, and giant kelp recolonized rapidly, especially on new rock surfaces exposed by the storm. As Ebeling et al. (1985) concluded, storm effects can clearly vary depending on community structure prior to a disturbance.

The loss and recovery of giant kelp associated with the 1982–1983 El Niño in California indicated that effects varied on regional scales, probably due to large-scale variation in the types and severity of disturbances among forests in Baja California, southern California, and central California. The exceptionally broadscale, multisite surveys of Edwards (2004) before and after the 1997–1998 El Niño thoroughly documented these differences and showed that the regional effects of El Niños on giant kelp abundance can supersede processes affecting local variation. Recovery, however, was more variable at smaller scales, presumably a reflection of variation in local physical conditions, maturity of the giant kelp stand prior to disturbance, sources of propagules, and interactions with other members of the community. Perhaps the most remarkable result of this study, as well as previous site-specific investigations, is that regardless of region and amount of loss, giant kelp is generally able to recover within 2 years.

The oceanography from northern Chile to the coast of southern Peru is generally similar to that from southern California into Pacific Baja California where *Macrocystis* is concentrated in areas of upwelling within gradually warming coastal currents. Like its counterpart in the northeast Pacific, this area of the southeast Pacific is greatly affected by El Niños. The 1982–1983 El Niño resulted in the elimination of many *Macrocystis* 'integrifolia' forests and a severe reduction in the large kelp *Lessonia nigrescens*, which often dominates the low intertidal above giant kelp (Camus 1994, Fernández et al. 1999, Vásquez et al. 2006). Although some *Macrocystis* recruitment occurred the year following the 1982–1983 El Niño, recovery of giant kelp took up to 10 years, and *L. nigrescens* longer, probably due to differences in dispersal from remnant populations (Martínez et al. 2003). The effects of the 1997–1998 El Niño were not as severe, but revealed complex dynamics related to upwelling and changes in grazers and predators (figure 5.7). The surveys of kelps and large benthic invertebrates by Vásquez et al. (2006) in a kelp forest at Caleta Constitución in northern Chile between 1996 and 2004 revealed little effect of the El Niño on *M. 'integrifolia'* but large increases in recruitment and survival of the sea urchin *Tetrapygus niger* during the La Niña that followed. Grazing by this sea urchin had a greater effect on giant kelp abundance than during the El Niño, when effects were mitigated by local upwelling. The greater survival of *T. niger* during the La Niña was attributed to reduced predation by sea stars, whose diet and distribution changed after the El Niño. Giant kelp re-established by 2003 after a decline in *T. niger*, an effect attributed to the return of predatory sea stars to their former distribution and diet. The abundance of the kelp *L. trabeculata* and the sea urchin *Loxechinus albus* did not vary greatly during the study period (figure 5.7), and overall species richness in the forest remained relatively constant during the study period.

## LONG-TERM PERSISTENCE

Most long-term studies have been done in southern California where scuba first became widely used in kelp forests, and where harvesting and the effects of sewage discharge stimulated research (Chapter 1, History of Research). There are also long records for Tasmania, particularly from aerial photographs from 1946–2007, which have shown a general decline in giant kelp (Johnson et al. 2011; discussed in Chapter 13).

In California, the pioneering, 7-year study by Rosenthal et al. (1974) of the Del Mar kelp forest north of San Diego was the first to examine longer-term variation in the kelp forest community. There were no El Niños during the study and the forest exhibited dynamic stability. Adult *Macrocystis* abundance varied as mortality of large, old plants from storms was followed by recruitment. With this short-term exception, along with those of a few sessile invertebrates that either declined, increased, or varied in abundance annually, the composition of the forest stayed remarkably stable over the years of the study.

The longest set of giant kelp forest observations was begun by Neushul and colleagues in a giant kelp forest on the southwest end of Anacapa Island, 20 km offshore from southern California (Neushul et al. 1967). They semiquantitatively documented the distribution and abundance of giant kelp and other conspicuous organisms along a 700 m long transect from the intertidal zone to a depth of 34 m in 1965–1966. The site was observed and described again from 1981 through 1987 by Ambrose et al. (1993) and then by Carroll et al. (2000) from 1988 through 1998. In the mid-1960s, *Macrocystis* was abundant at depths from 8 to 31 m, with different species of stipitate kelps above 8 m (*Eisenia arborea*) and below 31 m (*Agarum fimbriatum*). Red and purple sea urchins (*Strongylocentrotus* spp.) were common from the intertidal zone to 13 m depth, and the white urchin *Lytechinus pictus (=anamesus)* from 13 m to the end of the transect, but extensively grazed areas were not apparent. By 1981, giant kelp occurred only above 8 m, with the rest of the transect dominated by *Strongylocentrotus* spp. in shallower water and *L. pictus* in deeper water, and with much of the bottom extensively grazed and covered by nongeniculate coralline algae. Extensive recruitment of giant kelp was seen to at least 20 m depth in 1985 during the La Niña that followed the 1983–1984 El Niño, but by 1987 the composition of the site again became similar to its state in 1981. By 1998 the entire transect was mostly devoid of kelp, sea urchins were abundant above 18 m depth, and brittle stars (*Ophiothrix spiculata*) had become dominant below 18 m, covering nearly 100% of the bottom.

Surface waters around this part of Anacapa Island are warmer and more affected by El Niños than islands to the east, and the island is biogeographically more similar to islands to the south (Murray et al. 1980, Hamilton et al. 2010). Ambrose et al. (1993) and Carroll et al. (2000) suggested the large changes observed over 35 years were likely due to extended periods of warm water and low nutrients unfavorable to giant kelp, and possibly enhancing recruitment of some echinoderms, with occasional pulses of cooler, nutrient-rich water favorable to giant kelp. Furthermore, during periods when giant kelp can re-establish, its productivity may not be sufficient to offset losses from high grazer densities.

The information from these and other studies on the recovery from El Niños indicates that giant kelp and associated species in California recover from most natural disturbances, even large ones like El Niños, within 1–10 years, which is not unusual for marine communities (Foster et al. 1988, Jones and Schmitz 2009) composed of many species with rapid growth, relatively short life spans, and widely dispersed spores and larvae. Moreover, historical records from explorers and more modern records from some areas in California show that forests have persisted in the same locations for hundreds of years. Connell and Sousa (1983) argued that populations should be considered to have a stable equilibrium point only if they persist longer than their turnover times (i.e., when all adults are replaced) in an area large enough that replacement comes from within. Although red sea urchins (*Strongylocentrotus franciscanus*) can live over 100 years (Ebert and Southon 2003) and many nongeniculate corallines are probably long-lived (Frantz et al. 2005), the life span of *Macrocystis 'pyrifera'* is less than 10 years and the life spans of many longer-lived associated species are only in the order of tens of years, well within known persistence times of populations. This may not be the case in Tasmania where a persistent change in oceanographic conditions (Johnson et al. 2011) may prevent recovery.

Cavanaugh et al. (2013) combined satellite-derived biomass estimates of individual kelp stands, wave, and sea-surface temperature data (as a proxy for nutrients), sea urchin abundance, and giant kelp recruitment in an effort to link population and landscape ecology and reveal factors that may affect giant kelp population dynamics at different spatial scales. The data were obtained from repeated sampling of kelp forests over

550 km including the coast around Santa Barbara and northward, as well as offshore islands. Variables were measured several times within the years 2000–2011. Additional data were used to determine the relationship between satellite images of canopies and kelp canopy biomass (Cavanaugh et al. 2010, 2011). These showed the strong relationship between the biomass of giant kelp and significant wave height of winter storms north of Point Conception, as suggested by previous small-scale studies (e.g., Kimura and Foster 1984). Cavanaugh et al. (2013) found declines in population synchrony (similarities in giant kelp biomass within and among stands) at scales of 50 m to 1.3 km, correlated with variation in sea urchin abundance and giant kelp recruitment, at scales of 1.3–172 km, which was most strongly related to differences in wave disturbance. These studies are noteworthy not only for their innovative use of remote sensing and linking of population and landscape scales, but also because they provide a much needed spatial context for evaluating the relative importance of recruitment, grazing, and storm disturbance. It is likely that the findings may differ in other regions, depending on their particular biological and oceanographic settings (see Argentinean and some Chilean giant kelp forests discussed previously), but the approach of Cavanaugh and colleagues, applied to other regions, holds great promise for providing the information needed for a global understanding of the causes of giant kelp population dynamics and, eventually, the effects of these on giant kelp communities.

## DIVERSITY

The causes of differences in diversity among giant kelp forests are best considered at different spatial and temporal scales. On a large spatial scale, Darwin (1839) observed that "we see the fucus [*Macrocystis*] possessing a wider range than the animals that use it as an abode," referring to differences in the species occurring with giant kelp among the now recognized biogeographic provinces in Chile. It is clear from the community descriptions in the previous section that differences in diversity or species richness of giant kelp forests among provinces is a function of isolation, how long they were isolated, ocean climate, frequency of large-scale disturbances (such as glaciation), and evolutionary history of each province as well as some locations within it. An example of such influences is the 50 or so species of fishes in southern California kelp forests compared to three at the subantarctic Marion Island.

The causes of variation in species composition among forests within a region are also implicit in the community descriptions. In this case, assuming all possible species that occur in forests within a region could occur in a particular forest, the question becomes: in what ways are some local forests more diverse than others? Total diversity of macroorganisms alone has never been rigorously assessed at multiple sites within a region, but available quantitative information and qualitative observations indicate that at a primary level, habitat diversity and persistence related to substratum hardness and topography are particularly important, as is water quality. For instance, sites with high-relief and topographically complex hard rock in relatively clear water with low sedimentation generally appear to be the most diverse (e.g., the examples above of southern California islands vs. the mainland and Stillwater Cove vs. Sand Hill Bluff). At a secondary level within a site, diversity is no doubt affected by the structure of some of the community members that can occupy the site (e.g., coralline mats) and disturbance at various scales (Cowen et al. 1982, Dayton et al. 1984), including disturbance caused by episodes of sea urchin grazing (Byrnes et al. 2013). These studies, observations, and experiments, and those discussed in Chapter 7, show that sessile animals and understory algae can compete for space, and that surface canopies of giant kelp can reduce understory algae via reduction in light. Connell's insightful 1978 paper argued that intermediate disturbance is essential to maintaining diversity in tropical rain forests and coral reefs as it removes dominant competitors, thereby reducing competitive exclusion. The argument seems quite applicable to giant kelp forests.

————————

Community studies have made a major contribution to our knowledge of many local giant kelp forests and, when viewed in aggregate within and across regions, have revealed important general processes affecting species throughout their distributional range. Such studies have established that giant kelp occurs with many different assemblages of associated species in areas with different ages, evolutionary histories, and disturbances. There is no universal giant kelp community, and the most important structuring processes are those of storms and high temperatures / low nutrients on giant kelp, and the direct and indirect effects of these on associated species. Even the presence of giant kelp communities in similar coastal locations can be caused by different oceanographic and coastal features, such as being in the lee of headlands (e.g., Sand Hill Bluff) or associated with localized upwelling (e.g., Baja California). Appreciating such wide variation in what we understand as "giant kelp communities" is truly a case of celebrating diversity.

# 6

# DETACHED GIANT KELP COMMUNITIES, PRODUCTION, AND FOOD / CONTROL WEBS

In regions with high marine macrophyte production and / or
kelp forests the community structure of sandy beach macrofauna
is closely linked with the input and fate of macrophyte wrack.

—Dugan et al. (2003)

We still face a dilemma: is it appropriate to say that herbivory during early
successional stages is irrelevant, or should we present estimates of per capita
interaction strength and expect them to be general?

—Sala and Dayton (2011)

## DETACHED GIANT KELP COMMUNITIES

Kelp populations are exceptionally productive and an estimated 80% of the productivity (excluding dissolved organic carbon) ends up as detached detritus (review in Krumhansl and Scheibling 2012). Giant kelp is no exception, and it is therefore not surprising that detritus from *Macrocystis*, moved by currents and wind, can be an important source of habitat, food, and nutrients for other communities and components within kelp forests. Large size and distribution throughout the water column make fronds and entire *Macrocystis* plants especially vulnerable to detachment from waves and surge, and grazing concentrated in and above the holdfast results in further losses (e.g., Tegner et al. 1995—"Sea urchin cavitation"). These processes, plus loss of tissue from natural senescence (Rodriguez et al. 2013), can release massive amounts of giant kelp detritus into the water. Hobday (2000b), for example, estimated that 20–40% of adult plants were lost from the Point Loma kelp forest during winter storms, and Gerard (1976) estimated over 60% of the giant kelp production in a relatively protected kelp forest in Monterey Bay became detritus. Much of this remains buoyant by floats on the blades and ends up either onshore or else drifting in currents and wind offshore where it may lose buoyancy and sink (figure 6.1). Pieces of kelp generated within the kelp forest often

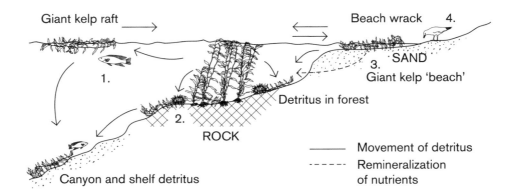

1. Fish feeding on associated invertebrates.
2. Sea urchin feeding on detritus.
3. Microfauna feeding on detritus and microbes.
4. Bird feeding on amphipods and insects.

FIGURE 6.1

Stylized depiction of the movement and use of giant kelp detritus. Floating rafts are occupied by numerous species. As kelp gets washed up onshore as wrack or sinks to the sea floor, it is eaten by a wide range of species.

lack floats, or have floats that have lost their buoyancy, and end up on the bottom where they may remain or be transported to other benthic areas. The amounts and transport of floating detritus were commonly observed by early explorers including Hooker (1847) who noted giant kelp's "wandering habits," that some waters "literally swarm with *Macrocystis*," and a Falkland Island beach "lined for miles with entangled masses." Such inputs, common in marine communities where transport via various water motions can blur boundaries, are currently referred to as "subsidies" (review in Polis et al. 1997), but detached giant kelp can have very large effects on other habitats, effects that suggest it could be considered a foundation species as it is when attached within a forest.

GIANT KELP RAFTS

Floating masses of giant kelp, often referred to as "rafts" or "paddies," can constitute a large proportion of objects that drift on the sea surface where giant kelp occurs (review in Thiel and Gutow 2005a). In the extreme, storms detach entire plants that may become entangled with other plants and cause their detachment (Rosenthal et al. 1974). When transported offshore, these entwined plants can form rafts with surface areas exceeding hundreds of square meters (Harrold and Lisin 1989) with denuded fronds and holdfasts hanging several meters below the surface (Kingsford 1995). Depending on how long they persist, rafts can be composed of a mix of their original associates from the kelp forest and new pelagic colonizers. The review of Thiel and Gutow (2005b) noted over 1200 species associated with rafting objects of all types, and Hobday (2000c) reported 150 macrofaunal species associated with giant kelp rafts in southern California. The initial floating giant kelp community is composed of giant kelp and other plants, including some without floats that are attached to the holdfast or wrapped up in the raft. It also includes animals living in the holdfast and on the fronds. Even the odd terrestrial animal can go rafting; Prescott (1959) reported a jackrabbit on a large raft of giant kelp offshore in southern California. Vásquez (1993) examined changes after dislodgement in the southern California holdfast fauna by removing holdfasts from the benthos and mooring them in the water column. The abundance of individuals in the most abundant faunal groups, such as polychaetes and ophiuroids, greatly declined within 20 days, but a few individuals from all groups remained until around 100 days when the holdfasts began to disintegrate. Isopods and amphipods were the exception, increasing later in the experiment, probably because they are brooders and reproduce inside holdfasts. In contrast, the fauna in the very diverse (172 species) Tasmanian kelp forest holdfasts studied by Edgar (1987) did not greatly decline after dislodgement and holdfasts survived over 190 days, eventually succumbing to grazing by isopods and amphipods. Unattached giant kelp can grow in protected inlets (Gerard and Kirkman 1984) and also while floating in the open sea, as indicated by their physiological performance (Rothäusler et al. 2011) and elongated haptera. However, the relatively low nutrients in the surface waters of the open ocean may reduce growth below the ability to compensate for grazing losses or may contribute to frond deterioration.

Depending on their time at sea, rafts can be colonized by numerous pelagic invertebrates (e.g., gooseneck barnacles, *Lepas* spp.) as well as larval, juvenile, and adult fishes (Mitchell and Hunter 1970, Kingsford 1995, review in Thiel and Gutow 2005b). The sessile animals and smaller fishes feed on plankton that might be considered a subsidy from the surrounding pelagic community. Hobday (2000c) considered giant kelp rafts in southern California waters as floating islands, but, contrary to predictions from island biogeography, species richness varied little with raft age because as kelp forest species declined, recruitment of pelagic species increased. Diversity did increase with raft weight.

Rafts of giant kelp and other floating seaweeds contribute to the transport of larval and juvenile fishes to nearshore habitats (Kingsford and Choat 1985, Kingsford 1995). Rafting giant kelp and associated invertebrates and small fishes are also a source of food for larger, grazing, and predatory fishes that accumulate around rafts (Mitchell and Hunter 1970; figure 6.2). Large predatory jacks, such as the yellowtail *Seriola lalandi* in southern California, may be attracted to giant kelp rafts because of their structure (similar to FADs) or to feed on smaller fishes there. These highly prized sport fishes make giant kelp rafts productive fishing sites (Waters 2002).

Life span, currents, winds, and landing sites interact to affect how far giant kelp rafts travel. Like Vásquez (1993), Hobday (2000a) estimated a raft life span of 100 days, based on changes in the lengths of blades while floating. He also found that artificial drifters designed to mimic kelp raft movement were transported a maximum of around 1000 km in 40 days in southern California and therefore could be transported to all of the mainland

FIGURE 6.2

School of halfmoon (*Medialuna californiensis*) around a raft of detached giant
kelp adrift off southern California. Fishes are 20–30 cm long.

SOURCE: Photo courtesy of Ron J. McPeak, UCSB Library Digital Collection.

and offshore islands. The distribution and genetics of giant kelp populations in the Southern Ocean suggest that once giant kelp migrated from the northeast Pacific to southern South America, initial colonization of the rest of the southern hemisphere occurred via very long distance dispersal in the Antarctic Circumpolar and related currents (see Chapter 1, Biogeography). While the points of origin may vary, this scenario is also likely for the large, buoyant fucoid *Durvillaea antarctica* (Fraser et al. 2009) and attached invertebrates like the brooding bivalve *Gaimardia trapsina* (Helmuth et al. 1994). Pearse et al. (2009) argued that long-distance dispersal of brooding invertebrates is the most likely reason for their abundance in Antarctica, and kelp rafts may have provided the dispersal vector. The probability of long-distance dispersal also seems reasonable based on raft life span and current velocities. The surface waters of the Antarctic Circumpolar Current move at around 0.5 m sec$^{-1}$ (Fyfe and Saenko 2005) or about 43 km day$^{-1}$. Therefore, this current alone (neglecting wind which also moves rafts; e.g., Harrold and Lisin 1989) could transport a kelp raft over 4000 km in 100 days, well within the distances between most island groups in the Southern Ocean. With a longer life, as suggested by Edgar (1987), and/or some help from the wind, rafts could also drift over the 5700 km stretch of open ocean from Kerguelen Island to Tasmania. Giant kelp rafts can clearly support a diverse fauna and are an important, long-distance dispersal mechanism for giant kelp and its associated flora and fauna.

GIANT KELP "BEACHES"

Giant kelp rafts deposited onshore become wrack (figure 6.1). With the exception of pieces of wrack observed being eaten by intertidal sea urchins, turban snails, and other grazers, the effects of giant kelp wrack on rocky intertidal shores are poorly known. They may be large in some areas, as suggested by Bustamante et al. (1995) who found other species of drift kelp to be a major contributor to grazer populations on South African rocky shores. Enormous piles of wrack can accumulate on sandy beaches (figure 2.2), where they can be an important part of sandy beach communities. Sandy beaches are complex, very dynamic systems (McLachlan and Brown 2006), and kelp detritus can be an important source of food (review in Krumhansl and Scheibling 2012).

We will focus here on the animal assemblages associated with giant kelp wrack based primarily on studies in California where most research has been done. Zobell's (1971) thorough, 12-year study of wrack abundance on southern California beaches near giant kelp forests found that in winter, masses of wrack composed primarily of giant kelp along 150 m of beach could attain volumes of 190 m³. Abundance was highly variable, however, with little to none in late spring through early autumn. This seasonal cycle of beach-cast kelp related to storms and thinning of giant kelp is like that of raft production, and appears typical (Yaninek 1980, Marsden 1991), as does local variation among beaches in the amount of wrack deposited at any one time (Zobell 1971, Dugan et al. 2011). Moreover, wrack abundance can change dramatically after first arriving on a beach, with the bulk of it returning to the sea during the next high tide unless stranded high on the beach when the tides are decreasing. Zobell (1971) estimated an average of less than 10% remained "beyond the reach of high water, buried by sand, or otherwise entrapped." Material transported off the beach could be returned to the beach or spend some time as a raft and wash up on another beach.

Beach wrack has been characterized as one of the subsidies to sandy beaches, contributing to the "bottom-up" regulation of the community (Dugan et al. 2011). This could probably be surmised by any observant beach goer who turns over a pile of fairly fresh giant kelp wrack and sees the grazed blades with a fascinating combination of terrestrial fly larvae, plus marine amphipods that rapidly burrow into the sand (figure 2.2C). Macrofauna in and around wrack piles can reach extraordinary densities of nearly 100,000 m⁻² with a richness of over 60 species that well exceeds that found in areas without wrack (Yaninek 1980, Dugan et al. 2003). Dugan et al. (2003) found that species richness and the total number of individuals increased as wrack cover increased. Lavoie (1985) repeatedly sampled cohorts of wrack piles soon after they were stranded as fresh kelp until they were almost completely degraded at around 36 days. He characterized them as "islands of plenty in a sea of sand" with a succession of macrofaunal consumers starting with amphipods, adult flies, and fly eggs, followed by beetles, fly larvae, and pseudoscorpions, and ending with another suite of beetles different from those in mid-succession as well as fly pupae. Litterbag (fresh kelp placed in mesh containers) experiments deployed over 18 days by Inglis (1989) found a similar early succession in the first three days followed by an assemblage of meiofauna, fly larvae, and mites. As found by Lavoie (1985), the wrack was rapidly consumed, with only 36–59% remaining at the end of the experiments. Lastra et al. (2008) reported the preferential consumption of brown algal wrack on beaches near Santa Barbara, California, where the amphipod *Megalorchestia corniculata* may eat 55% of palatable giant kelp. Wrack buried for various lengths of time may exist longer than 36 days (Yaninek 1980), and very little wrack is consumed if it is transported further inland and dries out (Lavoie 1985). In these situations, however, dried holdfasts may serve as the nuclei for the formation of embryonic sand dunes (figure 2.2D). Amphipods and other primary consumers of wrack are eaten by shore birds, fishes that move onto the beach as the tide rises, and terrestrial animals that forage on the beach. Bradley and Bradley (1993) found that the abundance of shore birds, especially sanderlings (*Calidris alba*), black turnstones (*Arenaria melanocephala*), and black-bellied plovers (*Pluvialis squatarola*) increased along the Palos Verdes Peninsula in southern California as the abundance of offshore giant kelp increased, and Dugan et al. (2003) found that the abundance of plovers was positively correlated with the abundance of wrack on beaches near Santa Barbara. The relationships between beach wrack and the abundance of other foragers remain to be investigated.

The rapid colonization and consumption of giant kelp wrack by primary consumers as well as microbial decomposers (ZoBell 1971, Delille and Perret 1991), combined with further remineralization within the beach sand, recycles considerable quantities of nutrients that are undoubtedly used by microalgae inhabiting the sand as well as organisms living offshore. Dugan et al. (2011) examined the relationship between nutrients in beach pore water and wrack biomass on beaches along the mainland near Santa Barbara, California. They estimated an average wrack biomass input of 50 kg dry weight per year per linear meter of beach. Dissolved inorganic nitrogen concentrations ranged from 1 to over 6000 µmol $L^{-1}$, with concentrations positively correlated with wrack biomass. These nutrients may represent significant inputs to offshore communities as tides and wave action flush nutrients from the permeable beach sands.

Giant kelp beaches, unlike kelp forests, are not universally appreciated, especially when natural beaches and nearshore ecological processes conflict with tourist perceptions and recreational desires for beaches without rotting kelp and flies, beaches that from the point of view of the tourism industry are "pristine." Many beaches are groomed and raked to remove wrack and produce this distorted notion of what "pristine" is (according to the Merriam-Webster dictionary, pristine is defined as *not spoiled, corrupted, or polluted [as by civilization]*). As pointed out by Martin (2008), over 160 km of beaches are estimated to be groomed in southern California alone. This, no doubt, subsidizes rubbish dumps rather than beaches, but also affects beach invertebrates and shore birds. The large-scale disturbance of sand may affect fishes like grunion (*Leuresthes tenuis*) that lay their eggs in the beach. Perhaps educating beach managers about the importance of wrack and tourists about interesting beach animals and processes on naturally pristine beaches will lead to a decline in beach cleaning. After all, who knows how far ecotourism will go in the future, and if rotting kelp, replete with its aroma, and its associated communities and ecological connections will be as valued as sunbathing and marine mammal watching?

Giant kelp detritus that sinks (figure 6.1), whether deposited directly on the bottom or indirectly after drift plants have spent some time on the shore or as rafts, can be abundant offshore where, as on sandy beaches, it provides food and habitat (review in Krumhansl and Scheibling 2012). Soft bottoms composed of sand, mud, or gravel often border the lower limit of kelp forests where exported detritus is a primary source of food for animals like the large, tube-dwelling, omnivorous polychaete *Diopatra ornata* (Kim 1992). On offshore rocky reefs without macrophytes but near algal stands, red sea urchins (*Strongylocentrotus franciscanus*) also depend on algal drift for food (Britton-Simmons et al. 2009). Giant kelp detritus is increasingly recognized as an important source of food and habitat in deep water, especially when near kelp forests and concentrated by water motion and topography. Vetter (1998) and Vetter and Dayton (1999) used a combination of diving, ROVs, and submarines to document the accumulation, distribution, and use of drift algae in submarine canyons near kelp forests in southern California. They found "mats" of drift with up to 100% cover of the bottom in depths of 15–65 m at the head of the Scripps Submarine Canyon. The mats supported an abundant macrofaunal assemblage with extraordinarily high densities ($10^6$ m$^{-2}$) of small crustaceans (Vetter 1995). These were consumed by a wide variety of fishes that congregated around the mats, with *Merluccius productus* (hake) and *Pleuronichthys* sp. (turbot) being especially common. The abundance of accumulated drift kelp declined with depth, with <1% cover below 700 m. Macrofauna are also abundant in the giant kelp and green algal drift that accumulates at the head of the Monterey Submarine Canyon in central California (Okey 2003). Amphipods and polychaetes dominated detritus at this site. Relative abundances and species composition were correlated with changes in the age and composition of the detritus, and the dispersal characteristics and dissolved oxygen tolerance of the fauna. Drift giant kelp and other macrophytes also provide habitat for species such as the persimmon eelpout, *Maynea californica*, which was thought to be rare until collected with detritus in trawls and caught in traps filled with surfgrass (Cailliet and Lea 1977). Harrold et al. (1998) also found large accumulations of drift in the Carmel Submarine Canyon in central California, the head of which is surrounded by giant kelp forests. As in the Scripps Canyon, drift abundance declined with depth. They estimated that *Macrocystis* may account for 20–80% of the particulate organic carbon (POC) reaching the canyon floor at depths between 150 and 450 m. Surveys in southern and central California have noted a very sparse distribution of detritus on shelves near canyons, which is likely the result of the lack of concentration mechanisms and increased consumption by the deep-sea echinoid *S. (= Allocentrotus) fragilis* (Harrold et al. 1998, Vetter and Dayton 1999).

Giant kelp can clearly be an important contributor to the secondary productivity of benthic communities at depths to several hundred meters if kelp forests occur in the vicinity. Like giant kelp on beaches, however, detrital export to these communities varies with water motion that causes detachment, and the abundance of detritus is a function of transport time and decomposition rate. Transport times have not been determined but could be rapid, particularly down the axes of submarine canyons (Vetter and Dayton 1999). Decomposition rates have been investigated in experiments using replicate litter bags retrieved at various time intervals. Albright et al. (1982) placed litterbags with a 1.5 mm mesh and filled with *Macrocystis 'integrifolia'* fronds at 10 m depth on the bottom of a bay in British Columbia. The fronds degraded to a residue of small particles in 50–60 days, largely due to decomposition by microbes. Gerard (1976) used 5 mm mesh bags of *M. 'pyrifera'* placed at 10 m in a Monterey Bay kelp forest and found that 75% of the biomass was lost in 60 days. Foster and colleagues (unpublished) deployed larger-holed (50 mm) mesh bags of *M. 'pyrifera'* on the bottom at a depth of 15 m within the Stillwater Cove kelp forest, at 30 m at the lower edge of the forest, and at 300 m depth offshore. The biomass at 15 and 30 m declined by 75% in 70–90 days, with more rapid loss at 30 m, while less than 1% of the biomass remained after 40 days at 300 m. Amphipods, gastropods, and decapods were most abundant in the shallow bags, while annelid worms dominated at 300 m. Bernardino et al. (2010) placed litter bags of giant kelp on the bottom at a depth of 1670 m offshore from Santa Barbara, and sampled them using an ROV at 100 and 200 days. Macrofaunal abundances increased within 0.5 m of the bag, but diversity decreased, most likely because of increased sulfide and organic content in the adjacent sediment. The time to disintegration was not determined, but photos taken at 100 days indicated that considerable biomass of kelp remained. The investigators argued that while shorter-lived than a whale carcass or wood falls to the sea floor in deep waters, drift kelp contributes "stepping stones" for dispersal in the deep sea.

These studies indicate that at depths to 300 m detritus is completely degraded after around 100 days, and degradation rates may increase with depth in this range but decline in deeper water. The similar results obtained for different mesh sizes indicate that degradation is caused primarily by microbial decomposition, with amphipods and other larger consumers feeding on the partially decomposed and nutritionally enhanced kelp / microbe mix (review in Krumhansl and Scheibling 2012) rather than on fresh material. These relatively rapid decomposition rates probably contribute to reduced detrital abundance on shelves near kelp forests and in canyons below 700 m. Decomposition plus episodic drift generation caused by storms result in drift abundance being highly variable even in canyons, but with large effects on the diversity, abundances, and population dynamics of associated species.

## KELP FOREST PRODUCTIVITY
### GIANT KELP

Given the dominance of giant kelp, it is not surprising that *Macrocystis* has been the focus of most studies of giant kelp forest productivity. Kelp forests are, however, usually inhabited by numerous other species of macroalgae, and they are open systems with transport much affected by water motion. All or parts of macroalgae and sea grasses can senesce or be dislodged by storms and grazers, and can be transported out of the community or in from other communities. Phytoplankton production also occurs within kelp forests and biomass produced by plankton outside the kelp forest is transported into it. Consequently, primary production within a forest is not necessarily a measure of what is available to primary consumers within it. Moreover, primary production available to consumers can vary from live, attached plants, and suspended plankton to drift to particulate organic matter (POM) and dissolved organic matter (DOM). This combination of different sources and forms of production and its use by varied consumers makes unraveling production and energy flow within a forest a formidable and continuing challenge.

Previous chapters have discussed how several factors contribute to the very high growth rates of adult giant kelp sporophytes: having a surface canopy in full sunlight, the ability to translocate the products of photosynthesis, having a high percentage of photosynthetic versus non-photosynthetic biomass, and often having high nutrients in coastal waters. High growth rates indicate high productivity. Unfortunately, the size and complex morphology of giant kelp make measuring productivity difficult. A variety of methods have been used (Table 6.1), but many of the results are questionable because of methodological problems and short-term nature of measurements (review in Foster et al. 2013). The most accurate but most difficult method was pioneered by Gerard (1976), based on changes in standing crop from measurements of frond density, growth, and loss. She measured net primary production (NPP), which is gross primary production minus the production lost in plant respiration. She determined the NPP of giant kelp in a central California forest to be 2.23 kg dry weight $m^{-2}$ $year^{-1}$ (production is integrated over a $m^2$ column from the surface to the sea floor). Similar methods in kelp forests along the southern California mainland have yielded similar results (Reed et al. 2008, Rassweiler et al. 2008; Table 6.1). This method underestimates NPP because it does not account for production in the form of dissolved organic exudates. These exudates may be around 20% of NPP at sites around Santa Barbara (D.C. Reed, pers. comm.). Even without considering DOM, the production of *Macrocystis* alone ranks giant kelp forests among the most productive ecosystems on earth (review in Reed and Brzezinski 2009).

In contrast to NPP of giant kelp, the standing crop of *Macrocystis* (biomass present at any one time) measured in the studies cited above was relatively low over the area occupied by kelp forests, averaging only around 0.4 kg dry weight $m^{-2}$. This seems low given the visual dominance of giant kelp, but the plant is composed of around 90% water. The productivity / biomass ratio illustrates the dynamic nature of giant kelp production and loss, with turnover times of 6–7 times per year (Gerard 1976, Reed et al. 2008). The high NPP of giant kelp therefore results from a relatively small standing crop that turns over many times per year, which indicates that variation in biomass can have large effects on production. In a study at three southern California kelp forests over 4.5 years, Reed et al. (2008) developed a measure of "foliar standing crop" (FSC) based on all plants greater than 1 m tall in several permanent plots. Each plant was measured

TABLE 6.1    Studies of net primary production (NPP) in giant kelp forests.

| Method | Location | Duration of study | NPP (dry kg m$^{-2}$ year$^{-1}$) | Reference |
|---|---|---|---|---|
| GIANT KELP (*MACROCYSTIS PYRIFERA* UNLESS NOTED OTHERWISE) | | | | |
| Product of blade standing crop and net carbon assimilation as measured by in situ incorporation of $^{14}$C by individual kelp blades during 3 hours of incubations. | Monterey, CA | Multiple days during 1 month | 8.68 | Towle and Pearse (1973) |
| Difference between measured dissolved $O_2$ concentration and the concentration predicted by temperature for midday in the middle of Pt. Loma kelp forest at 3–6 m depth. | San Diego, CA | 1 day | 12.12 | Jackson (1977) |
| Simulation results from a model of whole plant growth as a function of environmental parameters that affect the flux of light. | 33 N (San Diego) | NA | 1.88 | Jackson (1987) |
| Standing crop of harvested plots coupled with rates of frond initiation and loss. | Falkland Islands | Spring and fall of 1 year | 2.80 | van Tussenbroek (1989) |
| Product of the density of growing fronds and the mean monthly increase in frond mass. | Monterey, CA | Monthly for 21 months | 2.23 | Gerard (1976) |
| Changes in standing crop based on allometric measurements of fronds and plants in fixed plots combined with independent estimates of biomass loss from tagged plants and fronds. | Santa Barbara, CA | Monthly for 54 months, data collection is ongoing | 2.46 | Rassweiler et al. (2008), Reed et al. (2008) |
| Product of leaf area index (LAI) from *M. 'integrifolia'* harvested from Grappler kelp forest, and net photosynthetic rates measured in the laboratory | British Columbia | Bimonthly for 12 months | 4.33 | Wheeler and Druehl (1986) |
| GIANT KELP FOREST ECOSYSTEM | | | | |
| Dissolved $O_2$ concentrations in diurnal water samples collected in a kelp forest at discrete depths | Paradise Cove, CA | 2 days | 6.39 | McFarland and Prescott (1959) |
| Allometric measurements of changes in standing crop combined with independent estimates of biomass loss for giant kelp. $O_2$ evolution in enclosed chambers for intact assemblages of understory algae. $C^{13}$ incorporation for phytoplankton | Santa Barbara, CA | Monthly for 17 months | 2.73 | Miller, R. J. et al. (2011) |

SOURCE: From Foster et al. (2013).

NOTES: Values of NPP were converted to kg dry weight m$^{-2}$ year$^{-1}$ using a wet/dry ratio of 10.31 and a carbon/dry ratio of 0.286 (Rassweiler et al. 2008). NPP estimated in units of oxygen were converted to carbon using a photosynthetic quotient of 1 (following Rosenberg et al. 1995).

in three sections: subsurface fronds that did not reach the sea surface, water column fronds that did reach the surface, and the portion of fronds in the canopy at the sea surface. By combining the counts with the water depth above the holdfast and the length of the fronds in the canopy, they derived the wet mass of each plant and converted this to dry mass. By revisiting tagged plants and measuring changes in the FSC, they were able to derive a mean growth rate over time. Integrating this growth rate with FSC over time, they derived a measure of NPP (detailed methods in Rassweiler et al. 2008). They found relatively high values of NPP up to 2.4 kg dry weight m$^{-2}$ year$^{-1}$ and that FSC and recruitment collectively accounted for 84% of the variation in annual NPP. Reed et al. (2009) went on to explore a simpler method for estimating annual NPP and found that frond density during summer explained 80% of the variation in NPP from year to year. Reed et al. (2011) argued this relationship was also true for central California given similarities in frond morphology and seasonal growth, mortality, and recruitment. A more general understanding of these relationships could greatly simplify productivity determinations in other regions.

Giant kelp productivity can vary as a result of bottom-up (producer-driven) and top-down (consumer-driven) processes, and abiotic disturbances such as storms. Top-down effects have been highly publicized and highlighted by Jackson et al. (2001), who reviewed the literature supporting their belief in the ubiquity of top-down control of California giant kelp forests. However, given the relationship between productivity and frond density, productivity can also vary depending on wave disturbance that removes fronds individually or as entire plants. Reed et al. (2011) tested the relative importance of these influences by comparing productivity between central and southern California. Central California experiences large, annual wave disturbances, but low-nutrient conditions are rare. Large sea urchins are also rare, in large part because of sea otter foraging. In southern California, wave disturbance is much lower, but episodes of low nutrients are common, sea urchins are more abundant, and sea otters are absent from most of the region. The top-down and bottom-up hypotheses would both predict that central California kelp forests should be more productive but for different reasons: more sea otters, fewer sea urchins, and more nutrients in central California versus few otters, more sea urchins, and less nutrients in southern California. Reed et al. (2011) rejected both of these hypotheses. Instead, they found that the annual reduction in frond density from storms was most important, accounting for a southern California productivity twice that of central California. As they concluded, their study illustrates "the potential for physical disturbance to overwhelm the effects of top-down and bottom-up forces in determining NPP."

Total production of giant kelp at a particular location most likely varied over geologic time as changes in sea level and climate caused variation in the distribution and area of habitat suitable for giant kelp growth. Graham et al. (2010), using bathymetry and palaeoceanographic information, estimated that giant kelp forest biomass in southern California has decreased 40–70% relative to its biomass 14,000 years ago.

Understory algae can be abundant within giant kelp forests, but it seems reasonable to expect that their productivity would be much lower than that of *Macrocystis* because of its seemingly high biomass, surface canopy in full sunlight, and consequent shading of the bottom. This is suggested by Heine's (1983) in situ measurement of the productivity of two understory red algae in a central California kelp forest. On a per-gram of tissue basis, red algal productivity was one-third to one-tenth that of giant kelp. Recent research by Miller, R. J. et al. (2011) comparing giant kelp, understory algae, and phytoplankton production in southern California kelp forests with and without giant kelp produced a surprising result. By removing the giant kelp canopy and maintaining its absence for around 1.5 years, they found that the combined production of understory algae and phytoplankton in the absence of giant kelp can, at times, equal that of giant kelp. They determined giant kelp productivity according to Rassweiler et al. (2008), understory algal productivity using oxygen production in chambers placed in situ over understory algal assemblages and sealed to the bottom, and phytoplankton productivity via $C^{13}$ tracer incubations in light and dark bottles suspended in the kelp forests. Measurements were made monthly for 17 months. Understory algal productivity increased by 500% relative to the understory in control plots with giant kelp, and phytoplankton production increased by 200%. Phytoplankton production increased rapidly following the removal of giant kelp, but there was a lag in understory production related to a gradual increase in understory biomass resulting from recruitment and growth in the higher-light environment. Harrer et al. (2013) found that as for giant kelp, annual NPP by understory algae could be predicted from understory biomass (greatly simplifying production estimates; details in Miller, R. J. et al. 2011, Harrer et al. 2013), but only in late spring and summer after opportunistic species colonized and grew. These results are intriguing as they suggest compensatory production if giant kelp is removed. They also beg the questions of how a relatively constant amount of production but, from different sources, affects the dynamics of the community, and how these effects compare with those that result from differences in vertical structure and habitat among the algal taxa.

## FOOD WEBS AND CONTROL WEBS

The preceding descriptions of the physical and chemical environment, structure, and dynamics of giant kelp forest communities highlight the diverse biotic and abiotic environments within which *Macrocystis* occurs. We define "community structure" as the composition, abundance, and distribution of species within a community, and "causes of structure" as processes including productivity, species interactions (e.g., competition and facilitation), recruitment, disturbances that remove organisms, and substratum characteristics. Communities are composed of populations of numerous species, so causes of community structure could hypothetically be understood by linking together the causes of the demographic features of particular species within the community. This would require a considerable amount of coordinated research. Other approaches involve aggregating species into various categories, such as trophic levels, functional groups, and guilds, looking for patterns within and among categories and the abiotic environment, building causal models to explain the patterns, and then testing hypotheses derived from the models.

The most common of the aggregation approaches is via food webs. Foods webs have been of interest since being broadly introduced by Elton (1927) as "food cycles." Elton emphasized their utility as a means of illustrating feeding relationships. He pointed out and considered the causes of patterns in "key-industry animals," "food niches," and the "pyramid of numbers." As reviewed by Menge (2008), food web studies have expanded into searches for other patterns, such as food chain lengths, connectance, and trophic cascades, stimulated by the results of experimental studies of feeding interactions and interest in the causes and consequences of the diversity and abundance of species at different trophic levels.

Giant kelp forest food webs have been used as illustrations of feeding relationships and as grist for the mill of model generation and hypothesis testing. Rosenthal et al. (1974) produced the first illustration of feeding relationships among the larger species in the kelp forest at Del Mar in southern California. We have used their web plus others (e.g., Castilla 1985, Foster and Schiel 1985, Graham et al. 2008) and feeding relationships described at various locations to illustrate a generalized giant kelp forest food web organized by trophic level and feeding type (figure 6.3), bearing in mind not only that there is huge variation among the kelp forests of the world, but also that these food webs are greatly simplified versions of the feeding relationships among the diverse organisms that can occur in kelp forests (for a more complete web, still under construction, see http://kelpforest.ucsc.edu/visualize/Network_marker2.html). The food web illustrated in figure 6.3 is based on feeding observations and gut content analyses. The trophic levels in the web have recently been evaluated in giant kelp forests around Santa Barbara, California, using $\delta^{15}N$ analyses assuming a +3.4‰ increase in $\delta^{15}N$ per tropic level, and estimates of the $\delta^{15}N$ producer baseline. The analyses of Page et al. (2013) found feeding relationships similar to those in classic food webs, with the most notable exception being that many more of the common adult kelp forest fishes (e.g., rockfish, *Sebastes* spp.) are secondary consumers that feed on invertebrates rather than on small fishes. Their results also show a more continuous distribution of $\delta^{15}N$ in consumers (figure 6.4), indicating food types are more diverse than implied by the usual trophic partitions and links in food webs (figure 6.3).

Perhaps the most informative feature of the food web is the importance of detritus. The detrital pathway has been emphasized by numerous investigators (e.g., Castilla 1985) in many of the kelp forests discussed previously, placing detritus / decomposition with living, attached primary producers, not as in terrestrial forest webs where it is placed after consumers as a source of nutrients and carbon for below-ground

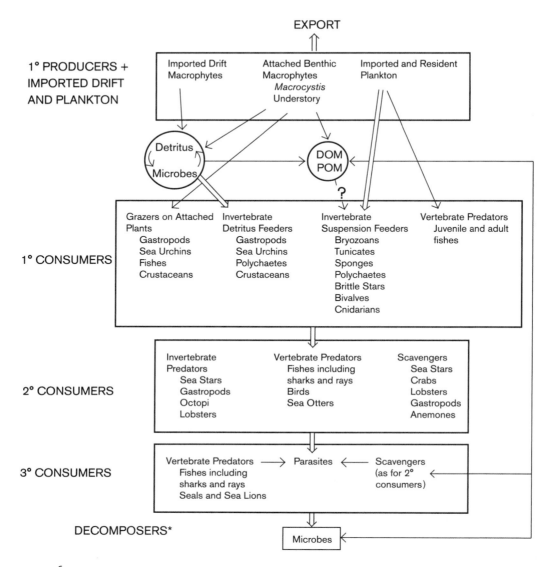

FIGURE 6.3

Generalized food web of a giant kelp forest, showing primary producers (top) to tertiary consumers and decomposers (bottom). Arrows indicate the direction of energy flow and double arrows the most likely large flows. Microbes include bacteria and fungi.

DOM = dissolved organic matter; POM = particulate organic matter; ? = flow uncertain; *(Decomposers) indicates decomposition also occurs in the detritus pathway.

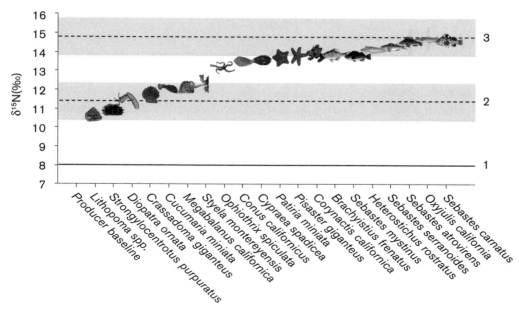

FIGURE 6.4

Nitrogen isotope values of invertebrates and fishes sampled in a kelp forest near Santa Barbara, California. The producer baseline is 1 (solid line), and position of trophic levels 2 and 3 (dashed lines; assuming an enrichment factor of +3.4‰) are shown for comparison to values from the animals. Shading indicates ±1‰ standard deviation for the trophic levels. Standard errors for the animals are covered by the icons. Note, for example, that most suspension feeders fall within the SD of the mean for trophic level 2, and most fish are at level 3.

SOURCE: From Page et al. (2013).

communities. Gerard (1976) found that 70% of giant kelp production in a central California kelp forest became detritus. Detritus biomass could exceed 20 g dry weight m$^{-2}$ on the bottom, and over half of it was consumed by benthic herbivores. Consumption by animals ranging from small crustaceans to sea urchins, abalone, and sea stars is well documented, as is secondary consumption of these invertebrates by larger invertebrates, fishes, birds, and mammals (see the "Subtidal Soft Benthos" section; Tutschulte and Connell 1988, Basch and Tegner 2007). Consumption of live giant kelp may actually be rare in some kelp forests, as Attwood et al. (1991) noted at Marion Island where the only two abundant grazers, isopods and sea urchins, appeared to consume almost nothing but detritus. Assessments of within-forest drift algae in the northeast Pacific have shown it to be highly seasonal. Abundance is high during summer and early autumn when water motion is reduced, but lower during winter when storm-induced water motion disperses it out of the kelp forest (Gerard 1976, Harrold and Reed 1985, Druehl and Breen 1986, Tutschulte and Connell 1988, Basch and Tegner 2007; see the "Detached Giant Kelp Communities" section above). With the exceptions of Harrold and Reed's (1985) and

Kenner's (1992) finding that small sea urchins in coralline mats eat more live geniculate coralline algae when drift is low, and Basch and Tegner's (2007) observation that sea urchin reproductive output was positively correlated with drift algal abundance, the direct consequences of this variation in drift abundance on drift consumers and food web structure have not been determined.

Another feature highlighted by the food web is the inputs or subsidies of production from adjacent habitats, especially plankton from the surrounding water but also kelp detritus (Vanderklift and Wernberg 2008): kelp forests are sinks as well as sources. Numerous suspension feeders (animals that extract POM, live or dead, from the water) live in kelp forests and can feed on plankton. They may also feed on POM derived from the senescence of living algae and decomposition of dead giant kelp and other macroalgae. Early studies based on differences in carbon stable isotope composition between phytoplankton and kelp suggested that nearly 70% of the fixed carbon eaten by suspension feeders in or very near kelp forests came from kelp (review in Miller and Page 2012). A more comprehensive evaluation by Page et al. (2008), however, comparing reefs with varying amounts of giant kelp found that although large pieces of kelp detritus were the primary source of food for grazers such as sea urchins, phytoplankton were the primary source for suspension feeders. Yorke et al. (2013) measured the production of small (<1 cm) detritus particles by giant kelp and coupled these measurements with a 5-year record of giant kelp biomass and POM in a giant kelp forest near Santa Barbara. They concluded that giant kelp contributed <0.2% of total suspended POC and particulate organic nitrogen (PON) in the forest. Moreover, Miller and Page (2012) pointed out that the carbon isotope composition of phytoplankton varies widely in coastal waters and can be similar to that of kelp. This was not accounted for in many early studies, leading to overestimates of the importance of kelp POM. Therefore, the main food of benthic suspension feeders appears to be plankton, and the contribution of kelp-derived POM remains to be determined. Cross-shelf surveys by Halewood et al. (2012) found increased amounts of DOM inshore of kelp forests in the Santa Barbara Channel, suggesting input from macroalgae that are used by heterotrophic bacterioplankton. The fate of the bacterioplankton is unknown (is it eaten, e.g., by sponges and tunicates?), but there is some suggestion that DOM may be taken up directly by invertebrates (e.g., Fankboner and Druehl 1976). In hindsight, it is not surprising that plankton can be the main source of food for suspension feeders in kelp forests given that the abundance and diversity of these animals on nearshore reefs with few large brown algae can be very high, especially in areas where water motion enhances plankton delivery (e.g., Pequegnat 1964).

Zooplankton are well documented in the diets of kelp forest fishes in California (e.g., Quast 1971c, Bray 1981). Consumption by fishes can be high enough to affect adult populations of species like intertidal barnacles whose planktonic larvae pass through the kelp forest (Gaines and Roughgarden 1987). Interestingly, large zooplankton like salps (a pelagic tunicate) imported into kelp beds can become food for sea urchins, reducing their active foraging and indirectly increasing macroalgal recruitment (Duggins 1981). Zooplankton such as opossum shrimp or mysids are produced within the forest. These omnivores often occur around concentrations of drift algae on which they forage. They can occur at densities so high that their swarms appear as "clouds" within a kelp forest, which are consumed by several species of fish (Clarke 1971, Turpen et al. 1994). A final important feature of kelp forest food webs, and perhaps most important functionally to the maintenance of macroalgal populations, is that many of the common grazers are "double" generalists, feeding on a wide variety of macroalgal species, both alive and as detritus.

Of all the possible patterns that might be discerned (real or not; see Peters 1988, Polis 1991) from food webs, the effects of grazers and predators or "top-down control" have been the primary pattern of interest, based on some kelp bed and forest sites where trophic cascades have been shown or suggested to have widespread effects (e.g., North and Pearse 1970, Mann and Breen 1972, Estes and Palmisano 1974). In our assessment, many attempts to investigate these patterns have, for a variety of reasons, not been rigorous enough to be considered sufficiently informative to derive meaningful relationships. For example, Halpern et al. (2006) concluded that giant kelp forests show top-down control, based on correlative relationships among species in food webs constructed from data obtained by others around the Channel Islands in southern California. The study did not include the influence of variation in ocean climate, misinterpreted some important feeding relationships (e.g., Foster et al. 2006) and did not consider alternative explanations for the results of their correlative statistical analyses (e.g., Steele et al. 2006). Halpern et al. (2006) also came up with the rather startling conclusion that Kellet's whelk, which feeds on invertebrates on kelp forest floors and is never particularly abundant, is the key controlling species in kelp forest trophic control. This conclusion is contrary to just about everything we know about kelp forest ecology, perhaps highlighting the caution that must be exercised in meta-analyses remotely applied on complex data and relationships. Similar problems, plus a lack of consideration of among-site variation in ocean climate, depth, substratum, and topography, make questionable Halpern and Cottenie's (2007) conclusion that there was little evidence for climate effects on forest structure and dynamics at these sites. Classification of kelp forest organisms into trophic and other categories labeled functional groups was used by Micheli and Halpern (2005) to investigate the relationship between species diversity and functional diversity, again using the Channel Islands data set. Although admitting that the results are sensitive to the functions chosen and without an operational definition of "function" they proceeded. Their functions included feeding types further split into invertebrates and fishes because of general differences in mobility, which they felt should be a function. Giant kelp was placed in its own functional group because of its "unique role" in kelp forests. The result that functional diversity increased with species diversity is neither surprising nor informative; species are often functionally unique and their classification arbitrary and highly subjective.

Steinberg et al. (1995) explored the possibility that the number of trophic levels affects whether or not macroalgal abundance is controlled by top-down effects of predators on grazers by contrasting North Pacific kelp communities with an odd number of tropic levels (otters–sea urchins–kelp) to Australasian communities with an even number of levels (because they had no sea otters). They concluded the theory was correct based on differences in consumption and phlorotannins (putative chemical defense against grazing) between regions. An alternative interpretation is that given the opportunity, high densities of hungry sea urchins will eat lots of algae with or without phlorotannins. Moreover, macroalgae can do quite well in giant kelp communities where predators appear to have little effect on grazers (e.g., Castilla 1985), and Estes et al. (1998) suggest that at least in parts of the North Pacific killer whales eat sea otters, giving the food web an even number of trophic levels. In any case, the simplification of kelp forest systems to "odd and even" trophic levels has several pitfalls and caveats, not least of which are the facts that other predators can exert strong trophic linkages on adult urchins at some austral sites (e.g., Johnson et al. 2011), kelp species and their diversity are greatly different in the northern and southern hemispheres, and site-specific trophic control by benthic-feeding fishes on juvenile sea urchins (e.g., Andrew and Choat 1982) can influence kelp forest dynamics.

Others have used field correlations and laboratory experiments to determine the effects of interaction strengths and predator diversity. Byrnes et al. (2006) used these to investigate the effects of predator diversity on trophic cascades. They found positive correlations between predator diversity and kelp abundance in the data set used by Halpern and colleagues above. They tested if these were causal via effects on herbivores by placing varying numbers of different predators in laboratory tanks with a suite of herbivores plus giant kelp fronds, and determining how much kelp was consumed. In a similar experiment, Sala and Graham (2002) determined grazer–plant interaction strengths because they might help test the theory that weak interactions increase community stability and resistance to invasion. They determined the interaction strengths of eight grazers by placing each into separate containers with monocultures of microscopic giant kelp sporophytes and then measuring consumption. These experiments and those of Byrnes et al. could be useful conceptually, but their relevance to natural communities is unknown, not the least because grazers (and predators) have numerous food choices in the field, do not graze in isolation from other animals, and are not naturally confined in small spaces with few choices. Again, a cautionary note is warranted. As Connell (1974) pointed out in his argument for using field experiments in ecological research, there is every reason to believe organisms in laboratory experiments will not behave naturally. Sala and Dayton (2011) assessed grazer–giant kelp interaction strengths in laboratory experiments similar to those of Sala and Graham (2002), and then examined the relevance of these to the field by using a long-term data set from field sampling of grazer densities and kelp recruitment in the Point Loma kelp forest near San Diego, California. Grazer densities that eliminated sporophyte recruitment in the laboratory could eliminate recruitment in the field, but not always. As Sala and Dayton (2011) pointed out, the ecological significance of *per capita* interaction strengths would be expected only under specific assumptions and conditions, in their case including when giant kelp is removed by storms or some other abiotic phenomenon over a large area and recruitment after removal comes from a single new cohort. Ecological significance also depended on the behavior of sea urchins, the distribution of giant kelp, and whether recruits managed to reach a size refuge from particular grazers.

A larger-scale approach was used by Byrnes et al. (2011) to understand kelp forest dynamics using food webs as part of a model. They combined a long-term data set on the abundance of organisms from coastal forests near Santa Barbara, well-vetted feeding relationships, and satellite imagery to assess kelp canopy dynamics, and site-specific wave data to examine the effects of storms on food web structure using structural equation modeling (SEM). Moreover, since the main direct effect of storms is to remove giant kelp, they evaluated the model with a multiyear field experiment at four sites where *Macrocystis* was removed from one of two paired 2000 m$^{-2}$ plots. They concluded that a single severe storm helped maintain the complexity of kelp forest food webs, primarily by altering light that affects competitive interactions between sessile invertebrates and understory algae (Arkema et al. 2009). An increase in storm frequency, however, reduced food web diversity and complexity due to local species loss, particularly among higher trophic levels. The results beg the question: how much of the change in food webs resulted from changes in feeding relationships versus structural alterations that can affect species presence and abundance? After all, the loss of giant kelp during storms removes most of the three-dimensional structure of the kelp forest and therefore much of its biogenic habitat. Therefore, how much of the change is due to the consequences of altering the food web, such as through reduced sessile invertebrate abundance resulting in the reduction of predators that feed on them, compared to how much is due to the direct effects of the disturbance itself, such as fewer sessile invertebrate predators due to loss of habitat, or some combination of many factors? Such questions cannot be investigated with current models, but perhaps future modeling will become more realistic by incorporating the relative importance of these alternatives. Some hint of the answers might be found in a discussion of species that changed and when they did so, but, unfortunately, such natural history information is usually lacking in modeling papers.

We concur with Menge (2008) that the trend toward understanding the causes of community structure by incorporating the direct and indirect effects of competitive and facilitative interactions, of inputs of food and larvae from surrounding communities, attributes of the substratum, and disturbances caused by variation in the chemical and physical environment is refreshing. Figure 6.5 is an attempt to cast these interactions and processes in a "control web" for a giant kelp forest, a web of factors and interactions that can control or have large effects on populations with the community. Such a web de-emphasizes the assumed primacy of feeding relationships inherent in food webs, and

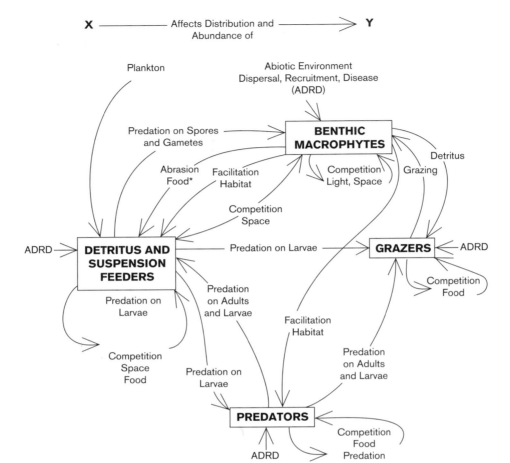

FIGURE 6.5

Generalized control web of factors and interactions affecting the distribution and abundance of organisms within a giant kelp forest. Arrows indicate the direction of effects; those that circle back indicate interactions within groups. ADRD refers to abiotic environment, dispersal, recruitment and disease. * indicates food reduced by flow reduction and interference of fronds. Note that as in figure 6.3, this web includes feeding relationships (predation and grazing) but also recognizes numerous other important factors and interactions.

recognizes the variety of factors and their interactions that have been identified from natural history observations and community descriptions, experiments, and observations and are known to affect kelp forests. Reed et al. (2011) exemplify this approach, and if food webs are useful in this regard, then Byrnes et al. (2011) is a step in the right direction. However, we think a control web that incorporates this variety is the conceptual model that better approximates relationships and processes that can affect the composition of giant kelp forests, and emphasizes that hypotheses based on a particular relationship or process need to be tested relative to alternatives. Perhaps this web can eventually be modeled, eliminating the need for food web-based modeling.

---

It seems to us that there has been somewhat of a disconnection between some efforts to model kelp forest community dynamics or sort out important interactions via elaborate and often opaque correlation analyses, and the hard-won observational and experimental research that has gone on for over 70 years. This disconnection can be both practical and philosophical. The practical side is that many known relationships and interactions can be subsumed as averages or noise in models, leading to nonsensical conclusions (at least to an empirical ecologist). The philosophical differences may lie in how empiricists and modelers or theoreticians view the world. Are there truly nontrivial, underlying relationships that can be discerned across systems, forest types, biogeographic zones, and the myriad species that inhabit them? Or are we forever to work out the intricacies of nature on a case-by-case and site-by-site basis, perhaps like the "stamp collecting" referred to in the early 20th century by Earnest Rutherford (in Birks 1962; "All science is either physics or stamp collecting"; see also Todd 2012) These issues have been debated for some time (e.g., Lawton 1999, Simberloff 2004), but their resolution is still in the future. A better understanding of the relative effects of variation in production and structure on the community is needed. In our opinion, food web-based models have so far made only marginal contributions to understanding giant kelp communities in nature. We look forward to more realistic modeling approaches that incorporate important processes in addition to feeding relationships that are well anchored by field observations and experiments.

# 7

# FACILITATIVE AND COMPETITIVE INTERACTIONS IN GIANT KELP FORESTS

The dynamics of this system are driven by variability in the abundance of a single structure-forming species (Macrocystis) that has indirect positive, as well as direct negative, effects on associated species.

—Arkema et al. (2009)

Of the myriad interactions that occur in giant kelp forests, most emphasis has been on processes that remove kelp through trophic relationships. Relatively few studies have explored the ways in which nonconsumptive interactions affect overall forest composition and the array of patches encountered within a kelp forest. These interactions can be facilitative or competitive. To many, these direct and indirect interactions are made possible by higher-order trophic dynamics, such as reduced grazing by sea urchins because top predators such as sea otters are present. To others, these interactions are key relationships involving fundamental processes of settlement, recruitment dynamics, light adaptations, food provision, and habitat use, which together provide the biogenic requirements supporting the vast array of species within kelp communities. In this chapter, we review what is known about facilitative and competitive interactions and their importance relative to other interactions in structuring giant kelp communities.

## FACILITATION

Broadly defined, facilitation denotes positive interactions between species where at least one species benefits and no harm is done to the other. The subcategory of mutualism refers to interactions where both species benefit (review in Bruno et al. 2003). The clearest examples of facilitation in giant kelp forests are the numerous invertebrates that use the fronds and holdfasts of giant kelp as a place to live, a refuge from predators, and perhaps an enhanced food supply in the form of plankton and the small epiphytes on fronds or detritus in holdfasts (e.g., Ghelardi 1971, Coyer 1984; see Chapter 5, Within Communities: Habitats and Patches). Similar facilitation is provided by dense mats of coralline algae (Dearn 1987, Kenner 1992). Not all habitat effects are facilitative, however. Some crustaceans (e.g., Cerda et al. 2009) and gastropods (Watanabe 1984b) graze on giant kelp fronds, sea urchins (Kenner 1992, Tegner et al. 1995—"Sea urchin cavitation") and isopods (Barrales and Lobban 1975) graze on holdfasts and coralline fronds, and encrusting bryozoans may contribute to blade loss (Dixon et al. 1981). *Macrocystis* can also facilitate fishes that use it as recruitment habitat, as a refuge from predators, and that feed on plankton or associated invertebrates (Quast 1971b, 1971c, DeMartini and Roberts 1990). The relationships with fishes can be mutalistic. Bernstein and Jung (1979) found that the señorita *Oxyjulis californica* reduced grazing as well as frond sinking by preying on crustaceans and bivalves. Field experiments by Davenport and Anderson (2007) that excluded fishes from stands of juvenile and adult giant kelp showed that in the absence of predators, grazing invertebrates increased, thereby reducing the number of kelp fronds and frond growth rates (figure 7.1). Predation by *O. californica* and the kelp perch *Brachyistius frenatus* had especially large effects. These authors argue that the indirect, facilitative effect of fish predation on mesograzers that eat giant kelp may be one of the most important interactions in giant kelp forests. The positive effects of predation by fishes, however, may become negative if in the pursuit of their prey the fishes remove giant kelp tissue (Bernstein and Jung 1979, Dixon et al. 1981).

Giant kelp may also enhance recruitment of juvenile fishes through the provision of recruitment and nursery areas. This has been shown for various rockfishes (*Sebastes* spp.) in central California (Carr 1991), and kelp bass (*Paralabrax clathratus*), kelp surfperch, the giant kelp fish (*Heterostichus rostratus*), and señorita in southern California (Carr 1989, 1994, O'Connor and Anderson 2010). Facilitative interactions are therefore clear for some giant kelp associates but not for others. Facilitation can also be complex, indirect, and even circular, such as giant kelp facilitating fishes, which eat grazers, thereby facilitating kelp. Many, and probably most, of these relationships are not obligate. Invertebrates and fish associated with giant kelp can recruit elsewhere within the forest, especially in areas with high-relief substrata. It is important to remember that interactions described in California may not hold in other areas that have very different fish and invertebrate faunas (see Chapter 5). On a geographic scale, however, the communities described in Chapter 5 allow the probability of most of these types of interactions, although the intensity and magnitude of most of them are yet to be explored.

Although not strictly an "interaction," the role of microhabitats in kelp recruitment dynamics is probably important. These small habitats can provide a refuge from grazing for the early life history stages of kelps. One somewhat bizarre relationship was observed by Garbary et al. (1999) of kelp gametophytes occurring endophytically in various filamentous and foliose red algae. It could be expected that facilitation is important to the recruitment of *Macrocystis* because spores arrive in numerous microhabitats and microscopic stages may be particularly vulnerable to grazing. Under suitable conditions

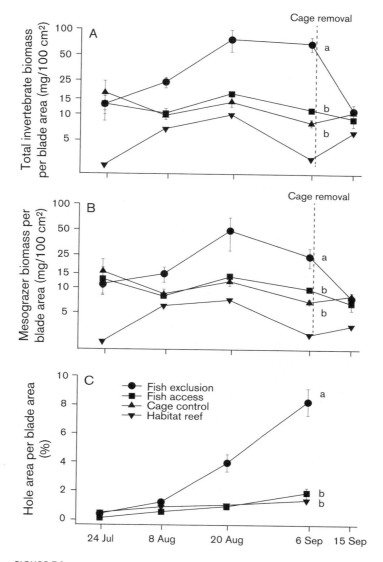

FIGURE 7.1

The effects of fish inclusion and exclusion on small grazing invertebrates and kelp fronds, from an experiment at Santa Catalina Island, southern California. (A) Total invertebrate biomass per blade; (B) mesograzer biomass per blade; and (C) the percentage of hole area (i.e., grazed through) per area of blade for three treatments (±1 standard error). Treatments were fish exclusion, fish access, and cage control, plus open reef areas near experiments, which were done over 45 days. Treatments with the same lowercase letter are not significantly different from each other. Note the log scale in panels A and B.
SOURCE: From Davenport and Anderson (2007).

of light, nutrients, temperature, and lack of sedimentation, however, giant kelp can recruit to a wide variety of natural substrata (Reed and Foster 1984) as well as concrete blocks (Foster 1975b) and boulders (Reed et al. 2004). Although not precluding a role of microsites in spore and germling survival, this does suggest that at the scale of whole forests, facilitation may not be very important to giant kelp recruitment, growth, and survivorship.

Evidence from clearing experiments indicates that nongeniculate corallines and the crustose holdfasts of geniculate corallines may enhance recruitment relative to bare rock (Reed and Foster 1984). Whether the enhancement was due to facilitation or the prior occupation of corallines by microscopic stages of giant kelp was not determined. Other experiments have shown that corallines do facilitate the recruitment of the large, foliose brown alga *Desmarestia ligulata*, an annual understory species that opportunistically recruits when giant kelp canopies are reduced (Edwards 1998). In contrast, Okamoto et al. (2013) showed that fleshy red and brown algal crusts can inhibit kelp recruitment but not enough to significantly reduce adult densities relative to those in natural stands, and Breitburg (1984) found nonsignificant effects of nongeniculate corallines on macroalgal recruitment but significant negative effects on sessile invertebrate. It appears, therefore, that crustose algae can affect algal and sessile invertebrate recruitment, but it is not always clear if the effects are ecologically important. Future studies that include better identification of crusts and the organisms that might settle on them are needed; the "recruitment function" of crusts most likely varies among crusts and potential recruits.

In southern Chile, the slipper limpet *Crepipatella fecunda* can facilitate annual populations of giant kelp by enhancing recruitment success, possibly by removing the propagules of potential algal competitors (Henríquez et al. 2011). Harris et al. (1984) described an interesting facilitation of *Macrocystis* by algal turfs. They found enhanced recruitment of *Macrocystis* on Naples Reef (near Santa Barbara, California) within dense red and brown algal turfs that provided a refuge for juvenile sporophytes from grazing by herbivorous fishes. This facilitation may be unique as it occurred after a storm that removed most large algae from the study reef, thereby concentrating grazing by resident fish on small benthic plants.

Numerous other organisms besides kelp have facilitative relationships in a kelp forest. Nongeniculate coralline algae and noncalcified red algal crusts and turfs have been demonstrated to facilitate the recruitment of abalone and sea urchins. Abalone can preferentially settle and metamorphose on nongeniculate coralline algae and noncalcified red algal crusts in response to the presence of γ-aminobutyric acid (GABA) in the surface cells of the thalli (Morse et al. 1979). The response can vary depending on the species of abalone and algae (Daume et al. 1999a), and settlement may also occur in response to diatom and bacterial films (Daume et al. 1999b). Rowley (1989) found that newly settled larvae of the sea urchins *Strongylocentrotus franciscanus* and *S. purpuratus* were most abundant on nongeniculate coralline algae and foliose red algal turfs, and *S. purpuratus* larvae preferentially settled on these substrata in the laboratory. These relationships could be mutualistic because newly settled abalone (and probably sea urchins) feed primarily on diatom films and bacteria (review and data in Garland et al. 1985) that can potentially inhibit algal growth.

An associational interaction, another type of facilitation that occurs when consumption of one species is reduced by neighboring species, was delineated with field experiments by Levenbach (2009) in California kelp forests. The interaction is among the strawberry anemone *Corynactis californica*, algal turfs, and sea urchins. Levenbach found that algal turfs could be protected from sea urchin grazing by the strawberry anemone *C. californica*, but that this varied with grazing intensity. When urchin grazing was reduced, the anemone and algae compete for space, but when grazing was moderate the anemone facilitated algae nearby by inhibiting sea urchin foraging. When grazing was intense, inhibition by the anemone was insufficient to prevent grazing, so the turfs were consumed.

## COMPETITION

Competition, a negative interaction, occurs within and between species when individuals require the same resource that is insufficient for all and therefore compromises some combination of growth, reproduction, and survival. Without some other process such as storm disturbance that constrains populations below the point where a resource is limiting, the likely critical resources in giant kelp forests are light, space, and food.

Anyone who has dived beneath a thick surface canopy of giant kelp has experienced the drastic reduction in light that occurs (figure 7.2). Early investigators noted this, as well as a reduction in understory algae beneath thick canopies (e.g., Neushul 1971b), suggesting giant kelp outcompetes understory species for light. This is further indicated by a comparison of some sites at Catalina Island where understory kelps are abundant on the benthos above and below the margins of giant kelp distribution, but not under giant kelp (see Chapter 5), and by shallow offshore banks in the same region where, in the absence of giant kelp (most likely due to high water motion), understory kelps are abundant at depths where giant kelp usually grows (Lewbel et al. 1981, Lissner and Dorsey 1986).

Various canopies can act alone or in concert to alter the light environment on the benthos so light measurements are critical to unraveling responses to canopy removals. Neushul (1971b) measured the light reduction caused by giant kelp canopies using a custom-made, self-contained diver's photometer and found, as have subsequent investigators with more sophisticated instruments, that light attenuation by the water alone may reduce bottom light (at 20 m depth) to around 5–10% of light at the surface. The addition of a surface canopy of giant kelp and fronds in the water column can reduce this to 0.5–1%, which is at or below the limits for the growth of kelps (see Chapter 3).

Within kelp forest patches, the array of species may be partially dictated by intraspecific competition within kelp species. For example, Reed (1990b) experimentally thinned

FIGURE 7.2

Canopy effects on light and understory. Left: light reduction looking up under a giant kelp canopy.
Right: understory beneath a dense canopy of *Pterygophora californica* in Stillwater Cover, California.
The benthos is dominated by nongeniculate corallines, sponges, and bryozoans. Sea stars are *Patiria
miniata* (around 15 cm dia.). The sea star on the *P. californica* is probably digesting bryozoans growing
on the stipe.

SOURCE: Photos by M.S. Foster.

populations of the understory kelp *Pterygophora californica* to 2, 58, and 102 plants per
m². Recruits of *Pterygophora* and other macroalgae, as well as epiphytes on the kelp
plants, appeared only in the low-density stands. Furthermore, the grazers *Lytechinus pictus (=anamesus)* and *Aplysia californica* were common only at low *Pterygophora* density.
On a per capita basis, individuals at high density had lower growth and less reproductive
output than those at low density, probably a consequence of light competition. Reed and
others (e.g., Schiel and Choat 1980, Schiel 1985, Stewart et al. 2009) argue that species'
architecture, wave climate, and the depth at which they occur can affect the degree of
dominance and suppression of stand density, both within and between species.

The negative relationship between giant kelp abundance and understory canopy cover
or plant density has been documented by correlative studies using giant kelp surface
canopy cover (Foster 1982) or stipe and adult densities (Carr 1989, Tegner et al. 1997)
as a proxy for light reduction, and field experiments that removed giant kelp. Pearse
and Hines (1979) did the first such experiment using the kelp forest at Point Santa
Cruz, California. They removed all giant kelp from a 10 × 20 m plot at 8 m depth and
followed changes in the understory vegetation relative to a nearby control at the same
depth. The removal resulted in an increase in light to the benthos from 0.2% to 4%
of surface irradiance. Juvenile kelps and foliose red algae increased rapidly, exceeding
abundances in the control plot within 3 months. In another much-cited study, Reed and
Foster (1984) examined the effects of multiple canopies at 15 m depth in the kelp forest

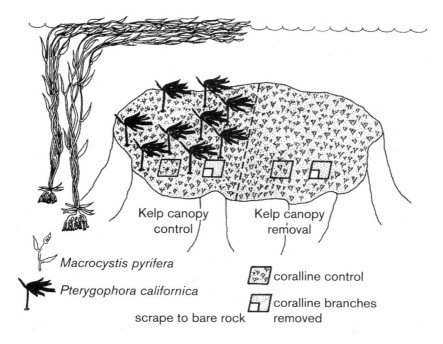

Kelp canopy
control

Kelp canopy
removal

*Macrocystis pyrifera*

*Pterygophora californica*

scrape to bare rock

coralline control

coralline branches
removed

FIGURE 7.3

Schematic of a removal experiment testing the direct and interacting effects of various canopy layers and benthic cover on algal recruitment in Stillwater Cover, central California. The experiment involved having plus and minus canopies of the kelps *Macrocystis pyrifera* and *P. californica*, the lower canopy formed by branches of geniculate corallines, and benthic cover of coralline crusts. Large quadrats were 1 m².

SOURCE: From Reed and Foster (1984).

at Stillwater Cove, California, over a 2-year period by removing various combinations of giant kelp, dense stands of the understory kelp *P. californica*, a bottom canopy composed primarily of the geniculate coralline *Calliarthron tuberculosum*, and a crustose layer of nongeniculate corallines and geniculate coralline holdfasts (figure 7.3). Removal of the dense *Macrocystis* canopy alone resulted in increased kelp recruitment and increased abundances of other understory brown (especially *Desmarestia ligulata*) and red algae, but only if *P. californica* was also removed; *P. californica* alone was capable of reducing light to below 1% of surface irradiance and suppressing recruitment. Recruitment was further enhanced by the removal of *C. tuberculosum* and coralline crusts but only in areas were kelp canopies were also removed. Recruitment was also greater on crusts than on areas scraped to bare rock, perhaps because of microscopic stages remaining on the crusts. These experiments, like those of Dayton et al. (1984), showed that all canopies except the crustose layer can inhibit recruitment and that competition for light, as it affects recruitment, increases as the number of canopy levels increase. The strength of this competition most likely varies with the density of the canopies and with depth, as suggested by the reduced growth and reproduction of the understory fucoid *Stephanocystis (=Cystoseira) osmundacea* as depth increased within kelp forests in central California (Schiel 1985).

An interesting contrast is that although large increases in foliose red algae occurred after the removal of giant kelp by Pearse and Hines (1979), the kelp removals by Reed and Foster (1984) resulted in only very small increases in red algae. Clark et al. (2004) examined the red algal response to canopy removals in more detail in Stillwater Cove, including the effects of removing the thick bottom canopy of the foliose brown alga *D. ligulata* that develops when kelp canopies are removed. After kelp canopy removal, the seasonal *D. ligulata* canopy alone reduced light to <1% of surface levels. Foliose red algae as a group did not increase significantly until 2 years after the clearings were initiated. Nine common species of foliose red algae occur in the cove, and further analyses indicated that these responded differently in space and time. Some species ("light-flexible"; e.g., *Plocamium cartilagineum*) changed little in response to canopy removals, while others ("light-adapted"; e.g., *Fauchea laciniata*) increased within a year in the canopy removal treatments. These groups could be further subdivided into high- and intermediate-light species depending on their response to the removal of one or more canopies. The difference in red algal responses between Point Santa Cruz and Stillwater Cove is probably due, therefore, to differences in understory species composition. Point Santa Cruz was and remains dominated by very thin-bladed, foliose red algae such as *Cryptopleura violacea* and *Polyneura latissima* (Breda and Foster 1985) that grow rapidly in response to the increased light.

These results of canopy removals are in contrast to those at Puerto Toro, Chile, by Santelices and Ojeda (1984b) (see figure 5.3B), who removed the upper 1 m of the giant kelp canopy in a 5 × 50 m transect perpendicular to shore, and compared the result to an unmanipulated transect. *Macrocystis* recruitment increased in the removal transect, but the understory kelp *Lessonia flavicans* decreased in biomass and fleshy red algal biomass remained similar to the control. The authors argued that these effects of the *Macrocystis* canopy were much less than those found in North American giant kelp forests, and questioned the designation of giant kelp as a foundation species. However, the provision of habitat and other effects of giant kelp qualify it as a foundation species, and the different effects of canopy removal at Puerto Toro may well be due to the way giant kelp was manipulated; only the canopy was removed at Puerto Toro leaving fronds in the water column, whereas entire plants were removed in the California experiments discussed above. Other studies involving the removal of only part of the surface canopy of giant kelp yielded similar responses. For example, where the surface canopy was removed by a kelp-harvesting ship and effects compared to nonharvested areas in central California, Kimura and Foster (1984) also found the small changes similar to those in the Puerto Toro experiment. Moreover, Reed and Foster (1984) and Clark et al. (2004) showed that dense understory kelp canopies can have the same effect as giant kelp. The response to the giant kelp removal of Pearse and Hines (1979) may have reflected the sparse canopy of understory kelps at their sites. The results of Santelices and Ojeda (1984b), therefore, may have been a consequence of the remaining shade produced by giant kelp fronds in the water column plus shade produced by a dense canopy of *L. flavicans*. The different results are difficult to reconcile without appropriate light measurements and may also be related to differences in understory composition in the different areas, as between Point Santa Cruz and Stillwater Cove.

It is tempting to infer from *Macrocystis* canopy removals that species living below the lower limit of giant kelp are prevented from moving into the forest due to competition for light. This has rarely been tested, but for at least one species, the elk kelp, *Pelagophycus porra* (figure 1.1), it is not the case. This large, float-bearing kelp commonly occurs outside the deeper edge of giant kelp forests from San Diego into Baja California. Using a combination of field transplants, light measurements, and lab experiments, Fejtek et al. (2011) found that elk kelp does not usually occur in giant kelp forests because their sporophytes take nearly 100 days to develop from spores to macroscopic size, and these microscopic stages are intolerant of high light. Giant kelp canopies can reduce benthic light to levels suitable for the small stages, but Fejtek et al. argued that canopies often vary over 100 days, so it is likely there will be one or more high-light events sufficient to kill the microscopic stages. Therefore, elk kelp is not excluded by competition for light, but does not occur in giant kelp forests or shallower water because it cannot tolerate the high-light conditions there, even if these conditions occur only occasionally. The study by Fejtek et al. (2011) shows well that a full understanding of the distribution and abundance of a kelp species requires understanding the ecology of all life history stages, as we have argued previously (e.g., Schiel and Foster 2006).

Given that giant kelp can suppress understory development, how does the understory persist? Most likely this is because giant kelp canopies vary in extent and thickness, which affects the light environment below. These differences are largely due to storms (Arkema et al. 2009). For example, the giant kelp canopy in Stillwater Cove is greatly reduced during winter storms, with a burst of understory growth in the following spring before the canopy recovers (Reed and Foster 1984). Giant kelp canopies are more persistent within more protected sites in Monterey Bay and, interestingly, the understory kelp *P. californica* does not occur at these sites (Harrold et al. 1988). As suggested by the effects of shading by giant kelp on the physiology of *P. californica* (Watanabe et al. 1992), and in contrast to the study of Fejtek et al. (2011) above, perhaps this kelp is absent because there is not enough light at some time during the year to sustain populations.

Competition also occurs among the microscopic stages of giant kelp and probably between these and other small seaweeds and invertebrates, but is not well understood given the difficulties of field investigations. As Reed (1990a) pointed out, there is undoubtedly competition-based density-dependent mortality among developing *Macrocystis* and other kelp sporophytes given that successful reproduction requires spore settlement densities of over $1 \text{ mm}^{-2}$, but the size and spacing of adult sporophytes are orders of magnitude larger. There may also be competition among gametophytes of different kelp species. Because all female kelp gametophytes produce the same pheromone, lamoxirene, that causes release and attraction of sperm to eggs, Müller (1981) (see Chapter 2) suggested "chemical warfare" may occur among gametophytes of different species if they co-occur. The species producing mature female gametophytes (and thus the pheromone) first could cause premature sperm release and consequent reduced fertilization in other species. This form of interference competition was suggested in experiments by Reed (1990a), but its importance relative to other causes of differences in the relative abundance of kelps in mixed stands in the field is unknown.

The competitive effects of giant kelp on understory seaweeds have indirect effects on sessile animals. Various small-scale studies have suggested understory algae and sessile invertebrates compete for space in kelp communities and the competition is often mediated by light. Low light favors sessile invertebrates that can overgrow algae (Breda and Foster 1985) or possibly inhibit their recruitment, while high light favors algae that can negatively affect sessile invertebrates by increasing sediment loadings (Duggins et al. 1990) and by overgrowing, abrading, and inhibiting their recruitment (review in Arkema et al. 2009). These interactions may vary with life history stage. For example, Breitburg (1984) found that nongeniculate coralline algae outcompete various sessile invertebrates for space by inhibiting larval settlement, but as adults the invertebrates can outcompete algae by growing over them. Predators can enhance algal abundance in low light by removing invertebrates (e.g., Foster 1972, Foster 1975b), and grazers can enhance invertebrate abundance in high light by removing algae that can reduce invertebrate abundance via abrasion (e.g., Coyer et al. 1993). In a rigorous and thorough study, Arkema et al. (2009) examined the relationship between giant kelp shading and the abundances of understory algae and sessile invertebrates in southern California using a combination of removal experiments and multisite surveys over 8 years, and structural equation modeling (SEM) to evaluate direct and indirect interactions. Removal of giant kelp resulted in an increase in light from 0.7% to 1.7% of surface levels, and quite large increases of 15–60% cover of understory algae in areas where invertebrates were also removed. Multisite surveys showed that giant kelp abundance was negatively correlated with understory cover but positively correlated with sessile invertebrate cover. Overall, however, the indirect effect of giant kelp on sessile invertebrates was six times greater than the direct effect (figure 7.4). Arkema et al. (2009) argued that coexistence of understory algae and sessile invertebrates occurs because of variation in giant kelp abundance, primarily due to storms, that causes variation in the light effect. The responses of benthic organisms occurred in less than a year, such that neither understory algae nor sessile invertebrates could dominate for long periods. They concluded that the main driver of community dynamics is giant kelp abundance (see introductory quote to this chapter). Like the direct competitive effects of kelp canopies discussed previously, it appears that the relative abundances of sessile species in habitats with potentially high light are greatly affected by the direct and indirect effects of kelp canopy dynamics. The abundance of plankton at particular sites may also affect

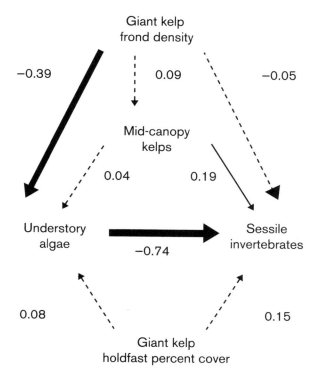

FIGURE 7.4
Fitted competition and facilitation model to estimate the strength of direct and indirect effects of giant kelp frond density, a surrogate for light reduction. Numbers are path coefficients indicating the magnitude of the interaction and whether it is negative (–) or positive (+). Solid lines indicate significant paths ($p < 0.05$). The magnitude of the indirect effect of giant kelp (via understory algae) on sessile invertebrates (–0.39 × –0.74 = 0.29) is nearly six times greater than the direct effect (–0.05).
SOURCE: From Arkema et al. (2009).

the outcome of interactions related to light and space because food availability, like light for algae, may alter the ability of sessile invertebrates to overgrow algae (see figure 6.5).

Competition for light between giant kelp and understory algae can also have indirect effects on fishes that feed on invertebrates in the understory. Schmitt and Holbrook (1990) found that the abundance of "turf" (small algae, sessile invertebrates and debris) was positively correlated with the abundance of giant kelp at Santa Cruz Island, southern California, and that the abundance of foliose algae (especially the red alga *Gelidium robustum*) was negatively correlated with giant kelp. These differences in understory composition affected the abundances of two species of surfperch. Black surfperch (*Embiotoca jacksoni*) forages for crustaceans associated with turf, and striped surfperch (*E. lateralis*) forages for similar food associated with foliose algae. The relative abundances of these surfperches may also be affected by direct competition between them (Hixon 1980), as are the abundances of some species of rockfishes along depth gradients in California (Larson 1980). These and other examples of competitive interactions in the diverse fish fauna of California giant kelp forests are discussed in detail by Hixon (2006).

The relationships discussed above show that understory algae can be key species in kelp forest dynamics. They can be greatly affected by competition for light with giant kelp and can also compete with each other, most likely for light. Perhaps the best examples of this are the interactions among the kelp *Eisenia arborea*, the fucoid *Stephanocystis (= Halidrys) dioica*, and the foliose red alga *Pterocladiella (=Pterocladia) capillacea* elucidated with field experiments by Kastendiek (1982) at Catalina Island, California. He found that canopy of *E. arborea* acts as a refuge for *P. capillacea* by excluding *S. dioica*. When the kelp canopy was removed, *S. dioica* spread adventitiously into the removal area, excluding *P. capillacea* and covering all space on the bottom. The effect of kelp canopy removal on *P. capillacea* was not physiological because *P. capillacea* persisted after canopy removal if *S. dioica* was experimentally prevented from entering the canopy removal area. These sorts of effects can also influence the success of introduced species, as demonstrated in Tasmania by Valentine and Johnson (2003) and Edgar et al. (2004), and in New Zealand by Schiel and Thompson (2012) who showed that the distribution of the introduced kelp *Undaria pinnatifida* was inhibited by native algal canopies. Perhaps surprisingly, understory canopies can also affect their own abundances as well as that of associated species by inhibiting invasion by grazers that react negatively to fronds that sweep over and touch them (Velimirov and Griffiths 1979, Konar 2000). This effect has been reproduced using plastic strips to inhibit sea urchin grazing on transplanted giant kelp (see Chapter 11).

Few studies have examined competitive effects among invertebrates in giant kelp forests. Three species of trochid gastropods occur in central California kelp forests and are segregated by depth. Their segregation, however, is probably related to recruitment patterns because competition among them is weak (Watanabe 1984a). The combined effects of competition and predation have also been shown to affect the distribution and abundance of two other trochid species in southern California (e.g., Schmitt 1982, 1996). Sessile animals may compete for space, as well shown by the field experiments of Chadwick (1991) on cup corals (*Balanophyllia elegans* and others) and the corallimopharian, *Corynactis californica*, in central California. These animals are segregated into zones on hard substrata with large vertical relief, and Chadwick showed this was caused, in part, by the ability of *C. californica* to reduce the reproductive output, increase larval mortality, alter recruitment patterns, and in some cases kill the polyps of cup corals. Lowry and Pearse (1973) suggested that, based on their distribution, abalone and sea urchins may compete for space in crevice habitats, and Tegner and Levin (1982) argued they may compete for food. Neither suggestion has been investigated with field manipulations. Schmitt (1987) demonstrated "apparent competition" between gastropods and bivalves in a kelp forest at Catalina Island in southern California. The two groups naturally separate by habitat, suggesting they compete. However, experiments showed that predator (lobster, octopi, and whelks) densities increased when the two groups of prey co-occurred, causing higher mortalities of both than when they occurred separately. The separation results from predation, therefore, and not competition.

There are clearly numerous facilitative and competitive interactions that affect the structure and dynamics of giant kelp forests. Current knowledge, however, indicates that the most important of these, at least in California, are related to variation in the abundance of giant kelp and the extensive habitat it provides, and consequent effects on light resources. In dense stands that can exist as discrete forests or as patches within a forest, these effects can cascade, affecting other facilitative, competitive, and trophic interactions. Giant kelp plants are generally distributed nonuniformly, but this may not result in variation in the benthic light environment even within a stand because of the spread of the canopy and fronds in the water column. These attributes tend to homogenize effects on light such that the patch scale affected is that of individual stands rather than areas within them. Other interactions within a giant kelp stand, such as those among and between understory algae, invertebrates, and bottom-dwelling fishes, are more "patchy" in occurrence (see Chapter 5), primarily because these taxa tend to reside in relatively small patches that are affected by more than just the light environment and habitat produced by giant kelp. Judging the relative importance of these smaller-scale interactions to the structuring of the entire forest requires understanding the distribution of the various kinds of patches and interactions among them, knowledge that is currently far from complete.

# 8

# GRAZING IN KELP COMMUNITIES

During the course of this study some southern California kelp beds have
suffered depredation chiefly by sea urchins. Certain beds have been severely
damaged or even completely destroyed by grazing. Others, however, have never
shown evidence of excessive grazing pressures.

—Leighton (1971)

The major agent of loss of kelp other than water motion and natural senescence of
fronds is grazing. A rich and diverse array of species make their living feeding on kelp
and kelp-derived detritus within most kelp forests, both directly and indirectly through
the food web and trophic interactions. As illustrated in previous chapters, these interactions occur across many feeding types and up to five trophic levels. These include primary producers (kelp and other algae), primary consumers (the herbivores), secondary
consumers (smaller mobile predators), and tertiary consumers (or secondary predators;
figure 5.11), with some authors delineating a higher level of apex predators (large predators such as killer whales and sharks that may cause trophic cascades by having a strong
influence on important prey species or trophic levels; e.g., Carr and Reed, in press).
Among primary consumers, a distinction is often made that depends on the source
of food and mode of feeding. Primary consumers include grazers that feed directly
on living attached algae, detritivores feeding on detached drift or litter material, and
planktivores feeding on material produced in or imported into a kelp forest (Graham et
al. 2008). Most grazers and detritivores, however, will feed on all types of algae depending on availability and therefore those categories are not mutually exclusive. Here, we
simply refer to these herbivores as grazers, in the interest of simplicity.

Animals that feed on algae range in size from microscopic crustaceans and snails
to large herbivorous fishes. Historically, the now extinct dugong-like Steller's sea cow
(*Hydrodamalis gigas*), which reached 8 m or so in length and occurred along the west
coast of North America, was the largest grazing animal in kelp forests, feeding on giant

kelp canopies and inshore algae as it lolled around the sea surface nearshore (Domning 1972, Anderson 1995). The diversity of herbivorous taxa that can occur in giant kelp communities is exemplified by those in Californian kelp forests (Table 8.1). Amphipods and isopods can be extremely abundant on kelp fronds, where they graze epithelial tissue or epiphytes. Turban snails and crabs move up and down fronds, abalone feed on drift algal material, thousands of small crustaceans and snails inhabit kelp holdfasts and algal turfs, and chitons and sea hares graze along the sea floor. Numerous fishes have kelp as at least part of their diet and in New Zealand a large labroid fish is almost exclusively herbivorous within nearshore kelp stands (Clements and Choat 1993, Taylor and Schiel 2010). However, by far the most important grazers in almost all kelp forests worldwide are sea urchins.

Numerous publications and literature reviews have documented extensive feeding by sea urchins on kelp, removing attached plants, and creating unforested patches largely devoid of all kelps and other macroalgae (e.g., Lawrence 1975, Foster and Schiel 1985, Schiel and Foster 1986, Harrold and Pearse 1987, Graham, Vásquez, et al. 2007). Such "overgrazing" in kelp beds is a worldwide phenomenon and has been recorded along the west coast of North America from Baja California, Mexico (e.g., Beas-Luna and Ladah 2014), and through Alaska (e.g., Estes and Palmisano 1974), the northwest Atlantic (Hart and Scheibling 1988), the northeast Atlantic (Fagerli et al. 2013), Britain (Kitching and Ebling 1961, Jones and Kain 1967), New Zealand (Choat and Schiel 1982, Andrew 1988, Schiel 1990), eastern Australia (Andrew and O'Neill 2000), and southeastern Australia (Johnson et al. 2005). Some of the observations of such deforestation predated modern ecology. Lawrence (1975) reported that destructive grazing by sea urchins was noted in the mid-1800s by fishermen who saw how algae were eliminated from an area by large populations of *Strongylocentrotus droebachiensis*. Concern about expanding sea urchin populations and removal of kelp grew considerably when early scuba diving scientists noticed "feeding fronts" composed of thousands of individuals of *S. purpuratus* and *S. franciscanus* removing large tracts of kelp in southern California (Leighton 1971). Several causes of these outbreaks were hypothesized, including expanding urbanization effects and excessive fishing of top predators that might once have kept urchin populations in check (North and Pearse 1970). Similar observations of vast areas denuded of kelp along the coasts of Nova Scotia, Canada (e.g., Chapman 1981), New Zealand (Choat and Schiel 1982), and more recently Chile (Vásquez and Buschmann 1997), Australia (Johnson et al. 2005), and Norway (Fagerli et al. 2013), but with different species of urchins and kelp, highlighted that the importance of kelp–urchin interactions had a far wider context than just southern California. In Nova Scotia, however, there was little urbanization and the major suspected cause of urchin population explosions was the removal of top predators, especially the lobster *Homarus americanus*, through overfishing. Breen and Mann (1976) were the first to formalize this in a simple model involving lobster biomass, urchin densities, and kelp loss. This was essentially the start both of marine "trophic cascades" (Sala, Boudouresque et al. 1998), a term coined by Paine (1980; see Polis et al. 2000),

TABLE 8.1   Common grazing invertebrates in California giant kelp forests that remove tissue from kelps or affect recruitment.

| Species | Common name[a] | Grazing effects |
|---|---|---|
| PHYLUM ECHINODERMATA | | |
| *Strongylocentrotus franciscanus* | Red sea urchin | Directly removes plants; consumes all parts of plants |
| *S. purpuratus* | Purple sea urchins | As above |
| *Lytechinus pictus (=anamesus)* | White sea urchin | May graze juvenile kelp, portions of holdfasts, and lower fronds, weakening plant attachment |
| *Centrostephanus coronatus* | – | Grazes drift plants; possibly grazes holdfasts |
| *Patiria miniata* | Bat star | At high densities, may graze microscopic stages, affecting recruitment and early survival |
| PHYLUM MOLLUSCA | | |
| *Haliotis rufescens* | Red abalone | Feeds extensively on drift kelp; may graze attached stipes and sporophylls |
| *H. fulgens* | Green abalone | As above |
| *H. corrugata* | Pink abalone | As above |
| *Tegula brunnea* | Brown turban snail | Abundant on giant kelp, other kelp, *Stephanocystis (=Cystoseira)*. Grazes surface of blades and fronds, may weaken tissue |
| *T. funebralis* | Black turban snail | Same as above but only in low intertidal zone |
| *T. pulligo* | Dusky turban snail | As above |
| *T. eiseni* | Banded turban snail | As above |
| *T. montereyi* | Monterey turban snail | As above |
| *T. aureotincta* | Gilded turban snail | As above |
| *Norrisia norrisi* | Norris's top snail | As above |
| *Calliostoma annulatum* | Purple-ringed top snail | Eats kelp, but probably feeds mainly on bryozoans, diatoms, detritus |
| *Megastraea undosa* | Wavy top snail | Found on substratum; may graze lower stipes and sporophylls |
| *Pomaulax gibberosus (=Astrea gibberosa)* | Red top snail | As above. Often found grazing drift kelp at night. |
| *Alia (=Mitrella) carinata* | Carinated dove snail | Abundant on giant kelp blades; feeds mainly on diatoms and detritus |
| *Lacuna unifasciata* | Chinkshell | Feeds on stipes, producing pits |

*(continued)*

TABLE 8.1    *(Continued)*

| Species | Common name[a] | Grazing effects |
|---|---|---|
| *Megathura crenulata* | Giant keyhole limpet | Minimally affects kelps; feeds on understory algae and ascidians |
| *Discurria (=Lottia) insessa* | Seaweed limpet | Almost exclusively on *Egregia menziesii*; grazes fronds and weakens them |
| *Lottia instabilis* | Unstable seaweed limpet | On stipes of *laminarian* spp. and *Pterygophora*; no evidence of damage to plants |
| *Ischnochiton intersinctus* | – | Only indirect effects; may graze algal spores |
| *Lepidozona cooperi* | – | As above |
| *Tonicella lineata* | Lined limpet | On encrusting corallines; as above |
| *Cryptochiton stelleri* | Gumboot chiton | Grazes on bottom; effects unknown |
| *Aplysia californica* | California brown seahare | Occasional grazing on lower portions of giant kelp; mostly feed along bottom |
| *A. vicarria* | California black seahare | Grazes on *Egregia*; effects unknown |

PHYLUM ARTHROPODA

| Species | Common name | Grazing effects |
|---|---|---|
| *Pentidotea (=Idotea) resecata* | Kelp isopod | On giant kelp and *Pelagophycus*; eats holes in blades, causing weakening and providing centers for infection |
| *Paracerceis cordata* | Pillbug | May derive nourishment from kelp, but no visible damage |
| *Parampithoe (=Ampithoe) humeralis* | Kelp curler | Rolls and cements edges of blades to form a sticky web; probably feeds on blades |
| *A. rubricata* | – | As above |
| *Cymadusa uncinata* | – | As above |
| *Limnoria algarum* | Gribble | Burrows into holdfasts and may cause weakening |
| *Pugettia producta* | Kelp crab | Mainly herbivorous; eats kelp and other algae |
| *Taliepus nuttalli* | Southern kelp crab | As above |

a. Common names are from North (1971a) and Morris et al. (1980).

and an influential and pervasive literature on trophic control of kelp forest food webs. This early work of Breen and Mann, based on observations, field and lab experiments, and empirically based modeling led to one of the most hotly debated topics in the marine ecological literature about the causes of urchin population explosions and the role of predators in them (review in Elner and Vadas 1990). Urchin–kelp relationships, their causes and effects, have probably spawned more research and speculation than any other interactions in kelp forests worldwide (e.g., Dayton 1985b, Andrew 1988, Vásquez and Buschmann 1997, Graham, Vásquez, et al. 2007, for reviews), and the literature on the topic is more like a torrent than a cascade.

Although the direct role of urchin grazing on kelp can be fairly unequivocal and readily observed, it has proven to be far more problematical and elusive to tie down definitive relationships between predators and sea urchins. Numerous species of fish, lobsters, other crustaceans, octopus, sea stars, predatory snails, and some mammals (including humans) eat sea urchins at various stages of their life history, and their larvae are no doubt eaten by predatory plankton and nekton. Whether or not they "control," or are even capable of controlling, urchin populations is a topic of continued debate. Predator–urchin relationships tend to be far more diffuse than those of urchin–kelp interactions, predatory effects are rarely seen directly, and historical reconstructions of putative past relationships, such as before modern fishing and (in some cases) the extirpation of marine mammals, are largely surmised and speculative, leaving ample room for plausible alternative explanations.

In this chapter, we discuss these relationships and interactions with respect to giant kelp forests. The evidence is skewed with respect to case studies, which come mostly from the coast of California. We call on the literature from other areas to gain a wider perspective on predators–urchins–kelp relationships. Furthermore, there are other grazing and predator interactions not involving sea urchins that can be at least locally important in giant kelp forests.

## KELP–SEA URCHIN INTERACTIONS

Sea urchins are generally the most obvious grazers in *Macrocystis* forests and kelp beds worldwide, and their extensive grazing effects have been recorded in various habitats in tropical, temperate, and boreal regions (Lawrence 1975). From the extensive literature on this topic, a few general patterns have emerged relating to urchin feeding activities and effects in kelp-dominated habitats. Sea urchins within kelp forests feed mostly on drift algal material (Mattison et al. 1977, Vadas 1977, Duggins 1981, Foster and Schiel 1985, Harrold and Reed 1985) and even when quite abundant they often have little effect on attached kelp (Lowry and Pearse 1973, Foster 1975b, Cowen et al. 1982). This is at least partially due to them being widely dispersed in cracks and crevices where drift algae accumulate. Extensive feeding on attached plants appears to be initiated by a shortage of drift algae that triggers behavioral changes to active foraging (Dean et al. 1984, Harrold and Reed 1985), and which may also be intensified by the dispersion of individuals and their increasing densities within patches of substratum (e.g., Schiel 1982, Lauzon-Guay and Scheibling 2007a). Once a dense aggregation of urchins is formed, often containing >90 animals per square meter (figure 8.1), a "feeding front" emerges that can advance through a kelp forest. Urchins in the vanguard are often large and tightly packed together (Leighton 1971, Dean et al. 1984) and most or all plants in their path are consumed or

FIGURE 8.1

Two modes of sea urchin grazing. A: active front of *S. purpuratus* (gray in photo) and *S. franciscanus* (larger urchins, dark to black in appearance) moving into a kelp forest in Baja California, with a Garibaldi (*Hypsypops rubicundus*, center) and Señorita (*Oxyjulis californica*; upper right) in the water. B: stationary aggregation of *S. franciscanus* surrounded by foliose red algae in central California. *S. franciscanus* are 15–20 cm diameter. SOURCE: (A) Photo courtesy of Ron J. McPeak, UCSB Library Digital Collection. (B) Photo by M. Foster.

removed. Because urchins graze mostly through holdfasts and the lower fronds of kelp and detach often large plants, much material can be made available on which urchins nearby can feed, although urchins further back in aggregations can be food limited and resort to active foraging (Mattison et al. 1977). Urchin-dominated areas with little or no kelp can persist for many years, even decades (Chapman 1981, Andrew and Choat 1982). These areas were called "barren grounds" by Pearse et al. (1970), a term that persists to the present in various forms (e.g., urchin barrens; Hughes et al. 2012). Although usually devoid of kelp, these areas can have a rich array of other macroalgae and sessile invertebrates (Choat and Andrew 1986, Graham 2004, Vásquez et al. 2006) and support a different suite of fishes (Choat and Ayling 1987) and large invertebrates (Graham 2004, Johnson et al. 2005) than found in nearby kelp beds. Harrold and Pearse (1987) therefore recommended that rather than "barren grounds" such areas should simply be referred to as "deforested," a term used by many. In a study from the Channel Islands National Park in southern California, for example, Graham (2004) found there was a 36% reduction in the number of species (by 96, of 274 taxa) from forested to deforested areas, but that the vast majority of species were more abundant in kelp forests. Larger mobile invertebrates such as abalone moved out of deforested areas and, of course, there was a loss of the canopy fish assemblage. Similarly, in Tasmania, southeast Australia, Ling (2008) found that of the 296 taxa recorded in an area, 221 were present within intact kelp beds and 72 within deforested areas. In the deforested areas, these were mostly coralline algae, low-lying turfs, and numerous small sessile and mobile invertebrates. These deforested areas can be self-sustaining because the remaining algal turfs can be good recruitment grounds for sea urchins (Andrew and Choat 1982, Kenner 1992). For example, Rowley (1989) found in southern California that there was no significant difference in urchin recruitment into turfs within kelp forests and in deforested patches. He found up to 1000 recruit urchins per meter square in these turfs. Eventually, however, these areas usually are reinhabited by kelp, but only after urchins reach suitably low densities, usually through disease (Pearse et al. 1977, Scheibling 1986), recruitment failure, or storms (Ebeling et al. 1985; Scheibling and Lauzon-Guay 2010). The sequence can also be reversed with kelps re-establishing first due to changes in oceanographic conditions (Harrold and Reed 1985) or the availability of alternative food for sea urchins (Duggins 1981), causing the urchins to shift back to passive feeding.

Given the general phenomena of intensive grazing by sea urchins on kelp, however, there are many variations on the theme because of the large compositional differences in giant kelp forests worldwide, and in the diversity of grazers, types of urchins, and their behaviors (Schiel and Foster 1986, Graham, Vásquez, et al. 2007). Although most of the descriptions of effects of sea urchins on *Macrocystis* come from the west coast of North America where extensive deforestation has occurred from time to time, there have also been extensive episodes in Tasmania, Australia (Johnson et al. 2011), and some less extensive episodes in southern South America (Graham, Vásquez, et al. 2007). Although other areas are known to have had extensive grazing by sea urchins in kelp beds, such as northern New Zealand (Andrew and Choat 1982, Andrew 1988), there have been no similarly recorded extensive effects in *Macrocystis* forests (Schiel 1990, Schiel et al. 1995), which is perhaps a reflection of fewer studies.

Particular conditions seem to be associated with overgrazing by sea urchins, but not all of these are the same between biogeographic regions. In northern Chile, for example, where the dominant kelp is *Lessonia trabeculata* and not *Macrocystis*, the urchin *Tetrapygus niger* can maintain deforested patches in areas of high water motion (Vásquez 1993). In southern Chile, the urchin *Loxechinus albus* appears to have few effects on the abundant *Macrocystis* stands, but there are some exceptions. It was initially found that *Loxechinus* was in low densities or absent in most places in southern Chile and that it did not affect *Macrocystis* populations (Castilla and Moreno 1982, Santelices and Ojeda 1984a, 1984b, Vásquez et al. 1984, Castilla 1985). In their review of herbivore–kelp interactions in Chile, Vásquez and Buschmann (1997) concluded that in central and southern Chile, *Macrocystis* beds were not controlled by sea urchins. However, from his observations and experiment in the 1970s, Dayton (1985b) concluded that *Macrocystis* populations seem to be limited by physical factors in the fiords and by *Loxechinus* grazing in most other places. Dayton recorded high densities of *Loxechinus* and low densities of *Macrocystis* in some of the more exposed sites. More recently, it was found that when *Loxechinus* densities were over 20 m$^{-2}$, *Macrocystis* sporophytes in an annual population were reduced from 24 to 2 m$^{-2}$ (cited in Graham, Vásquez, et al. 2007). Because of the long and heterogeneous coastline of southern Chile, and the wide array of *Macrocystis* populations, it appears the jury is still out on the ecological role of sea urchins in this region. Graham, Vásquez, et al. (2007) concluded, however, that large-scale overgrazing is a relatively rare occurrence there.

"Overgrazing" by sea urchins has elicited comment and concern from the early days of the marine ecological literature. To early diving scientists, the appearance of vast feeding fronts of sea urchins and destruction of large tracts of kelp must have been alarming. It would have been rare to see wholesale change in a giant kelp forest as it was happening and to some the notion that a whole forest could be removed did not seem to be the natural state of things. Indeed, Leighton (1971) recorded that "at that time Conrad Limbaugh was one of the few scientific diving personnel aware of the potential threats by grazing populations (Limbaugh (1955)." North (1971a) and colleagues therefore examined grazing effects as integral to the accumulating knowledge of kelp forest ecology. In the Point Loma kelp forest off San Diego, for example, Leighton (1971) recorded a "grazing front" and "frontal attack" of a dense aggregation of urchins on giant kelp. There was a gradation of percentages of different species of urchins from the front toward the back of the aggregation. Ninety-four percent of urchins at the front were the large red urchin *Strongylocentrotus franciscanus* and 6% were the smaller purple urchin *S. purpuratus*, with a combined weight of almost 2.5 kg m$^{-2}$. Eighteen meters back from the front, these percentages were reversed, and the biomass was halved because of the smaller sizes of *S. purpuratus*. Over a period of 2 months, the entire kelp stand of 91 × 183 m was destroyed, leaving only remnants of holdfasts and encrusting coralline algae. Leighton also recorded that small urchins may tunnel into holdfasts and destabilize them, making them prone to removal. He found a single holdfast with 52 *S. franciscanus* and 45 *S. purpuratus*.

This type of dramatic, real-time observation of wholesale destruction has been re-enacted more recently in southeastern Australia, where one of the most extensive incursions of sea urchins into kelp communities has been seen. The diadematid urchin *Centrostephanus rodgersii* was seen in limited numbers in sites along northeastern Tasmania in Bass Strait in 1974. Johnson et al. (2005) reported the expansion and spread of populations along the eastern coast of Tasmania. By 1981, the overall density of sea urchins had not changed, but they were more aggregated and restricted to depths greater than 10 m, where reefs were largely devoid of macroalgae. The progressive expansion of urchin populations affected several fucoids and the kelps *Ecklonia radiata* and *L. corrugata* and, later, *Macrocystis*. *C. rodgersii* is known to have a large effect on macroalgal assemblages around 700 km further north in central and southern New South Wales, where up to 50% of shallow reefs are barrens habitat (Andrew and O'Neill 2000). Along eastern Tasmania, up to 40% of boulder reefs at 20–25 m depth and around 20% between 5 and 30 m are barrens. Johnson et al. (2005) also recorded a high proportion of "incipient barrens" that were actively forming. It is interesting to note that two sea urchin species were present and can occur in high densities (*C. rodgersii* and *Heliocidaris erythrogramma*) but, unlike the case in California, these species were negatively correlated and *Heliocidaris* plays little role in the production and maintenance of barrens habitat along the coastline, although it may do so at a small scale in some sheltered bays (Johnson et al. 2005, Ling et al. 2010). *Macrocystis* has declined in almost all places, in many areas to near zero (Johnson et al. 2005, 2011). As elsewhere, there was a negative correlation between urchin numbers and those of kelp and large invertebrates, particularly abalone and lobsters.

## WHY DO URCHINS RAMPAGE?

The formation of urchin feeding fronts and initiation of destructive grazing are topics of considerable debate. The mechanisms underlying them may be proximal and more remote, and may vary in type and magnitude among biogeographic regions. Johnson et al. (2005) noted that "it is important to emphasise that the mechanisms underpinning the incursion of *C. rodgersii* into Tasmania may not be related to those underpinning the formation of barrens habitat." In general, the proximal causes seem to be massive recruitment of sea urchins, a shortage of drift algae and detrital material, a switch in behavioral mode from passive feeding to active foraging, and self-sustaining of barren habitats through continual recruitment. The more remote causes relate to oceanographic regimes and reduction or loss of predators. However, as in all things ecological, these factors are confounded and interact in various ways across the giant kelp forests and bioregions of the world.

Echinoderms have long been known to have sporadic and highly variable recruitment episodes. For example, Ebert (1968) found unusually good recruitment of *Strongylocentrotus purpuratus* in Oregon in 1963, but no or poor recruitment from 1964 through 1978. He found that recruitment to populations of the same species in southern California and Mexico was more frequent but still varied greatly among years (Ebert 1983). It is generally thought that such variable recruitment relates to temperature, food availability and quality, and oceanographic conditions affecting planktonic larvae. Hart and Scheibling (1988), in a study of the northern sea urchin *S. droebachiensis*, found experimentally that food quality and quantity affected survival at higher temperatures. They concluded the extensive barrens that formed along Nova Scotia were the result of intense sea urchin recruitment coinciding with a period of warm temperatures in one year, and that these were augmented by subsequent recruitment into altered habitats.

It is obvious that the density of sea urchins plays a critical role in feeding fronts. There seem to be critical threshold densities required for destructive grazing episodes to be initiated, but several factors can contribute to these. Dean et al. (1984) found that urchins reached densities of 47 m$^{-2}$ in a destructive feeding aggregation in a giant kelp forest in southern California but only 7 m$^{-2}$ in stationary gatherings. Feeding fronts in kelp beds in Nova Scotia were seen to reach hundreds per square meter (Lauzon-Guay and Scheibling 2007a) but when they had a mass less than around 2 kg m$^{-2}$ they did not destructively graze, a conclusion drawn earlier by Breen and Mann (1976). Observations worldwide are that the numbers necessary to maintain deforested areas are far less than those that formed them originally (e.g., Ling et al. 2009).

As Johnson et al. (2005) pointed out, however, the high numbers of urchins must act in tandem with other factors to be destructive. One of the main correlates of such

destructive grazing is the reduction or absence of algal drift. Many studies have identified drift algae and algal detritus as the main food of urchins in kelp forests (e.g., Mattison et al. 1977). Dean et al. (1984), for example, found that stationary aggregations of
the red urchin *S. franciscanus* subsisted on drift algae. Aggregations formed and active
foraging began after 2 years of declining kelp abundance. They concluded that the formation of feeding fronts appeared to entail a behavioral shift related to the availability
of drift algae. Harrold and Reed (1985) combined several of these factors in a qualitative
model that involved drift algae, behavioral shifts, hydrographic conditions, and kelp
abundance (figure 8.2). In their study at San Nicolas Island off southern California, they
found that brown algae, mostly *Macrocystis*, comprised 87–98% of drift algal biomass
across several sites. The drift biomass was lower in winter when storm activity is greatest, which transports drift out of kelp forests. When drift algae decreased, movement
of *S. franciscanus* increased. In barren areas, the urchins were poorly nourished and
they actively grazed on remnant algae, compared to inside kelp forests where drift was
plentiful. Harrold and Reed saw a gradual shift of substantial recruitment of kelp into
some barrens, a transformation of deforested areas back into kelp stands, the production of more algal drift, less movement of sea urchins, and an abandonment of open
habitats for protected cracks and crevices. Their model, therefore, shows that that the
transformation from kelp dominated to barrens can be triggered by a behavioral switch
in urchin feeding mode from passive feeding to active foraging without a change in
urchin numbers. This is a result of the availability of drift, which itself depends on the
amount of growing, attached kelp, which is regulated by prevailing hydrographic conditions. As in other areas, storms played a role both in the production and in the transport
of drift algae. Others have found that severe wave action can affect urchin aggregations
directly, especially in shallow water, and cause them to clump (Cowen et al. 1982) or
disperse (Lauzon-Guay and Scheibling 2007a, 2007b).

The Tasmanian urchin–kelp relationships are in many ways the most instructive
about underlying mechanisms and their interactions. In this case, kelp was known to
be historically abundant and the sea urchin responsible for destructive grazing was rare.
*Centrostephanus rodgersii* is an essentially warm-water species and of great ecological
importance along New South Wales to the north (Andrew and Underwood 1993, Ling
et al. 2009). The arrival in abundance of *C. rodgersii* to Tasmania was a range extension
brought about by the poleward penetration inshore of the East Australian Current (EAC)
(Johnson et al. 2005, Ling 2008). This caused ocean warming off eastern Tasmania of
2.3°C in the past century (Ridgway 2007). Ling et al. (2009) found that a lower threshold of 12°C for larval development was surpassed by this warming. The EAC therefore
provided both the necessary temperature and the conveyor belt for *C. rodgersii* larvae to
settle into Tasmanian waters. At the same time, the warmer, nutrient-poor conditions
coincided with a decline in giant kelp beds, especially since the 1980s (Edyvane 2003,
Johnson et al. 2011). These conditions combined to exacerbate the dramatic influence of
*C. rodgersii* on kelp stands. As in other areas, therefore, the interaction of oceanographic

## Barren Area

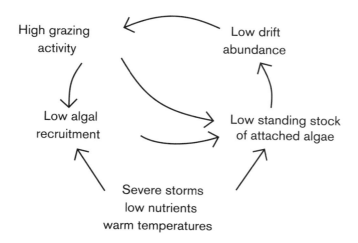

High grazing
activity

Low drift
abundance

Low algal
recruitment

Low standing stock
of attached algae

Severe storms
low nutrients
warm temperatures

## Hydrographic Conditions

Benign storms
high nutrients
low temperatures

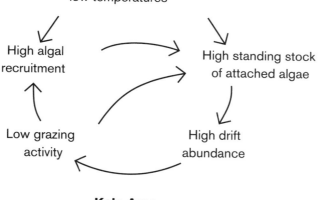

High algal
recruitment

High standing stock
of attached algae

Low grazing
activity

High drift
abundance

## Kelp Area

FIGURE 8.2

Qualitative model of formation and maintenance of areas dominated
by sea urchins and giant kelp, based on work at San Nicolas Island, off
southern California. Two states or areas of kelp forests are depicted, one
with sparse kelp and dominated by storms, low nutrients, warms water,
and heavy grazing (upper panel), and the other with lush kelp and associated with benign conditions, high nutrients, low temperature, and reduced
grazing.

SOURCE: From Harrold and Reed (1985).

conditions, species life histories, giant kelp abundance, and the production of algal drift and detritus affected the formation of barrens habitat. Rather than behavioral shifts, however, Johnson et al. (2011) and Ling et al. (2009) argued that the severe reduction of predators from fishing, particularly large rock lobsters, *Jasus edwardsii*, during the same period had a synergistic interaction between oceanographic conditions and its consequences on urchins and kelp, leading to decreased resilience of kelp beds and the formation of barrens habitat. They argued that this has been a "catastrophic phase shift," and the expansion of barrens habitat is the single largest immediate environmental threat to the rocky reef communities of eastern Tasmania.

Tasmania stands in considerable contrast to other areas of the world where barrens habitat has formed in *Macrocystis* forests. In their review of the global ecology of giant kelp, Graham, Vásquez, et al. (2007) concluded that various instances of overgrazing in *Macrocystis*-based systems have been described, but they are generally short-lived and seen at local scales, although they can have conspicuous ecological consequences. It appears that the Tasmanian kelp forests are particularly conspicuous in this regard.

## REVERSION OF BARRENS TO KELP HABITAT

Other than the effects of altered predation pressure (discussed below), several factors may act in concert when intensively grazed habitats revert to kelp domination. One of the most dramatic of these has been disease. Invertebrate pathogens and parasites are common, and can have significant impacts on populations (e.g., Jangoux 1984). Pearse et al. (1977) reported a mass mortality of *Strongylocentrotus franciscanus* near Santa Cruz, California, that allowed a deforested area to become forested and reviewed other occurrences of this phenomenon in California, the first of which was seen in 1971. When affected, urchin spines are no longer held upright and are eventually lost, the epidermis degenerates, and the animal usually dies. These mortalities occurred in large but localized patches in California, and not all urchins in them were killed. Low incidences of urchins with bacterial-infested lesions often occur in otherwise healthy-appearing populations, but it is unclear whether they are related to incidences of mass mortality (Gilles and Pearse 1986). Similar symptoms and widespread mortality have occurred in sea urchin populations in Nova Scotia (Miller and Colodey 1983). There, mortality was so great in 1982 that urchin tests washed up in windrows on beaches (Scheibling and Stephenson 1984). The likely agent was infection by a *Paramoeba* species (Jones and Scheibling 1985). Mortality of *S. droebachiensis* reached 70% in shallow inshore areas, equating to a loss of around 2 kg m$^{-2}$. As elsewhere, mortalities were temperature related. They began near the peak of seawater temperatures around 15°C and declined when temperatures dropped to around 7°C. Urchins in deeper and cooler waters offshore had far less mortality (around 6%). Scheibling (1986) found recurrent outbreaks of the disease between 1980 and 1983, and the almost complete elimination of urchins at many places along 280 km of coastline.

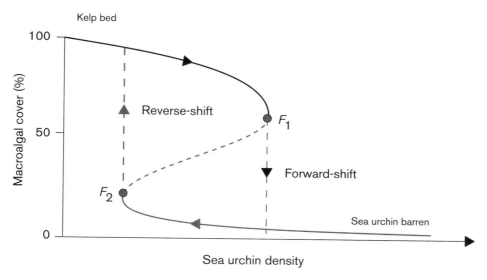

FIGURE 8.3

Phase shifts from domination by kelp and sea urchins, based on work in Tasmania. If a reef is dominated by kelp (on the upper arrow path) but close to a density threshold of sea urchins $F_1$, a slight increase in sea urchin density may induce a "forward-shift" to a state dominated by sea urchin barrens. A "reverse-shift" (lower arrow path) occurs only if the density of sea urchins is reduced below a return threshold ($F_2$). The broken gray line indicates the region of instability between the phase states. See text for further explanations relating to altered predator abundances and oceanographic changes.

SOURCE: Data from Scheffer et al. (2001), redrawn in Ling et al. (2009).

In other areas, such as those recorded by Harrold and Reed (1985), severe storm activity was involved in the transition between kelp and urchin domination. Perhaps the most unusual case was on Naples Reef off Santa Barbara, California, which underwent two transitions within 5 years. A large storm in 1980 removed canopies of giant kelp but left the understory kelps largely intact. Ebeling et al. (1985) found that this caused a decline in drift kelp, triggering active grazing by *S. franciscanus* onto open areas of reef and creating barrens habitat. A second storm in the winter of 1983 eliminated exposed urchins, which were no longer hiding in cracks. Remaining turfs and the arrival of fast-growing turfing algae facilitated survival of subsequent kelp recruits (Harris et al. 1984). Storms, therefore, had quite different effects depending on the state of the community.

"Phase" or "regime" shifts have been seen in many ecosystems (Scheffer et al. 2001, Scheffer and Carpenter 2003). In Tasmania, Ling et al. (2009) put the shift between forested and "barrens" states into the context of discontinuous phase shifts (figure 8.3). Their conceptual model identified a threshold of sea urchin density beyond which overgrazing of kelp occurs and there is a shift to sea urchin barrens. A shift back to kelp domination is difficult, and occurs only if sea urchin densities are returned to a much lower threshold. There is then a region of instability between states as kelp come back to dominance. A similar model has been used for urchin–kelp relationships in Nova Scotia and Alaska (Filbee-Dexter and Scheibling 2014).

One consistent feature of all biogeographic regions where urchins have changed behavior from active to passive foraging, or have been severely reduced in numbers or eliminated from barrens habitat, is a relatively quick recruitment of kelps and reversion back to kelp beds or kelp forests. In going through the quite large literature on urchin barrens, however, it is interesting to note the context dependency of how processes operate, both in the formation and in the reversal of barrens habitat. Harrold and Reed (1985) noted that the sequence of processes they identified were intrinsic to the communities they studied and that "because the sequences are cyclic, each habitat persists in the absence of input from outside the system." In summary, the conclusions of many studies of the depredation of kelp forests by sea urchins and the particular conditions that trigger phase or regime shifts between urchin and kelp domination are akin to saying that "every family is unhappy in its own way" (L. Tolstoy, *Anna Karenina*). These examples illustrate both the variability of different systems and processes and the futility of using single models across biogeographic regions in which the species, communities, processes, oceanographic regimes, and magnitude and ecological severity of responses are quite different.

## OTHER GRAZERS IN GIANT KELP COMMUNITIES

Seaweeds, and especially giant kelp, support the highly diverse food webs of *Macrocystis* forests and are eaten directly or as drift by myriad invertebrates. Few grazers other than sea urchins are capable of removing large attached kelp but many species live on kelp, feed on their tissues, and indirectly cause the removal of all or parts of the plants. For example, grazing of kelp tissue may provide centers for fungal and bacterial infections that can sever blades, fronds, or holdfasts. It may also provide sites for epiphyte growth or can weaken parts of the plants, rendering them vulnerable to removal by increased water motion or entanglement with other plants.

Across the giant kelp forests of the world, numerous species of echinoderms (other than the sentinel species discussed above) feed directly on attached, drifting rafts, and detrital kelp but few have more than highly localized effects. In southern California to Baja California, Mexico, for example, the small white sea urchin *Lytechinus pictus* (=*anamesus)* feeds extensively on small algae, particularly foliose reds, but it generally has much less an effect on large kelps than do purple and red urchins, although in very high densities it is known to remove adult *Macrocystis* by grazing through holdfasts and lower fronds (Clarke and Neushul 1967, Dean et al. 1984). Another sea urchin, the diadematid *Centrostephanus coronatus*, which is closely related taxonomically to the very destructive Tasmanian *C. rodgersii*, may eat kelp (Vance 1979), but its recorded effects are highly localized. It is not an aggregating species and it is not widely distributed in giant kelp communities except at some of the Channel Islands, off southern California, and islands offshore of Baja California. Also in North American giant kelp forests, the bat star *Patiria miniata* may affect algal recruitment by digesting spores and small plants when it everts its stomach over the substratum. It is found on rocky substrata up to densities of 4–5 m$^{-2}$ in many places throughout the range of giant kelp. Bat stars are omnivores and scavengers, and also eat tunicates and other encrusting animals. Leonard (1994) used lab and field experiments to test the effects of *P. miniata* on kelp recruitment. He found that bat stars grazed over 80% of the substratum over 90 days when at densities greater than around 4 m$^{-2}$. They locally caused mortality of *Macrocystis* gametophytes and young sporophytes in lab experiments but did not affect kelp recruitment in the field. He argued that giant kelp quickly reaches a size beyond which bat stars do not affect it, and that bat stars generally do not reach high enough densities to affect kelp recruitment.

In the giant kelp forests of southern New Zealand, the echinometrid urchin *Evechinus chloroticus* is commonly found, but it has not been recorded as having more than highly localized grazing effects on giant kelp along mainland and offshore island sites (Schiel 1990, Schiel et al. 1995, Schiel and Hickford 2001). This urchin is ubiquitous on rocky reefs throughout New Zealand. Although capable of creating urchin barrens in stipitate kelp stands in northeastern New Zealand (e.g., Choat and Schiel 1982), *Evechinus* appears not to do this in the *Macrocystis* stands further south. Similarly, the urchin *Heliocidaris erythrogramma* can form coralline-dominated barren grounds in stands of stipitate kelp along much of southeastern Australia (Connell and Irving 2008), but contributes to the formation of barren grounds only in some protected waters of Tasmania (Ling et al. 2010). In Chilean giant kelp forests, Castilla and Moreno (1982) tested the effects of four sea urchins on giant kelp with exclusion and addition experiments. They found that *Loxechinus albus*, *Pseudoechinus magellanicus*, *Arbacia dufresnei*, and *Austrocidaris canaliculata* had no significant effects on recruitment or persistence of adult *Macrocystis*. These sea urchins appear to feed mostly on drift *Macrocystis* (Vásquez et al. 1984).

Many species of molluscs feed on kelp. Worldwide, the species would number in the thousands. Particularly prominent in most *Macrocystis* forests are the abundant trochid and turbinid snails that graze on attached kelp as well as on understory algae and the substratum below. It is highly unusual for the grazing activities of these molluscs to result directly in the removal of adult kelp, but indirect damage to fronds that results in severing portions of plants is probably common. In Stillwater Cove, central California, for example, *Tegula* spp. reached very high densities on the bottom following the removal of most of the giant kelp by storms during the 1982–1983 El Niño. The snails crawled up the stipes of the remaining understory kelp *Pterygophora californica* and, at densities of 20 or so per plant, their grazing retarded the growth of new sporophylls and fronds. They also attacked the bases of the few remaining fronds on giant kelp, severing them and killing the plants (Schiel and Foster observation). Riedman et al. (1981) and Watanabe (1984a) recorded that the abundances of three species of *Tegula* were stratified with depth in a *Macrocystis* forest near Pacific Grove, central California. *Tegula brunnea* was the most abundant turban snail at 4 m depth and *T. pulligo* was more abundant at 11 m, with little overlap in their distribution. A third species, *T. montereyi*, was the least abundant of the three turban snails, and tended to be most common at 6 m depth. In addition to living and feeding on large brown algae, these species can be found at densities of up to 40 m$^{-2}$ on the substratum (Watanabe 1984a) where they graze small plants and spores. Many other grazing gastropods are present in *Macrocystis* forests. In California, for example, several species of *Calliostoma* occur on kelp plants where they feed on surface tissue, algal films, and sessile animals (Morris et al. 1980), but their effects on kelp tissue are probably minimal. *Alia (=Mitrella) carinata* and *Lacuna unifasciata* are small species, and can be the most abundant gastropods on *Macrocystis* in California (Leighton 1971, Abbott and Haderlie 1980), but their grazing effects are also probably negligible. *Norrisia norrisi* and turbinid snails can be abundant in kelp forests, particularly on the fronds of *Macrocystis* and *Eisenia* (Schmitt et al. 1983). *N. norrisi* feeds on sporophylls, stipes, and young kelp fronds, and Leighton (1971) reported that stipe breakage may result from this grazing. The giant keyhole limpet *Megathura crenulata* and several species of chitons may graze algal spores from the substratum, but no significant effects on kelp have been reported. *Tonicella* spp. are often abundant on encrusting corallines in central California kelp forests, and may be responsible for keeping small areas free of other algae. The California brown sea hare, *Aplysia californica*, may be locally common in kelp forests and grazes on a variety of algae (Morris et al. 1980). Individuals can occasionally be found in the tops of *P. californica*, grazing on blades (Reed 1990b). *A. vaccaria*, the California black sea hare, is most common in southern California and Baja California, Mexico. This species may be the world's largest gastropod, with individuals over 0.5 m long and weighing nearly 16 kg. The feather boa kelp *Egregia menziesii* may be its primary food (Morris et al. 1980).

Abalone comprise one of the most important molluscan groups commercially within kelp forests of California, New Zealand, and Australia. These large gastropods are almost entirely passive feeders reliant on a supply of drift algae, which they trap under a powerful foot. In New Zealand, large densities of *Haliotis iris* occur in shallow areas inshore of *Macrocystis* forests, where drift algae are plentiful, on mainland sites and offshore islands (Schiel 1990, Schiel et al. 1995, Schiel and Hickford 2001). Great densities of this species occurred historically, but their reduction through fishing has produced no discernible wider effects (cf. Schiel 2006). It appears their ecological role was mostly in recycling kelp and not in affecting its distribution and abundance. In Tasmania, Johnson et al. (2011) found that an increasing seawater temperature significantly reduces size at maturity and therefore maximum shell length in *H. rubra*, a commercially important species worth $100 million annually. Of great concern in Tasmania is the effect of expanding sea urchin populations and declining kelp on this species. They found increased densities of the sea urchin *C. rodgersii* and expanding sea urchin barrens are associated with a significant decline in numbers of *H. rubra*. Along California, several species of abalone have been serially depleted (Karpov et al. 2000) to the extent that there is now no commercial fishery for any abalone in California, and a very limited recreational fishery north of San Francisco. Disease has also played a role in the severe decrease in some abalone species, such as the black abalone *H. cracherodii* (e.g., Lafferty and Kuris 1993), as has predation by an expanding sea otter population (e.g., Lowry and Pearse 1973, Watson 2000). In laboratory experiments, *H. rufescens* and the red sea urchin *Strongylocentrotus franciscanus* compete for food, but this may not be the case in natural conditions (Tegner and Levin 1982). There is no evidence that abalone have any effect on attached *Macrocystis*.

In any given kelp forest, thousands of isopods, amphipods, and other small crustacean scuttle along kelp plants and dwell among every form of algae. Some of them have ephemeral effects on giant kelp, which occasionally can lead to extensive damage, probably during periods when kelp are temperature or nutrient stressed. Along California, the kelp isopod *Pentidotea resecata* dwells on the upper fronds and blades of *Macrocystis*, and can heavily graze them (Jones 1971). North (1966) reported that this feeding activity once extensively damaged the canopy of *Macrocystis* in a wide area of the Point Loma (San Diego, California) kelp forest. The holes they create in blades from their feeding may also act as sites for fungal and bacterial infections. Another isopod, the pillbug *Paracerceis cordata*, probably causes little damage to kelp plants.

The gribble *Limnoria algarum* can be abundant on *Macrocystis* holdfasts (Andrews 1945) and may occasionally cause adult *Macrocystis* to be detached from the substratum. It burrows into the haptera, forming tunnels that may severely weaken holdfasts (Jones 1971), which then can be dislodged by water motion. A related isopod can cause considerable weakening of giant kelp holdfasts in Argentina. The kelp curling amphipods *Ampithoe humeralis*, *A. rubricata*, and *Cymadusa uncinata* build tubes in kelp laminae by curling the edges of blades and sticking them together. They eat kelp and can puncture blades with their spines and hooks (North and Schaefer 1964). Similarly, the kelp curler *Peramphithoe femorata* grazes fronds of *Macrocystis* in north-central Chile. Cerda et al. (2009) found that this isopod grazes apical blades of *Macrocystis* but this was compensated by growth in other parts of the plant. They speculated that this was a mechanism for *Macrocystis* to tolerate low-level grazing. In southern California, Graham (2002) found that intensive grazing by amphipods could result in sterility in giant kelp, but this effect was sublethal. Other crustaceans known to feed on kelp forest seaweeds are various spider crabs (Hines 1982), especially *Taliepus nuttalli*, and the kelp crab *Pugettia producta* in California. Although often abundant, their grazing does not appear to have a great effect on plants. Similarly in Chile, the kelp crab *T. marginatus* grazes on *Macrocystis* fronds but uses giant kelp mostly as a nursery (Madariaga et al. 2013). Encrusting invertebrates, although not grazers, can heavily infest *Macrocystis* blades. Bernstein and Jung (1979) found that *Macrocystis* blades in the Point Loma kelp forest had a rich fauna of epiphytes consisting mostly of bryozoans (*Membranipora membranacea, Hippothoa hyalina* [now *Celleporella hyalina*], and *Lichenopora buskiana* [now *Disporella buski*]), a serpulid polychaete (*Spirorbis spirillum*, now *Circeis spirillum*), and hydroids (*Obelia* sp. and *Campanularia* sp.), which were mostly kept in check by their life histories and larval delivery through the kelp bed. They found that small kelp beds that were heavily infested with *Membranipora* could be affected by predation on encrusted blades by the small labrid fish, *Oxyjulis californica* (see also figure 7.1).

Determining how grazing species work in tandem can be a particular challenge both conceptually and logistically because of the great number of consumers, their differing densities, and abilities to switch food. Calculating interaction strengths is one way of categorizing herbivory. For example, Sala and Dayton (2011) estimated the per capita interaction strength of a suite of herbivores on the microscopic stages of *Macrocystis*. These included *A. humeralis, A. carinata,* the wavy top snail *Megastraea undosa, N. norrisi,* the white urchin *L. pictus,* and the urchins *S. franciscanus* and *S. purpuratus*. In aquarium experiments, they found that all of these grazers could significantly reduce the survival of microscopic sporophytes. The strongest per capita interactions were by *L. undosum, S. franciscanus,* and *S. purpuratus,* and the weakest by *M. carinata* and *A. humeralis*. They modeled their results with reference to densities found in field sites, finding that the small herbivores produced no added impact to that of bigger species (particularly the sea urchins). Sala and Graham (2002) calculated the predator–prey interaction strengths of 45 species in a series of laboratory experiments. They found that the per capita interaction strengths of small grazers could be similar to those of larger herbivores, and surmised that the greater densities of small consumers in the field could increase their importance as grazers in the community. This is an important point because interaction strength does not take account of densities of organisms and can give a quite distorted view of the importance and magnitude of interactions. More studies of the effects of small grazers would be especially informative given the low survivorship of giant kelp microscopic stages.

Another phenomenon seen occasionally in kelp forests is when "good grazers go bad." Normally benign grazers, such as the turban snails mentioned earlier, can become destructive when giant kelp abundance is reduced. In another case, restoration attempts of transplanting giant kelp to the Palos Verdes area near Los Angeles were unsuccessful because of severe grazing of sporophytes by the fishes halfmoon (*Medialuna californiensis*) and opaleye (*Girella nigricans*), which led to the elimination of plants (North 1976). These also exemplify the problems with using interaction strengths to predict "importance" because strength metrics like per capita consumption clearly vary with the relative abundance of the interactors that cause changes in behavior.

---

Given the great biomass, complex structure, production, and turnover of *Macrocystis*, it is hardly surprising that this single species can provide the fuel for such a diverse array of consumers, and that environmental changes increasing or reducing its abundance can lead to changes in herbivore behavior that may feed back to giant kelp. As more becomes known about giant kelp forests, particularly in the less-studied areas of the southern hemisphere, our perspective on the interrelationships between *Macrocystis* and grazers is bound to change.

# 9

# PREDATION AND TROPHIC CASCADES IN KELP COMMUNITIES

Phase shifts between forested and deforested states (the latter known as "sea urchin barrens") result from intense grazing due to increased abundance and altered foraging patterns of sea urchins made possible in turn by human removal of their predators and competitors.

—Jackson et al. (2001)

Based on these numerous field studies spanning the global range of Macrocystis, simple trophic cascades do not seem to exist in Macrocystis-based systems.
—Graham, Vásquez, et al. (2007)

The overwhelming focus of predation and trophic cascades in kelp communities has been the relationships between sea urchins and their predators. From the earliest days of observations of the formation of barren ground habitats, it was conjectured that the reduction of predators through hunting and overfishing played a key role in the expansion of sea urchin populations (North and Pearse 1970, Breen and Mann 1976, Estes et al. 1978). There are arguments both for and against the pervasive role of predators and their controlling influence on sea urchin populations, and the literature has somewhat of a split personality on this topic. However, the paradigm of top-down control has become so entrenched in the scientific literature, textbooks and popular media that it is almost codified and taken as a base condition of kelp forest ecosystems (e.g., Jackson et al 2001, Cousteau and Schiefelbein 2007, Estes et al. 2011). It often seems to be assumed that the natural state of kelp forests is tight trophic control such that when top predators are reduced in abundances by humans, there is a "relaxation in top-down control" on sea urchins and a subsequent decline in macroalgal abundance (Guenther et al. 2012), often referred to as a trophic cascade. This is a corollary to the belief in the "balance of nature" (review in Connell and Sousa 1983); it has the value of simplicity but the unfortunate downside of being uncertain or untrue in many kelp forests worldwide.

As discussed in Chapter 8, it is unequivocal that sea urchins and some other grazers can greatly affect kelp community structure, mostly in localized patches but sometimes on a much greater scale. Despite large differences in processes and their magnitude, it is also unequivocal that across biogeographic regions a reduction or removal of dense aggregations of grazers leads to relatively rapid recolonization by kelps. The strong linkages across the bottom part of the trophic chain have therefore been well tested and demonstrated. The link between predators and control of grazers as a generality, however, is more tenuous and uncertain. That predators can influence the distribution and abundance of prey is both obvious and demonstrated in myriad publications; after all, predators make their living eating prey and clearly have an effect on them. The questions are whether the strength or magnitude of such interactions are sufficient to control prey populations and, if so, whether this occurs over wide enough spatial and temporal scales to matter in terms of "phase shifts" or "alternate stable states" of communities, and whether other factors are equally or more plausible, or act in concert with predation (e.g., Elner and Vadas 1990). It is another large step to predator-mediated trophic processes that cascade through the food web.

To understand the prevalence of predator effects in the ecological literature on kelp forests, some background is warranted. The term "trophic cascade" came from Paine (1980), but the concept is said to have originated in Hairston's green world hypothesis (Hairston et al. 1960; discussed in Polis et al. 2000). Trophic cascades are probably the most studied indirect effect in communities (Strong 1997) and are particularly common in freshwater and marine systems (Strong 1992). As they were originally proposed, these cascades had three levels of interactions involving predators eating herbivores, which reduced grazing pressure and allowed plants to thrive. Polis (1999) and Polis et al. (2000) differentiated "species-level" and "community-level" cascades. In a species-level cascade, changes in predator numbers affect one or a few plant species in a compartment of the food web. Community-level cascades may apply to any multilevel linear food web interaction and alter the distribution of plant biomass through an entire system. Such community-level cascades tend to occur when the affected plant species is dominant, there is relatively low herbivore diversity, and there are strong trophic linkages (Polis et al. 2000). As an example, Paine (1980) described a keystone predator (a sea star) eating mussels, thereby releasing space on rocks that other species could occupy, thereby increasing diversity of attached organisms.

The best-known and most documented indirect effects in marine communities involve interactions between keystone predators, sea urchins and kelp, in which the overharvesting of predators has led to increased abundances of sea urchins and a reduction in kelp. The first papers examining these species interactions were by Mann and Breen (1972) and Breen and Mann (1976), and involved overfishing of the lobster *Homarus americanus*, dense aggregations of *Strongylocentrotus droebachiensis*, and severe reductions of stipitate kelp in Nova Scotia. Breen and Mann (1976) noted there was a decrease in the abundance of lobsters by around 50% in 14 years and that during the last six of these, sea urchins destroyed around 70% of the kelp *Laminaria longicruris*. Using urchin transplant experiments, they derived a threshold abundance necessary for destructive grazing to occur (2 kg m$^{-2}$). They also used underwater cages to examine feeding rates on several invertebrates, concluding that urchins were the major item in lobster diets (Breen and Mann 1976). They then constructed a model relating sea urchin settlement rates and lobster biomass, which produced areas where urchins were "controlled" and where they were not controlled. This model came at a time when there was an increasing awareness that extensive fishing of top predators may be taking a great toll on the structure of nearshore communities (e.g., North 1971b). Nevertheless, the direct linkages between predators and prey and consequent indirect effects on kelp were highly controversial, even in Nova Scotia where they were originally proposed (e.g., see review by Elner and Vadas 1990). Of particular note were arguments about unusually high but episodic recruitment of urchins (Scheibling and Lauzon-Guay 2010), the role of diseases in controlling urchin populations (Scheibling et al. 2010, Feehan et al. 2012), and the weak inference about the role of predators at relevant spatial and temporal scales (Elner and Vadas 1990).

The kelp forests of Alaska and the west coast of North America have been focal areas of top-down regulation (Estes and Palmisano 1974, Estes et al. 1978, Estes et al. 1998), but again some controversy remains about the extent of the roles of top predators on kelp community structure, especially in California (e.g., Foster and Schiel 1988, Foster and Schiel 2010). This topic and controversy has transferred across biogeographic regions. In their review, Sala, Boudouresque et al. (1998) focused on why many algal assemblages had been transformed to coralline "barrens" in the western Mediterranean Sea. In over half of the studies that postulated predation as the major controlling factor of the sea urchin *Paracentrotus lividus*, they found that some evidence was "completely speculative" and that there was a lack of long-term studies on the relationship between predator depletion and increases in urchin populations in the area. However, in reviewing the numerous factors affecting the relationships between predators, urchins, and macroalgae, they constructed a model depicting two states: one with fleshy, erect algal communities and the other with urchin-dominated barrens of low complexity. This model in various forms is still prevalent (e.g., Steneck 1998, Pinnegar et al. 2000, Jackson et al. 2001).

In any discussion of this topic, it is worth bearing in mind Strong's (1992) statement that all trophic interactions do not cascade, and single top-down dominance is not the norm of communities or ecosystems. Menge (1992) stated that it was more appropriate to ask how top-down and bottom-up processes affect each other and what are the mechanisms of variation. Others who highlighted the potentially serious effects of overfishing on the structure and functioning of marine ecosystems (e.g., Worm et al. 2005) noted that "our emphasis herein on top-down control is not meant to imply its predominance" (Baum and Worm 2009), thereby highlighting the context dependency of such control.

With this perspective in mind, we discuss the effects of predators and potential effects of trophic cascades within giant kelp communities.

Fish, lobsters, and sea otters have been identified as key predators of ecologically important grazers within kelp systems. Many factors may influence their effects, depending on the situation and the types of predators and prey involved. In the well-studied *Macrocystis* forests of California, these are primarily the sheephead wrasse *Semicossyphus pulcher*, the lobster *Panulirus interruptus*, and the sea otter *Enhydra lutris*. These and other predators of grazers are discussed below.

## FISH PREDATION ON GRAZERS

The Californian sheephead *Semicossyphus pulcher* is a labrid fish common in nearshore waters of southern California that feeds on a wide variety of prey, including sea urchins and lobsters (review in Cowen 1983). Sheephead populations are generally more abundant in areas with vertical relief provided by kelp or rock, possibly because relief provides sheltered habitat. Spiny lobsters (*Panulirus interruptus*) have a similar geographic distribution, a broad diet including sea urchins, and they shelter during daytime. They also migrate into deeper water in winter (review in Barsky 2001).

Tegner (1980) proposed that sheephead and lobsters might control urchin populations because they eat sea urchins and because of patterns of sea urchin abundance and distribution at offshore locations where sheephead were abundant. She noted that deforestation by sea urchins had also been observed at these offshore locations. Tegner and Dayton (1981) examined urchin populations in the Point Loma kelp forest, at two sites with sheephead and spiny lobsters and one site without them, to determine effects of these predators. They found differences in urchin size-frequency distributions among sites with and without these predators, but at three different depths, and interpreted this as indicating predator control, an argument also made by Barry and Tegner (1990). These studies were suggestive of predator effects, but did not actually demonstrate control of urchins because of differences in depth, vertical relief, and *Macrocystis* stand structure between localities.

In one of the few large-scale manipulative experiments on predators, Cowen (1983) removed sheephead from a 13,000 m² site at San Nicolas Island off southern California, and compared changes in populations of red sea urchins, *Strongylocentrotus francisca-nus*, relative to a control site. The changes he found were mostly in the sheltering of urchins. Where sheephead were removed, there was an increase in the total number of sea urchins and in the number of those appearing outside of crevices. Urchins placed in the open at the control site quickly moved to shelter or were eaten by sheephead. Cowen's (1983) data, however, did not show that sheephead control red sea urchin abundance. He himself pointed out that due to limited sampling around crevices, the increased abundances of sea urchins could have resulted from their redistribution within the experimental area, with a net movement into the crevice areas sampled. He later determined sheephead densities at 18 additional sites in areas not extensively fished for them and found densities varied from 0 to 510 sheephead per hectare (Cowen 1985). One-third of the sites had densities below the threshold of 100–150 fish ha⁻¹ that his earlier work suggested was necessary to confine sea urchins to crevices. Cowen's work was done in kelp forests where sea urchins occurred mostly in and around crevices, presumably eating drift algae. There is also evidence to the contrary. Dean et al. (1984) noted at a site with high sheephead densities of 258 ha⁻¹ that the spatial distribution and movement of urchin aggregations "appeared to be unrelated to predation pressure from fishes or lobsters," concluding that urchins exacerbate kelp declines initiated by other causes such as storms and high temperatures / low nutrients. Furthermore, large sea urchins appear to have a size refuge from predation by fishes (Tegner and Dayton 1981, Cowen 1983). It is still possible, however, that the predation effects of sheephead are greater when drift algae are scarce, and urchins are actively foraging out in the open. This interaction could have been more important historically if larger sheephead were very abundant and were able to keep the number of small and intermediate-sized urchins in check during periods of low abundance of drift algae. The initiation of the fish effect on urchins, however, seems more likely to be due to kelp dynamics and the production of drift (i.e., urchins stay in cracks and crevices feeding passively or are out in the open when drift is scarce) than to a controlling influence of fish.

An intriguing recent clue about these potential effects has come from a comparison of data from Cowen's work, a later period of intensive commercial fishing of sheephead in the 1990s, and a period in the late 2000s after management intervention reduced the fishery at San Nicolas Island off southern California. Hamilton et al. (2014) used gut content and stable isotope analyses and found evidence of dietary expansion after the 10-year recovery period that included increased consumption of sea urchins and other mobile invertebrates by large fish, compared to small crabs and suspension / filter feeding invertebrates during the period of heavy fishing when fish were much smaller (see also Chapter 12). There was a 140 mm increase in average fish length and the largest fish were males, which had been severely reduced by the earlier fishery. Their surveys indicated these dietary changes were not due to altered prey availability but because of the number of large sheephead capable of handling prey, a shift in life history traits seen in other once-heavily fished areas of southern California (Hamilton et al. 2007). The resulting increase in sheephead biomass within the Channel Islands reserves was correlated with a decrease in sea urchin densities and an increase in fleshy algae, including *Macrocystis* stipe densities (Hamilton and Caselle 2015). The stable isotope analyses of Page et al. (2013) also showed that some rocky reef fish at higher trophic levels were feeding mostly on invertebrates rather than on fish, as was often assumed. How these changes might affect urchin populations and kelp, in tandem with climatic and other factors, remains to be seen.

There is little other evidence of fish predation effects on grazers that have a major impact on *Macrocystis*. Small animals on kelp fronds are one trophic pathway linking kelp and animals at higher trophic levels, such as fish (e.g., Jones 1988, Holbrook et al.

1990), and there are hundreds of species that feed on small crustaceans and snails (Carr and Reed, in press). Moreno and Jara (1984) reported that amphipods dominated fish diets within a kelp forest in Tierra del Fuego, Chile. Although grazing by the kelp curling amphipod *Peramphithoe femorata* can induce compensatory growth dynamics in *Macrocystis* fronds in central Chile (Cerda et al. 2009, 2010), there have been no reports of widespread effects of predators having much influence on them. Similarly, amphipods may have small-scale effects on *Macrocystis* fronds in central New Zealand. Pérez-Matus and Shima (2010) used mesocosm experiments to test the effects of two labrid fishes, *Notolabrus celidotus* and *N. fucicola*, on amphipods and their grazing. They found that *N. celidotus* had a large and significant effect on amphipod survival (as did Jones [1988] in another kelp system in New Zealand), whereas *N. fucicola* had no effect. They also found, however, that the presence of *N. fucicola* reduced damage to kelp fronds because amphipods sought shelter and reduced foraging. They concluded that these types of trait-mediated interactions on prey behavior may benefit giant kelp in diverse communities.

The most thorough field-based studies on the interactive effects of small grazers, fish, and kelp were done by Davenport and Anderson (2007), who did a series of experiments at Santa Catalina Island, southern California, in which they excluded or included fish using large cages. They then tested for the abundances of small grazers and their effects on kelp growth. The primary mesograzers were the isopod *Pentidothea resecta* and the gammarid amphipod *Ampithoe humeralis*, and the main micro-carnivorous fish were the kelp perch *Brachyistius frenatus* and the señorita *Oxyjulis californica*. They found there was a significant increase in mesograzers with fish exclusion, which led to increased grazing and poorer growth of *Macrocystis* (see figure 7.1). Removal of fish exclusion cages resulted in a quick return of mesograzers to ambient levels. They also found that there was a critical threshold of fish abundance that affected mesograzer biomass, but that there was no difference in effects between the fish species. Davenport and Anderson's study, therefore, provided direct field-based evidence for mesograzers reducing kelp performance, which was mediated by the abundances of fish. As in previous examples, these apparently diffuse effects can have specific and nonintuitive impacts on localized kelp forests.

## LOBSTER PREDATION ON GRAZERS

As for fish, the role of lobsters in controlling overgrazing by sea urchins has been greatly debated. The scene was very much set for this debate in eastern Canada, where lobsters were first considered to be a primary candidate in the great expansion of urchin populations, due to a large reduction in the numbers of these predators from fishing (Breen and Mann 1976). The debate continued for decades, sometimes getting quite vitriolic, at least in terms of the normally sedate writing of the scientific literature (Miller 1985, 1986, 1987, Keats 1986, Breen 1987). Scheibling's (1996) review concluded for eastern Canada that "our understanding of interactions between this species (sea urchins) and its predators is insufficient to support any generalizations about the role of predation in regulating populations." It seems to be mostly accepted now that disease, recruitment dynamics, and an active sea urchin fishery play critical roles in periods of urchin population expansion and demise there. However, an important caveat is that it is not possible to have a true understanding of ecological baselines because of historical overfishing of many species. There is considerable correlative evidence that overfishing of cod and several other commercially exploited fish species over several decades is associated with changes throughout the food web (Frank et al. 2005, Scheffer et al. 2005).

Tegner (1980) was the first to suggest that in southern California, where no sea otters had been present for over a century, sheephead and the California spiny lobster *Panulirus interruptus* might control sea urchin populations. In the Point Loma kelp forest, Tegner and Dayton (1981) examined variation in size-frequency distributions of live and dead sea urchins and made other observations at three sites, two with sheephead and spiny lobsters and one without, to determine the potential combined effects of these predators. Differences between sea urchin size-frequency distributions among sites with sheephead and lobsters and the site without these predators were interpreted as indicating predator control. This was further argued by Barry and Tegner (1990). These results were suggestive but equivocal; there was no replication of the area without predators, there were differences in depth and topography among sites, and others have shown that urchin size-frequency distributions similar to those found in these earlier studies can also result from sporadic recruitment (Ebert et al. 1993).

Tegner and Levin (1983) provide the primary data concerning lobster predation on sea urchins. They found that lobsters preferentially fed on *Strongylocentrotus purpuratus*, and ate all sizes of them, in laboratory choice experiments. They interpreted this as accounting for the unimodal size-frequency distribution of live animals observed in the field where lobsters and sheephead were common. They also found in their laboratory experiments that lobsters would eat *S. franciscanus*, but mostly consumed intermediate sizes. This was suggested to account for the bimodal size distribution of live red sea urchins in the field where the two predators were common. The laboratory experiments of Tegner and Levin (1983) also determined consumption rates of sea urchins by lobsters in the laboratory, which is a good assay of potential feeding rates if no other food is available, but it is not possible to relate these to natural populations because lobsters were only fed sea urchins, which comprise only part of the natural diet of lobsters. From these results and landing data for the lobster fishery, Tegner and Levin (1983) concluded lobsters and sheephead controlled sea urchin populations until fishing reduced their abundance in the 1950s, leading to increases in sea urchin populations and deforestation.

Lafferty (2004) and Behrens and Lafferty (2004) argued for lobster control of sea urchin populations based on field survey data from the northern Channel Islands off southern California. Data were the abundances of sea urchins and the size-frequency distributions of live and dead sea urchins at fished sites and at reserve sites where fishing was not permitted. There were significantly more lobsters and fewer urchins at reserve sites. Based on size-frequency distributions of live *S. franciscanus* and

*S. purpuratus*, Behrens and Lafferty (2004) stated that these matched those suggested by Tegner and Levin (1983) to be the result of lobster predation. Both studies concluded these relationships verified the hypothesis that spiny lobsters control sea urchin populations in the field.

These data, although frequently cited as evidence for lobster control of urchins, are not conclusive about the role of lobsters. There are plausible alternative hypotheses and inconsistencies in the data and interpretation. The two reserve sites were very close together (within about 1 km) within the same reserve on the western end of Anacapa Island, while the fished sites were more widely separated and positioned to the east of the reserve on other islands (Lafferty 2004). The lack of interspersion of sites can be problematic here because ocean circulation around the Channel Islands is complex, leading to small-scale differences in the distribution of organisms (e.g., Reed et al. 2000). Murray et al. (1980) indicate that Anacapa Island is a warm biogeographic region distinct from the other Channel Islands. It also should be recognized that differences in size-frequency distributions may result from sporadic recruitment rather than from predation (Ebert et al. 1993) and are affected by habitat quality.

These studies highlight many of the problems relating to interpretation of pattern, as we seek to understand wide-ranging and sometimes diffuse effects. In this case, the "experiment" was fished and unfished sites. Ideally, these sites should be as similar as possible, with the same topography and depth structures, in the same biogeographic regions (or interspersed among them), either nested within oceanographic features or interspersed among them, and sampled over a long enough time period to yield treatment effects instead of "noise" or natural fluctuations. Unfortunately, nature rarely comes arrayed in convenient packages and so researchers are constrained to gleaning what they can from the data available. It makes ecology interesting, but also can compromise inferential ability, and it requires scientists to consider alternative hypotheses that potentially explain the same data.

Guenther et al. (2012) asked whether "fishing can trigger trophic cascades that alter community structure and dynamics and thus modify ecosystem attributes," with respect to the lobster *Panulirus interruptus* and sea urchins in a kelp forest near Santa Barbara, California. They found the answer to be a qualified "no." From eight years of survey and fishing data at eight sites, they compared lobster abundances (as measured by the commercial catch in each of the eight areas), and sea urchin density, as well as sea urchin abundance versus kelp abundance. They then tested the extent to which these were related to the intensity of lobster fishing, as measured by the number of lobster traps lifted. Guenther et al. were unable to detect meaningful relationships between lobster and urchin abundance data, partially because they tended to be quite variable over the years. Algal biomass and urchin densities were unrelated across sites and years, but at two sites (Naples and Carpinteria Reefs) a decline in urchin densities coincided with increases in kelp biomass. Although they found no significant relationships between any of the measure relating to lobsters, urchins, and algae, they only suggested cautiously that "a trophic cascade caused by lobster fishing . . . is not ubiquitous in the Santa Barbara Channel marine ecosystem."

The general lack of significant relationships between metrics of urchins, predators, and giant kelp was noted by Foster and Schiel (2010). Their analyses were based mostly around the Point Loma (San Diego) kelp bed, which is a frequently cited focus of trophic cascades in giant kelp ecosystems (cf. Jackson et al. 2001, Steneck et al. 2002). Their examination of all available data on urchin and kelp abundances, and commercial landings of kelp, sheephead, urchins, and lobsters, showed that there was no widespread decline in kelp forests in the region between 1950 and 1970, as had been indicated (Jackson et al. 2001) but there were localized declines in mainland kelp forests near the rapidly growing cities of Los Angeles and San Diego. There were few correlations between any of the predator–urchin–kelp variables, except a positive relationship between urchin commercial landings and kelp canopy cover, a reflection of fishing practices and the reliance of urchin gonad condition on adequate kelp. They concluded that declines in some coastal kelp forest sites around major cities were caused primarily by degradation of water quality, increased sedimentation and contamination, and unfavorable oceanographic conditions, rather than by trophic cascades.

Further afield in Tasmania, the case for lobsters generating a trophic cascade is relatively strong, and yet Craig Johnson and colleagues have been sparing in their use of the term. Instead, they have marshaled considerable evidence that overfishing of lobsters has reduced the resilience of kelp beds to climatic and oceanographic changes, and how this has led to a "catastrophic phase shift" to widespread urchin barrens (Ling et al. 2009). From a series of experiments, they estimated that the minimum size of lobsters capable of preying on *Centrostephanus rodgersii* individuals in natural conditions was around 140 mm carapace length (figure 9.1). Survival of tagged urchins was much higher where lobsters were fished compared to areas where fishing was not allowed. Their analyses of catch data showed that it was the large lobsters, those capable of eating urchins, that were mostly eliminated by the fishery. Despite these mechanisms and effects, Ling et al. (2009) pointed out that the majority of reefs in the region remained kelp dominated, but argued their resilience is low because effective predators are relatively rare. Furthermore, as discussed earlier (Chapter 7), it was several processes and nonlinear responses working in concert that produced these changes, including the inshore movement of the East Australian Current, warming water and larval drift enabling *C. rodgersii* larvae to survive and drift down from warmer northern waters, a general decline in abundance of *Macrocystis* due initially most likely to warming and low nutrients over several decades, and the rise of the lobster fishery. Johnson and colleagues (e.g., Johnson et al. 2005, 2011, Ling et al. 2009) have very firmly put their arguments into a resilience-based ecosystem management approach. They argue this conceptual shift away from "traditional equilibrium-based thinking of top-down predation-driven vs. bottom-up environmentally driven control" is both necessary and urgent in the face of high levels of predator removal and a rapidly warming climate.

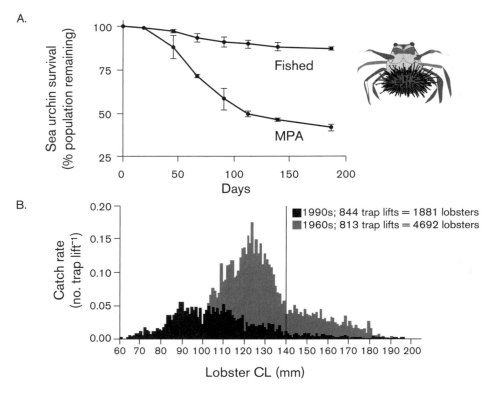

FIGURE 9.1

Effects of lobster (*Jasus edwardsii*) predation on sea urchins (*C. rodgersii*) in Tasmania. (A) Population trajectories of tagged *C. rodgersii* on reefs inside and outside marine protected areas (i.e., ±fishing); data are mean percentages (±1 standard error) of sea urchins surviving (initial *n* = 96 at each site). (B) Change in size-frequency distribution of lobsters before (1960s) and after intensive fishing (1990s) in northeastern Tasmania, showing the reduction in carapace length (CL) of lobsters. Lobsters must be around 140 mm CL to be effective predators of emergent sea urchins in the wild. SOURCES: Data from Frusher (1997) and Ling et al. (2009).

## SEA OTTERS AND PREDATION ON GRAZERS

The sea otter *Enhydra lutris* is the poster pinup of a keystone species. Publications on this species number in the hundreds, and citations relating to its effects on kelp-dominated coastal ecosystems in the thousands. The vast majority of this literature has been generated since 1985, when a discussion of otter effects on kelp communities could be relatively brief (Foster and Schiel 1985). What has changed since then is not only the "quest for keystone species" (cf. Power et al. 1996) that may be the lynchpins of trophic structure and ecosystem function, but also an awareness generally that the exploitation of top predators is taking a toll on many ecosystems worldwide. In this case, the poster pinup has mostly a deserved reputation. It appears to be cute and cuddly but lives up to its billing of being a dominant predator and having a wide-scale influence on the kelp systems it inhabits.

Sea otters are voracious consumers, especially of invertebrates, including sea urchins and abalone, and can severely reduce prey numbers. This was seen, for example, in the Monterey California region where there were enormous populations of large abalone, *Haliotis rufescens*, through the 1940s, and those that were not fished commercially or in cracks and crevices inaccessible to otters were quickly removed once otters reappeared in the region (McLean 1962). Nevertheless, their effects are not without controversy, especially as they relate to *Macrocystis* forests (vanBlaricom and Estes 1988). Sea otters are roving predators and their effects on large grazing invertebrates and kelp systems can sometimes take years to manifest themselves completely when new areas are inhabited (Lowry and Pearse 1973) and can be patchy (Watanabe and Harrold 1991). The controversy, however, relates more to putative historical effects of aboriginal and commercial exploitation that hunted otters to near extinction in the 1800s (described by Kenyon 1969) "with the attendant collapse of kelp forests grazed away by sea urchins released from sea otter predation" (Jackson et al. 2001). The corollary of this claim of collapse is that kelp forests where otters have not been present for considerable periods of time should be overgrazed by sea urchins and have little kelp. This is the case in many areas of the Aleutian Islands, where the sea urchin *Strongylocentrotus droebachiensis* forms barrens that dominate nearshore rocky reefs down to where otters do not forage and the virtual elimination of all large sea urchins where otters are present (Estes and Palmisano 1974, Estes et al. 1978, Estes and Duggins 1995). In California, in areas where otters had not been present since at least the early 20th century, however, only about 10% of the 224 giant kelp forest sites surveyed were deforested, and urchins were patchy within the other 90% of sites (Foster and Schiel 1988; figure 9.2). Clearly, there are numerous factors involved in determining the relative importance of dominant predators on grazers, including the relative influence of abiotic factors and disturbances on the standing stock of kelp and production of detrital material (Reed et al. 2011), the recruitment dynamics of sea urchins (Flowers et al. 2002), the species composition, and possibly the diversity of the system (e.g., Byrnes et al. 2006, Graham, Vásquez, et al. 2007).

Enactment of protective legislation for otters in 1911 was instrumental in the recovery of its populations. Before commercial exploitation, the range of sea otters extended in a continuous arc from the northern islands of Japan, along the Kamchatka coastline, across the Aleutian Islands chain, and southward along the west coast of North America into Baja California, Mexico (Kenyon 1969). The population southeast of the Aleutian chain was believed to be extinct, at least by the scientific community, until the discovery in 1938 of a raft of 50–90 individuals along an isolated stretch of coastline in central California (Bolin 1938, Woodhouse et al. 1977). After legal protection came into effect, the range of otters extended at the rate of approximately 4 km annually, with a total population increase of 5% per year through at least the 1970s (Wild and Ames 1974).

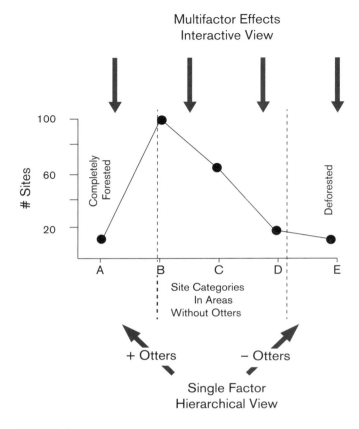

FIGURE 9.2

Categorization of sites in California without sea otters up to 1984.

Site categories:

(A) completely forested by giant kelp with no large sea urchins

(B) kelp dominated with sea urchins interspersed in crevices

(C) kelp present and abundant, but with patches of 1–3 m diameter dominated by sea urchins

(D) kelp present in patches, but with large patches (20–50 m diameter) dominated by sea urchins

(E) sea urchin 'barrens' dominated by sea urchins with no kelp. See text for more explanation

SOURCE: From Foster and Schiel (1988).

Sea otters currently occupy around 500 km of the California coastline from around Halfmoon Bay (just south of San Francisco) to around Gaviota (just southeast of Point Conception in southern California; Estes et al. 2003, Hatfield and Tinker 2012). The California population totals around 2700 otters, with slightly expanding numbers in the north and slightly declining numbers in the south portions of this range. The central area around Monterey, the source of the recolonization of otters in California, has had relatively stable numbers of around 1600 otters since at least 1990 (Hatfield and

Tinker 2012). There have been several successful translocations of otters into areas of their ancestral range such as southeast Alaska, Vancouver Island, Canada (Watson and Estes 2011), Washington state, and San Nicolas Island, off southern California (Estes and Harrold 1988).

The preferred habitat of otters in California is a lush kelp canopy in proximity to rocky substrata with deep crevices (Woodhouse et al. 1977). Kelp forests dampen wave action, creating areas of calm that serve as otter refuges from winter storms. The forests may also serve as protection from predators such as white sharks. Otters gradually move out into the offshore kelp canopies as these re-form in spring and summer. The diet of sea otters in kelp forests consists of epibenthic invertebrates commonly associated with low intertidal and subtidal rocky substrata with deep crevices (e.g., sea urchins, abalone), and with kelp fronds (e.g., *Tegula* spp., kelp crabs; Woodhouse et al. 1977, Estes et al. 1982). Otters can forage as deep as 40 m, securing their prey with their forepaws, and returning to the surface to eat (Kenyon 1969). California sea otters display a unique type of tool-using behavior when feeding on hard-shelled invertebrates: food items may be pounded against a rock held on the otter's chest while floating at the sea surface (Fisher 1939). Houk and Geibel (1974) described an incident of tool use by a sea otter pounding an attached abalone with a rock.

Numerous studies have recorded that otters colonizing kelp forests in central California prey preferentially on red sea urchins and abalone (Boolootian 1961, Ebert 1968, Benech 1980, Ostfeld 1982). In a colonizing group of otters in a *Macrocystis* forests off Point Santa Cruz, California, red sea urchins were initially the major prey item (Ostfeld 1982). Kelp crabs and clams replaced urchins as urchins became scarce. Abalone and cancer crabs were consistently exploited as dietary items at relatively low levels. Other prey items on rocky substrata include snails, mussels, octopus, chitons, tubeworms, limpets, barnacles, scallops, and starfish (Ebert 1968, Wild and Ames 1974, Shimek 1977, Woodhouse et al. 1977, Estes et al. 1982). VanWagenen et al. (1981) saw sea otters preying on seabirds. Otters are able to adapt to diverse environments (Woodhouse et al. 1977) and readily shift their diets as different prey items become available (Tinker et al. 2008). The spreading fronts of the California population of otters have successfully occupied sandy and silty bottoms in the coastal zone, exploiting Pismo clams, gaper clams, razor clams, mole crabs, and even echiuroid "worms" as prey (Wild and Ames 1974, Stephenson 1977, Hines and Loughlin 1980). They also inhabit bays and sloughs; one of the largest concentrations of sea otters in California is now in Elkhorn Slough where they have a profound positive effect on eelgrass (Hughes et al. 2013). This occurs because otters consume crabs, which releases grazing species from predation pressure. The grazers eat eelgrass epiphytes, which have increased due to eutrophication from agriculture. The reduction of epiphytes and consequent increase in light allows eelgrass to become more productive. Outside of the protection of kelp canopies and shallow inlets, sea otters are susceptible to attacks by great white sharks (Ames and Morejohn 1980). Other than man and parasites, this shark is its only known predator in California waters, although orcas eat otters in Alaskan waters (Estes et al. 1998).

Otters exert their large effects on invertebrate populations both by force of numbers and by satisfying their high energetic demand. Costa (1978) calculated that an average-sized otter must consume 20–25% of its body weight daily to meet its energy needs. Yeates et al. (2007) were the first to measure the energetic coasts of individual dives and metabolic rates from prolonged foraging sessions. They did a series of large tank experiments and radio tagging of wild animals, recording over 2000 foraging dives. They found that otters dive for up to 4 min, with an average dive of 1 min. The highest metabolic rates were after a foraging dive. Their metabolic rate was similar regardless of what prey they caught. Several features of otters contribute to their high energy demands. Otters, unlike other marine mammals, lack blubber and insulate themselves by trapping air against their skin with very dense fur (Williams et al. 1992). This insulation compresses with depth and the thermal energetic costs must be met by increasing food (Costa and Kooyman 1984). Air trapping and large lungs (Tarasoff and Kooyman 1973) make the animal buoyant. Yeates et al. (2007), in their fascinating study, concluded that otters have an energetically challenging existence because of their small size, exceptional buoyancy, and costly style of swimming, which are counterbalanced by them spending about half of their day resting.

The effects of otters in kelp forests are best known in central California, where it is now rare to find urchins other than in low densities in cracks and crevices apparently inaccessible to otters, a finding corroborated in numerous studies (e.g., McLean 1962, Lowry and Pearse 1973, Riedman and Estes 1988, Reed et al. 2011). Predation by otters is far from the only source of mortality of sea urchins in giant kelp forests with otters (e.g., Pearse and Hines 1977, 1979, Watanabe and Harrold 1991), but few larger urchins are able to survive for long outside of protective subhabitats. The clearest perspective on the effects of grazing, predation, and abiotic factors in the production of giant kelp forests of California is provided by Reed et al. (2011). They pointed out that the interplay and relative influences of top-down (predators), bottom-up (nutrients), and environmental forcing (wave disturbance) in central and southern California *Macrocystis* forests are mediated to some extent by the very low densities of urchins in central California due to predation by otters. Red and purple urchin densities averaged over 10 years at nine sites in southern California were 7.1 m$^{-2}$ compared to 0.06 m$^{-2}$ at eight sites in central California. The southern California sites showed considerable variation, up to 30-fold, in urchin densities among sites and years. Despite these differences in grazing and top-down control due to otters in central California, however, Reed et al. (2011) showed that net primary production (NPP) was consistently lower in central California, not greater as would be predicted by the absence of urchins, because of the intense wave disturbances that overwhelmed top-down and bottom-up processes. In both central and southern California, net primary production was best explained by interregional variability in wave disturbances. This is because the massive growth form of *Macrocystis* and its canopy on the sea surface creates considerable drag during storms and is susceptible to removal. Reed et al. (2011) noted, however, that the understory of perennial kelps may partially compensate for the loss of production of *Macrocystis*, but cannot compensate for its role as a foundation species. Furthermore, even though otters were absent from most kelp forests in southern California, the NPP of giant kelp in their southern sites was twice that of central California sites, despite seasonal nutrient limitation, strong localized grazing pressure, and some sites being repeatedly deforested over the years of the study.

The historical evidence for the effects of sea otters is interesting and suggestive, although not entirely conclusive. Simenstad et al. (1978) examined evidence from Aleut middens in the Aleutian Islands and concluded that alternate stable states existed in nearshore communities. They argued that the strata containing large quantities of sea urchin and limpet shells coincided with the absence of otters, due to hunting by Aleuts. The presence of fish remains coincided with times when otters were present, and macroalgae predominated. There is also evidence in the middens in Monterey (Gordon 1974) and on San Nicolas Island, off southern California, that prehistoric people hunted sea otters. Tegner and Dayton (2000) pointed out the large numbers of abalone shells seen on the Channel Islands, suggesting that aboriginal man had a significant impact on the nearshore community. A few studies have reviewed historical evidence and attempted to decipher temporal trends in middens to shed light on the use of shellfish populations and the possible effects of sea otters (Estes and Van Blaricom 1985, Erlandson et al. 2005, 2008). Erlandson et al. (2005), for example, found on San Miguel Island, 42 km off the coast of southern California that native Americans apparently coexisted with sea otters and productive shellfish populations for over 9000 years. They argued that based on modern knowledge of sea otter behavior and ecology, the domination in middens by large red abalone, *H. rufescens*, between 7300 and 3300 years ago was only likely if otters had been reduced by hunting or other causes because "an abundance of large abalone accessible to human foragers is fundamentally incompatible with the presence of significant numbers of sea otters." They speculated it was possible that native Americans intentionally controlled otter populations to enhance local productivity of fish and shellfish for harvesting. Erlandson et al. (2008) pointed out that some of the variation in sizes of shellfish in middens may have been the result of variation in water temperature, marine productivity, or other natural cycles, as well as an increase in the human population during the last 1000 years. An alternative explanation is that human foraging was not as intense as assumed, due to either small and/or variable human population sizes or the inability of humans to forage across all of the habitats occupied by shellfish. For example, many invertebrate species probably had a depth refuge because humans could not effectively forage below a few meters.

These historical reconstructions are interesting and relevant to understanding a broader perspective of human interactions with the marine environment, their reliance on sustenance from it, and the potential for local-scale impacts. As is the nature of most of this type of evidence, however, it is suggestive but inconclusive. The accumulating evidence is that the large-scale removal of sea otters from around the mid-1700s onward allowed urchins and abalone to increase in many areas (Erlandson et al. 2008). There also seems to be a general recognition that the diverse and complex giant kelp ecosystem of California provided considerably more buffering of predatory effects of otters so that they did not undergo the large changes seen in some Alaskan sites (Erlandson et al. 2005, Graham, Vásquez, et al. 2007, Foster and Schiel 2010).

## OTHER PREDATORS

The relatively high diversity and great biomass of giant kelp forests ensure there are multiple pathways for algal material to work its way through the trophic levels. Prominent predators of grazers and detritivores (not discussed in other sections here) are sea stars, crabs, whelks, octopi, harbor seals, and birds. The large sun star *Pycnopodia helianthoides*, for example, is an active forager in kelp forests along the west coast of North America, feeding on sea urchins, abalone, snails, and other sea stars (Moitoza and Phillips 1979, Duggins 1983, Watanabe 1984a, Harrold and Pearse 1987, Pearse and Hines 1987, Byrnes et al. 2006). Large Kellet's whelks (*Kelletia kelletii*) feed on a wide range of snails and other invertebrates. Octopi are highly mobile predators capable of squeezing into cracks and crevices, and feeding on snails, crabs, abalone, and other prey on the forest floor (e.g., Ambrose 1986). Further afield, a study in a *Macrocystis* stand in Tierra del Fuego in the Straits of Magellan showed that gulls could be an important high trophic level predator. Hockey (1988) found that the sea gull *Larus dominicanus* fed almost exclusively on small epifaunal bivalves, *Gaimardia* sp., on the floating fronds and upper stipes of giant kelp, consuming about 300 individuals per square meter daily.

None of these predators of grazing invertebrates appears to exert more than small-scale, highly localized effects within a kelp forest. In combination, however, there may be synergistic interactions among predators. Byrnes et al. (2006), for example, found in field surveys that as predator diversity increased, the number of herbivores decreased and kelp abundance increased. The importance of sea star predation to giant kelp community structure may be further elucidated as a result of the massive die-off of numerous sea star species at many sites in the Northeast Pacific, first observed in June 2013 (Stokstad 2014). The cause of the die-off and its ecological consequences are yet to be determined.

---

Any discussion of the role of predators in kelp forest biology is bound to generate lively debate in scientific circles. It is easy to find numerous examples of places and situations where grazers, especially sea urchins, have had a large impact. It is unequivocal that the sea otter *Enhydra lutris* can control urchin populations, but the evidence becomes more diffuse and patchy when it involves fish and lobsters. There are also numerous cases of snails and other invertebrates having localized impacts on giant kelp, particularly in ENSO years with large storms and low nutrients. The major question is how to separate the phenomenological from underlying mechanisms.

Essentially, there are two (mostly divergent) models of kelp forest structure, one involving trophic cascades and one involving multiple processes acting on kelp, sea

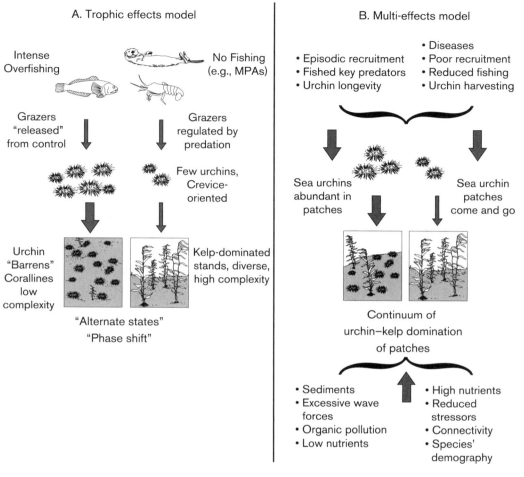

**A. Trophic effects model**

Intense
Overfishing

No Fishing
(e.g., MPAs)

Grazers
"released"
from control

Grazers
regulated by
predation

Few urchins,
Crevice-
oriented

Urchin
"Barrens"
Corallines
low
complexity

Kelp-dominated
stands, diverse,
high complexity

"Alternate states"
"Phase shift"

**B. Multi-effects model**

- Diseases
- Episodic recruitment
- Poor recruitment
- Fished key predators
- Reduced fishing
- Urchin longevity
- Urchin harvesting

Sea urchins
abundant in
patches

Sea urchin
patches
come and go

Continuum of
urchin–kelp domination
of patches

- Sediments
- High nutrients
- Excessive wave
- Reduced
  forces
  stressors
- Organic pollution
- Connectivity
- Low nutrients
- Species'
  demography

FIGURE 9.3
Conceptual models of dominant structuring processes of kelp beds. (A) Trophic cascades leading to different phases and (B) multi-effects model of a continuum of urchin to algal dominated patches (see text for explanations). Widths of arrows indicate strength of interaction effect. See also figure 6.5.
SOURCE: From Schiel (2013).

urchins, and predators (figure 9.3). In some conclusions of generality of predator control and "relaxation" of such control there is a logical flaw. Just because extensive grazing or significant predation effects on benthic grazers occur in most kelp forests of the world does not imply or mean they are generally of great magnitude or pervasive in space and time. For example, Reed et al. (2011) showed that urchin grazing is far more important in southern California, where there are no otters, than in central California, where otters are numerous, and yet kelp forests are abundant and functionally intact in both regions, largely responding to oceanographic conditions. Foster and Schiel (1988) calculated that around 10% of sites in southern California were dominated by urchins. In all cases where giant kelp occurs, it has thrived and its associated communities are functional, and not in a "collapsed" state from overfishing, except where topography and coastal sediment have been permanently altered. Interaction strengths vary, often depending on oceanic basin-wide conditions affecting kelp growth, survival, provision of detached drift material, and local-scale density-dependent effects of kelp. Even the severe case of overgrazing and overfishing in Tasmanian waters has several caveats, including the action of a current, transport of a warm-water urchin from the north, and long-term warming trends in SST.

Whether or not effects constitute a "trophic cascade" in kelp forests is only partially semantics. In many ways, overuse of this term trivializes most of the relevant processes in kelp forest ecosystems. Graham, Vásquez, et al. (2007) highlighted three features that buffer giant kelp forests from overexploitation of predators. All *Macrocystis*-based food webs are diverse, they are embedded within high-productivity oceanographic systems where production may simply override consumption, and the recruitment of sea urchins, the dominant grazer, tends to be highly variable in space and time. As discussed earlier (Part I), *Macrocystis* is also highly responsive to its environment in ways not found in terrestrial ecosystems. It is almost inevitable that we compare giant kelp, as a prominent foundation species in coastal waters, with terrestrial trees and forests. One of the many important ways they differ (cf. Carr et al. 2003), however, is that *Macrocystis* has a weedy life history, grows fast, does not usually live long, and an entire forest can regenerate quickly. It lends itself admirably to the physically and biologically dynamic waters in which it thrives. Predation and grazing can be important, and locally comprehensive, but they are just two of the numerous factors affecting the world's giant kelp forests.

# PART III

# HUMAN USAGE, MANAGEMENT, AND CONSERVATION

# 10

# ANTHROPOGENIC EFFECTS
# ON KELP FORESTS

Coincident with the increased discharge of sewage effluent,
harvestable kelp has virtually disappeared in the vicinity of Whites Point
for a distance of two or three miles along the coast.

—Revelle and Wheelock (1954)

Because giant kelp forests occur near the shore, they are immediately subjected to run-off and waste discharge from land. They can also be affected by shoreline modifications that result in loss of rocky habitat, increased turbidity, and changes in sediment distribution. Boat traffic and anchoring can thin canopies and remove entire plants, and be vectors for introducing non-native species. Shipping and oil extraction accidents can bathe the forest in petroleum products. Harvesting for food and chemicals removes *Macrocystis* and associated organisms. Of course, forests are also embedded in the larger ocean which is changing as a result of human activities. There are clearly ample opportunities for multiple, human-induced impacts on giant kelp forests, and all the above have occurred primarily along the densely populated shores of southern California. The effects related to ocean discharges, land use, boat traffic, toxic spills, giant kelp harvesting, and introduced species are considered below; fishing effects are discussed in Chapter 11 and climate-related changes in Chapter 13.

There are many anthropogenic effects on water quality and benthic habitats that negatively affect giant kelp growth and reproduction. Given the biological and physiological constraints on growth and reproduction reviewed in Chapters 2 and 3, it should be no surprise that activities that increase sedimentation, reduce light, and increase turbidity and temperature (if near upper thermal limit) will probably cause *Macrocystis* to decline. Although much of current coastal management is focused on sustaining and improving finfish and shellfish populations with marine protected areas (Chapter 12), maintenance of water and benthic quality is both fundamental and critical to ensure that diverse and productive forests persist (Schiel and Foster 1992). This can only be achieved through proper management of waste discharges and runoff.

## SEWAGE AND INDUSTRIAL WASTE DISCHARGE

The largest impacts of sewage discharge on giant kelp forests occurred from the 1940s through the early 1960s in southern California, coincident with rapidly increasing urbanization and industrialization (Foster and Schiel 2010). During this time the giant kelp forest at Palos Verdes adjacent to the Whites Point sewer outfall near Los Angeles declined from 12 km² to near zero, and the forest at Point Loma near San Diego from 6 km² to near zero (figure 10.1). The small remnants remaining by 1957 were further reduced by the 1957–1959 El Niño. These losses have also been attributed to increases in sea urchin abundance related to declines in populations of sea urchin predators due to fishing (e.g., Jackson et al. 2001), but sea urchin grazing became important only after the forests were reduced and urchins began to graze actively on the few remaining stands. Both forests recovered with increased regulation of sewage sources (source controls) and improved sewage treatment (figure 10.1), and when new construction extended the outfalls further offshore. These improvements also contributed to reduced sediment toxicity (Swartz et al. 1986, James et al. 1990). However, while the Point Loma forest has occasionally recovered to near its former size, the forest at Palos Verdes has not. Recovery may be inhibited by the continued discharge of high nutrient sewage effluent that may increase turbidity by stimulating phytoplankton growth; nitrogen inputs from sewage can exceed those from natural upwelling around large sewage outfalls (Howard et al. 2014). It may also be inhibited by runoff and sediment resuspension (Conversi and McGowan 1994) and the legacy of rocky habitat buried by sediment. The large-scale and long-lasting effects of these early discharges in southern California are in contrast to those that occurred in the Point Loma kelp forest from a broken outfall pipe. The break occurred in 1992, lasted 2 months, and released $7.1 \times 10^8$ L$^{-1}$ of treated effluent into the kelp forest. Increased ammonia concentrations and reduced light inhibited germination and growth of outplanted giant kelp microscopic stages, but these and other effects disappeared soon after the pipe was repaired, and the spill appeared to have no long-lasting effects on the community (Tegner et al. "Effects of a large" 1995).

The many potential waste materials in sewage effluent and their interactive effects make it difficult to determine which may be most important. Sewer effluent can contain a complex mix of potentially harmful materials, but the high, inverse correlation between canopy size and emissions of suspended solids (figure 3.1) and the accumulation of sediment associated with the outfall at Palos Verdes (Grigg and Kiwala 1970) suggest the most important effects are on microscopic stages and small juvenile sporophytes whose survival and reproduction is inhibited by light reduction, scour, and burial (see Chapter 3) as well as by toxic materials sorbed to particulate organic matter (James et al. 1990). The more advanced sewage treatment currently used removes

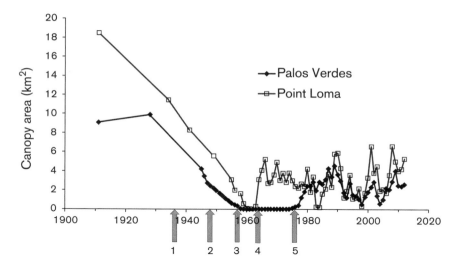

FIGURE 10.1

Sea-surface canopy areas of giant kelp (in square kilometers) at Palos Verdes and Point Loma kelp forests.

Events indicated by arrows: 1 = installation of shallow sewage outfall at Palos Verdes; 2 = sediment discharge from Mission Bay and sewage discharge from San Diego Bay at Point Loma; 3 = 1957–1959 El Niño; 4 = sewage discharge moved offshore at Point Loma; 5 = beginning of advanced sewage treatment at Palos Verdes and sea urchin fishery in southern California.

NOTE: Canopy variation after 1980 was primarily related to variation in oceanographic conditions and storms.

SOURCE: Modified from Foster and Schiel (2010), based on data in MBC (2013).

suspended solids but, depending on the level of treatment, nutrients may remain that reduce benthic light by stimulating phytoplankton growth, a problem that is reduced by moving outfalls offshore into deeper water (Conversi and McGowan 1994). As argued by Connell (2007) and Gorman et al. (2009) for kelp beds (*Ecklonia radiata*) in South Australia, excess nutrients may also encourage the growth of algal turfs that directly and indirectly (by accumulating sediment) inhibit recruitment of kelp and various large fucoids. Turf-sediment matrices have also been implicated in preventing recolonization of native algal species in Tasmanian kelp communities (Valentine and Johnson 2005). Ammonia can be toxic (Tegner et al. 1995—"Effects of a large sewage spill"), and domestic sewage can contain toxic metals and organic compounds that may be increased if the discharge also contains industrial wastes.

Offshore oil production produces very large waste streams of so-called "produced-waters" that can contain hydrocarbons, heavy metals, and other potentially toxic substances. Depending on the location of a discharge, this waste could affect kelp forests. Reed et al. (1994) used a combination of adult giant kelp transplants and outplanted zoospores, as well as laboratory toxicity tests, to evaluate the effects of a large ($2.64 \times 10^6$ L day$^{-1}$) produced-water discharge near Carpinteria, California. Diffusers on the discharge pipe diluted the wastes to 1% within 5 m, and the field experiments indicated discharge effects were probably limited to within 50 m of the outfall. Other studies at the same site indicated that effects on invertebrate reproduction and larvae extended to 100 m or more from the outfall (review in Reed et al. 1994).

Since the 1960s, regulation by the United States Environmental Protection Agency through the Clean Water Act and the National Pollutant Discharge Elimination System (www.cfpub.epa.gov/npdes) permitting process has led to the elimination of many discharges and improved treatment standards for those that remain. These improvements in the United States and elsewhere (e.g., Australia; Burridge and Bidwell 2002) have been informed by laboratory bioassays that expose life stages of organisms living in kelp forests, including giant kelp (Anderson and Hunt 1988), to variously treated effluent or particular toxins in the effluent, such as copper (Anderson et al. 1990). These assays can be used in conjunction with chemical analyses of effluent to determine potential toxicity. If the standards are exceeded, enforcement is supposed to bring the treatment plant into compliance. This path of regulation from the laboratory to enforcement may not always be reliable but in other than extreme cases, assessing effects by field sampling is difficult. Outplanting methods pioneered by Deysher and Dean (1986a) and applied to discharges of sewage (e.g., Tegner et al. 1995—"Effects of a large sewage spill") and produced waters (e.g., Reed et al. 1994) are an excellent alternative for determining the direct effects on macroalgae in the field. Of course the best solution, particularly given the growing shortage of clean freshwater, is not to discharge sewage wastes into the ocean at all but remove the wastes and nutrients and reuse the water. Depending on the nature of the wastes and the location and size of the outfall this may not be currently feasible or necessary, or may require trade-offs such as increased use of fossil fuel to power the recycling. In such cases it is precautionary to reduce the wastes as much as feasible and discharge the effluent into an environment where effects are minimized. Recent advances in floating bioreactor technology suggest treatment may eventually be feasible by using nutrients and water in wastes to produce biofuels (e.g., Trent et al. 2012).

## RUNOFF AND COASTAL MODIFICATION

Materials from land also enter coastal waters via runoff in natural waterways and storm drains. Landslides stimulated by coastal construction can directly introduce sediment (e.g., Konar and Roberts 1996) but are relatively rare and localized. Runoff is widespread and increasing, however, as coastlines worldwide become more urbanized and industrialized, and catchment land is converted to farms and ranches. Runoff most commonly includes sediment which, like suspended particulates in sewage effluents, can have large effects on kelp and other subtidal and intertidal reef habitats (review in Airoldi 2003). Extensive storm drain monitoring programs in the United States and elsewhere have documented numerous other constituents similar to those found in sewage effluent, including nitrates, phosphates, biocides, and heavy metals (review in

Maestre and Pitt 2005). The concentrations and amounts vary widely depending on catchment type and area, and this plus variation in volume, flow frequency, and amount of ocean mixing influence their effects in the ocean (e.g., Cox and Foster 2013). Moreover, unlike point source discharges such as sewers, storm drains are numerous and other sources of runoff are diffuse, making it difficult to attribute effects to a particular source. Nevertheless, there is considerable correlative evidence from around the world that runoff is degrading nearshore environments. Partial solutions have been obtained using upstream sediment basins, by directing flow through vegetated areas prior to discharge, and by better land-use planning and source control.

Coastal construction has the potential to degrade or eliminate kelp forests. Examples include dredging and breakwater construction for harbors and marinas, both of which probably contributed to kelp forest declines in southern California by eliminating habitat and increasing turbidity and sedimentation (Foster and Schiel 2010). Breakwaters are hard substrata that may be colonized by giant kelp, partially compensating for loss of natural habitat. Great impacts from sedimentation have been seen, for example, along the Baltic coast where increased sedimentation and turbidity have combined to reduce the depth of canopy-forming algae and impair their recruitment dynamics (Airoldi 2003, Schiel 2009). Upstream dams may reduce sediment transport but can have the unintended consequence of starving beaches of sand, leading to increased cliff erosion by waves (Bascom 1964). These and more subtle effects such as changes in current and wave patterns due to breakwater and groyne construction are, as yet, poorly documented for giant kelp forests; they are often difficult to detect in the midst of other changes such as sewage discharge and increasing runoff.

## INTAKES AND DISCHARGES FROM POWER AND DESALINATION PLANTS

Seawater is widely used to cool coastal power plants (called "once-through cooling") and growing shortages of freshwater have led to construction of desalination plants that produce freshwater from seawater. If seawater is withdrawn directly from the ocean, organisms in the water can cause fouling and clogging within the plant. Large mesh screens on the intake reduce these problems but, depending on intake velocity, can impinge and kill large organisms. Smaller organisms are entrained through the screens and into the plant where they are damaged or killed (figure 10.2). The discharge from power plants can exceed 20°C above ambient (Langford 1990) and the salinity of the brine discharged from desalination plants can be 75‰, or about twice ambient levels (Roberts et al. 2010). Depending on the amount of water affected, and devices and procedures used to reduce intake and discharge effects, power and desalination plants located near kelp forests can cause forest degradation.

Two nuclear power plants, the San Onofre Nuclear Generating Station (SONGS) in southern California and the Diablo Canyon Power Plant (DCPP) in central California,

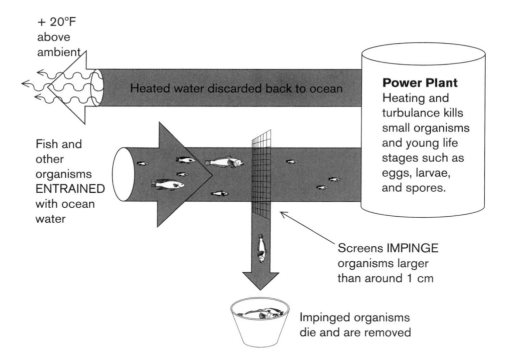

+ 20°F above ambient

Heated water discarded back to ocean

**Power Plant**
Heating and turbulence kills small organisms and young life stages such as eggs, larvae, and spores.

Fish and other organisms ENTRAINED with ocean water

Screens IMPINGE organisms larger than around 1 cm

Impinged organisms die and are removed

FIGURE 10.2

A typical once-through cooling system for a coastal power plant. Screen mesh size and discharge temperatures vary among plants. Desalination plants with surface seawater intakes are similar except entrainment mortality is primarily from turbulence and filtration, discharge temperatures are only slightly elevated above ambient levels, and the salinity of the discharge water (brine water) is typically twice that of the ambient water. Rapid dilution of heated and brine water to near-ambient temperatures and salinities can be achieved with diffusers on the end of the discharge pipe (see text).

are located near giant kelp forests and their intake and discharge effects have been extensively studied. Study results and regulatory actions are published primarily in reports by various state regulatory agencies. Some of these are cited below and can be obtained through the agency websites. DCPP uses over 8 billion liters of cooling water per day (York and Foster 2005). SONGS used a similar amount but there has been a recent large reduction in volume because the plant is no longer used to generate electricity. SONGS has a fish return system to reduce impingement, and discharges water through multi-port diffusers in the shallow subtidal zone. DCPP does not have a fish return system and discharges directly into the intertidal zone of a small bay. The amount of entrainment of organisms by the SONGS intake has been determined but the amount of habitat needed to mitigate for entrainment losses has not been assessed. The loss of organisms due to entrainment at DCPP has also been determined, and effects on rocky habitat estimated using "Habitat (= Area) of Production Foregone," a model that determines the amount of habitat needed to replace the organisms lost (Strange et al. 2004). Using this method, the amount of kelp forest habitat required to produce the organisms entrained by DCPP was estimated to be 85–200 ha (Raimondi et al. 2005).

The diffusers on the discharge at SONGS cause rapid dilution of the discharged water to minimize thermal effects. However, the intake is in shallower, more turbid water than the discharge, and turbidity is further increased around the discharge by the entrainment of a large volume of water to reduce discharge temperatures rapidly. These increase overall turbidity and sediment that affect the kelp forest (Murdoch et al. 1989). An artificial reef has been constructed to mitigate the loss (discussed in Chapter 11). Elevated temperatures from the discharge at DCPP have caused comprehensive changes in the intertidal and subtidal biota of the cove (Schiel et al. 2004) including a change in surface canopy from bull kelp (*Nereocystis luetkeana*) to giant kelp. Neither entrainment nor thermal effects have yet to be mitigated for at DCPP.

To our knowledge there are no desalination plants currently operating near giant kelp forests so effects in the field are unknown. If intakes are located in the water column, they have the potential to impinge and entrain organisms, although the impacts would most likely be less than those of power plants because desalination plants, at least those planned for California, will use much less water. Roberts et al. (2010) reviewed toxicological studies and the environmental effects of existing desalination plants elsewhere and concluded that the impacts of hypersaline discharges are commonly confined to within tens of meters of discharges unless the discharge is located in an area of limited circulation and mixing. In most cases there has been little damage if salinity is reduced to 3–4‰ above ambient. Entrainment may be reduced by using very fine mesh screens that exclude larger larvae, and eliminated by withdrawing water from beach wells or using saline ground water. Discharge effects can be reduced by the use of diffusers. They can also be reduced by increasing intake volumes and using the excess, unprocessed seawater to dilute the brine before it is discharged. This will reduce discharge impacts but increase those from entrainment at the intake.

New federal regulations make it doubtful that more once-through cooled power plants will be constructed in the Unites States, and California is requiring all such existing plants in the state (nuclear plants may be an exception) either to eliminate ocean impacts through the modification of existing cooling systems or to convert to cooling towers or dry cooling (large radiators). If this trend continues worldwide, the direct impacts of power plants will become minor. Indirect effects via $CO_2$, soot, and other emissions, however, may increase because, like advanced waste treatment, alternatives to once-through cooling require more energy. Desalination plants can be designed to minimize direct environmental impacts, but by potentially eliminating freshwater as a constraint on coastal populations and development, they may increase waste discharge and runoff. The trade-offs involved will no doubt be a matter of considerable debate if coastal populations and their consumption of resources continue to increase.

Powerboats motoring through kelp forests cut swaths through the surface canopy. Fortunately, small powerboats generally avoid the canopy because the fronds can foul propellers and clog cooling water intakes of engines. These are less of a problem for large boats, the repeated passage of which can cause the reduction of giant kelp canopies in the area traversed (M. Foster, pers. obs.). Sailboats usually avoid kelp canopies as fronds foul the rudder and frond drag on the keel can stop a boat altogether. Boat anchors can remove entire plants and scour the bottom but the difficulties of moving through a kelp canopy and retrieving a fouled anchor encourage most boaters to anchor at the edges of forests. Divers swimming through a kelp forest may remove plants and otherwise disturb the bottom (Schaeffer et al. 1999). Coral reefs are more sensitive to anchoring and diver disturbance and this has led to anchoring restrictions, use of mooring buoys, and restrictions on diver equipment and behavior (e.g., http://www.cabopulmopark.com /rules.html). We do not know of any such restrictions around giant kelp forests but they may be expected in some areas if boating and diving increase.

Spills of various kinds of toxic materials occur in nearshore waters, but oil spills have received the most attention because they occur more often, are unsightly, and stink. Large spills can occur from underwater pipelines, tanker groundings, and "blowouts" at offshore oil platforms (review in NRC 2003). To our knowledge, only two large oil spills have affected giant kelp forests in recent history, the 1957 diesel oil spill from the wreck of the tanker Tampico Maru in Baja California, Mexico, and the 1969 crude oil spill from the blowout of Union Oil Platform A off the coast of Santa Barbara, California. The Tampico Maru wrecked at the mouth of a small, shallow cove containing a stand of *Macrocystis*. As found in studies by North et al. (1964), the oil mixed into the entire water column causing massive mortality of invertebrates including sea urchins and abalone. Damage to seaweeds was less obvious. Vegetation in the cove, including juvenile giant kelp, increased within 5 months of the spill and *Macrocystis* covered much of the cove after 17 months. North and colleagues attributed the increased algal growth to a lack of grazing. Most animals recovered by 1961, but sea urchins and abalone had not recovered by the last sampling date in 1963, around 6 years after the spill.

Crude oil from the Santa Barbara spill polluted a large portion of the mainland coast and many of the Channel Islands (Foster et al. 1971—"The Santa Barbara oil spill. I"). Damage to kelp forests was reported by Foster et al. (1971—"The Santa Barbara oil spill. II") and overall damage by Foster and Holmes (1977). Assessment of spill effects was complicated by the simultaneous occurrence of record storms and associated runoff. Numerous sea birds were killed by contact with the oil, but other than a decline in mysid shrimp, little damage to kelp forest algae, invertebrates, or fishes was observed even though considerable quantities of oil fouled the surface canopies (figure 10.3). The partially weathered crude oil appeared to stay on the surface of the water and did not stick to the kelp fronds. Damage may have been more severe if the more volatile

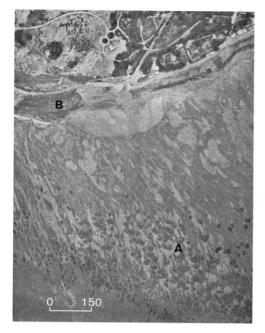

FIGURE 10.3

Aerial photo of the coast and nearshore kelp
forest, showing oil in a giant kelp canopy
near Santa Barbara, California, during
the 1969 Santa Barbara oil spill. The spill
was caused by a blowout on an offshore oil
platform.

A = oil streaming toward shore from the kelp
canopy; B = black area of oil on the beach. Scale
in meters.

SOURCE: Photo courtesy of Mark Hurd Aerial
Surveys.

components of the oil had not had time to evaporate prior to reaching shore, or if more
toxic, refined products were spilled and mixed into the water column as in the Tampico
spill. Dispersants are often used during oil spills. They can be toxic but their effects are
difficult to separate from those of the oil. Singer et al. (1995) tested the toxicity of Corexit
9554 on a suite of kelp forest organisms, including giant kelp zoospores, in laboratory
bioassays. Fifty percent of zoospores were killed at concentrations as low as 100 ppm
and abalone larvae were also very sensitive to the dispersant, suggesting impacts from
dispersants magnify those from the oil itself.

## GIANT KELP HARVESTING

Giant kelp has been harvested for food, fertilizer, and chemicals since the early 1900s
(see Chapter 1) and the effects of harvesting continue to be investigated. In the early
days of harvesting in California when there was little regard for environmental effects,
groups of entire plants were removed by putting a cable around a stand and dragging it
onto a ship (McPeak and Glantz 1984). This method was soon replaced by kelp harvesting ships with large cutting devices on the stern (see Chapter 11). The most modern
ships backed through the surface canopy and the cutting device severed fronds in an
8 m wide swath, much like a hedge trimmer. A metal conveyor moved the severed
fronds onto the boat where they were distributed by a grab that could move up and
down the deck on an overhead cable. Regulations evolved to limit the cut to no more

than 1.2 m below the surface. Harvesters also had to specify how they would "avoid" cutting bull kelp, *Nereocystis luetkeana*, when it occurs with giant kelp. In contrast to giant kelp, bull kelp is an annual species that produces spores on surface fronds. Harvesting these fronds can reduce the abundance of future generations of sporophytes (review in Springer et al. 2010).

Questions over possible impacts of harvesting on the environment and a desire for sustainability led to various harvesting investigations in California, beginning in the 1950s. Concerns included possible loss of entire kelp forests, reduction of canopy-dwelling fishes and invertebrates, reduction of fish populations due to loss of food and / or habitat, increased beach erosion due to less dampening of water motion, and an increase in the abundance of drift kelp on beaches from cut fronds and dislodged plants not captured by the harvester. Early studies of these potential problems were summarized by North and Hubbs (1968) who concluded that "No adverse influence of harvesting could be found among the statistics or field observations for the periods studied." A later review by Barilotti and Zertuche-González (1990) came to the same conclusion. Uncut, subsurface fronds grow to replace those cut, and it appeared that entire plants were only occasionally torn from the bottom during harvesting operations (Rosenthal et al. 1974). Harvesting the canopy increases light on the bottom, which can enhance recruitment of *Macrocystis* (Rosenthal et al. 1974; Kimura and Foster 1984), but Barilotti et al. (1985) found no adverse effects of harvesting on holdfast growth or plant survival if harvesting regulations and industry practices on frequency of harvest were followed. Numerous other organisms are removed along with the cut fronds (e.g., North and Hubbs 1968) but overall reductions within a forest have not been reported. Clendenning (1971b) estimated that 10% or less of giant kelp production was removed by harvesters in harvested forests. Possible changes in populations of consumers such as abalone and fishes in kelp forests, beaches, and offshore due to an indirect effect of removing the biomass and primary production of sea-surface kelp have not been investigated. At least for the present, harvesting impacts are of little concern in California because only small amounts of giant kelp are now harvested, primarily to feed abalone in aquaculture facilities (see Chapter 11).

Significant harvesting of *Macrocystis* (primarily *Macrocystis 'integrifolia'*) now occurs in Chile. Over 2500 metric tons are harvested in some areas of northern Chile, most of which is used in abalone aquaculture (review in Vásquez 2008). Some is harvested as drift on the beach, but most is removed by hand, either by cutting fronds or by removing entire plants. Kelp forests are not abundant in northern Chile, and there is concern that giant kelp is already being depleted. This could worsen as the demand of the aquaculture industry may eventually exceed supply, especially since supply also varies with oceanographic conditions. Harvesting regulations are being considered, including territorial user rights to encourage sustainability (Vásquez 2008). Cultivation of giant kelp is being explored as an alternative to harvesting natural stands (Buschmann, Varela, et al. 2008), and various methods of repopulating natural stands are being investigated (e.g., Westermeier et al. 2012). To our knowledge the effects of harvesting on associated Chilean invertebrates and fishes have not been investigated.

*M. 'integrifolia'* has also been harvested for the herring spawn-on-kelp fishery and other uses in Alaska and British Columbia, and various harvesting experiments have been done to determine effects, primarily on sustainability of the giant kelp resource (review in van Tamelen and Woodby 2001). Experimental harvesting results have been mixed, no doubt in part because different harvesting approaches and frequencies have been used. Druehl and Breen (1986) found increases in some understory algae and declines in herbivore abundance and reproduction in a plot harvested six times in 17 months. It seems clear from these studies and those done elsewhere that infrequent harvests of the canopy have little effect on giant kelp or its associates. However, more frequent canopy harvests and removal of entire plants can affect giant kelp and associated species. The growth and reproductive characteristics of *Macrocystis* make it the most suitable of all kelps for harvesting; plants are very productive, cutting the surface canopy is relatively easy, and uncut fronds grow to replace those that are removed. Best management practices will vary depending on ecomorph, location, and forest depth. There can be minimal impacts on the kelp forest community if harvesting techniques are based on knowledge derived from studies on the depth and frequency of cutting, as well as the proportion of the canopy harvested in particular stands that protects other species living in the canopy.

In New Zealand, kelp harvesting is limited to around 1500 metric tons annually. *Macrocystis* comes under the New Zealand Fisheries Quota Management System and can only be done with permits that allow particular harvests in set areas. As elsewhere, harvesting is limited to the surface canopy at no more than 1 m depth. This is not done by large harvesters but by cutters operated from a barge. No significant effects of harvesting on giant kelp communities have been found (Pirker 2002). The kelp is used to feed New Zealand abalone being raised for "blister pearl" production for jewelry (e.g., http://www.bluepearls.com/) and dried to be used as an additive in horticulture, agriculture, and as a stock food supplement. Its dried form is also marketed as "kelp pepper" and is used in cooking.

## INTRODUCED AND INVASIVE SPECIES
### MACROALGAE

We use the term "introduced species," also referred to as nonindigenous or non-native species, to indicate those not native to a community but transported into it by humans. Long-distance transport is usually via commercial shipping, aquaculture, or the aquarium trade. After arrival, introduced species may spread by recreational boating or natural dispersal. Introduced species may become invasive by spreading prolifically, although metrics of abundance that constitute an "invasion" are subjective. Invasions are considered to have negative impacts if they cause reductions in the abundance of

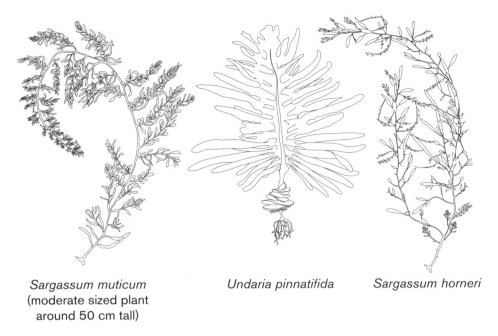

*Sargassum muticum*
(moderate sized plant
around 50 cm tall)

*Undaria pinnatifida*

*Sargassum horneri*

FIGURE 10.4

Invasive seaweeds found in giant kelp forests in California. Left: *S. muticum* (moderate sized plant around 50 cm tall). Center: *U. pinnatifida* (around 1 m tall with ruffled sporophylls above holdfast). Right: *S. horneri* (moderate sized plant around 75 cm tall; note distinctive, sausage-like floats).

native species or negatively affect economic activities such as aquaculture. Invasive species may also include native species that spread into and become dominant in habitats where they were not previously found. In this case the spread may not necessarily be caused directly, or even indirectly, by humans (review in Inderjit et al. 2006).

Relative to harbors, bays, and estuaries, giant kelp forests have experienced few introductions. In California and Baja California, for example, 27 non-native seaweeds have been found in the flora but only four of these are found in kelp forests and one, *Elachista nigra* (also non-native in Australia), is a rare epiphyte on the understory kelp *Eisenia arborea* (Miller, K. A. et al. 2011). Of the remaining three, one is so far confined to southern California and Baja California, and two have invaded worldwide. They are all large seaweeds native to Asia: the fucoids *Sargassum muticum* and *S. horneri* (=*filicinum*), and the kelp *Undaria pinnatifida* (figure 10.4). The earliest invader was *S. muticum*, arriving in the Northeast Pacific in the 1940s, most likely on oysters shipped from Japan for aquaculture (review and data in Critchley and Dijkema 1984). It was later introduced into the Northeast Atlantic and the Mediterranean Sea. The causes of these large-scale geographic distributional changes and concerns over their ecological and economic consequences resulted in numerous early and continuing studies on the biology and ecology of *S. muticum* (bibliography in Critchley et al. 1990, Britton-Simmons 2004).

In the Northeast Pacific, *S. muticum* spread north and south, eventually showing up in giant kelp forests in southern California. Fronds of this fucoid arise from a perennial holdfast, can be over 2 m tall, and have pneumatocysts that elevate them in the water column (Abbott and Hollenberg 1976). Ambrose and Nelson (1982) noted a severe decline in giant kelp around Catalina Island during a warm-water period in 1976, and *S. muticum* invaded some of these areas at densities of over 100 plants m$^{-2}$ at depths of 3–7 m. They experimentally removed the invader in three areas, and juvenile *Macrocystis* densities increased relative to controls in two of the areas, suggesting *S. muticum* inhibited giant kelp recruitment, probably by shading. Foster and Schiel (1993) also observed *S. muticum* invade two giant kelp forests at Catalina Island, an invasion that occurred in association with a very large decline of giant kelp and other native species during the 1982–1984 El Niño. A few *S. muticum* were observed in 1982 prior to the El Niño, but were not abundant enough to occur in samples. Abundance increased by 1984 after native species declined, and by June 1986 *S. muticum* plants up to 50 cm tall attained mean densities of over 25 m$^{-2}$. However, giant kelp and other native species recovered and *S. muticum* retreated to open areas within the giant kelp canopy and then became rare again by 1989 as adult giant kelp densities recovered to around 0.5 m$^{-2}$. This invasive fucoid may have inhibited giant kelp recruitment but did not eliminate it, and these studies show that negative effects on giant kelp and probably on other native macroalgae in these kelp forests were transitory. Miller and Engle (2009) suggested that *S. muticum* has now "equilibrated" with natural populations at Catalina Island. This invader can occur from the low intertidal zone well into subtidal areas. Its effects on intertidal communities are highly variable (reviews in Britton-Simmons 2004, Schaffelke and Hewitt 2007, Thomsen et al. 2009). Britton-Simmons (2004) found large negative effects on native understory species in shallow (2 m deep) areas without a surface canopy, and, based on this and the literature, concluded that "studies in the subtidal zone indicate relatively strong effects." The data from giant kelp forests, however, are more indicative that in this subtidal habitat and perhaps others with a thick surface canopy, the effects are weak.

The morphology, life history, and dispersal characters of the other fucoid, *S. horneri*, are similar to those of *S. muticum* except the holdfast may not be perennial. As of 2012, it has only been introduced in southern California and the Pacific Coast of Baja California, Mexico. Its invasion history and distribution were summarized by Miller and Engle (2009) in southern California and by Riosmena-Rodriguez et al. (2012) in Baja California. It was probably brought in by commercial shipping and was first observed in a mainland California harbor in 2003. Plants were discovered at the southern California Channel Islands (e.g., Catalina Island) in 2006, most likely transported from the mainland by recreational boaters. The invader rapidly spread around Catalina Island and south along the mainland and offshore islands into Baja California. Plants over 2 m tall have been found in several giant kelp forests around Catalina at depths of 4–12 m, and at Isla Natividad in Baja California at 3–8 m. *S. horneri* can form a dense band around 9 m deep at sites around Catalina Island (D. Kushner, pers. comm.), but its ecological effects have not been investigated. Densities were around 1 plant m⁻² at Isla Natividad in 2009, and Riosmena-Rodriguezet al. (2012) suggested it currently has no significant effects on native flora. The species spread north into the northern Channel Islands off southern California by 2009, where it can occur in dense patches to depths of 15 m, and may now also be at mainland sites. Plants are shorter than those at southern sites, and can reproduce when only 50 cm tall. Observations during annual surveys by the Channel Islands National Park staff (D. Kushner, pers. comm.) indicate no obvious, negative effects. Plants are eaten by sea urchins and abalone, and serve as habitat for fishes, especially juvenile kelp bass (*Paralabrax clathratus*).

*U. pinnatifida* (wakame) is native to temperate Asian waters where it occurs naturally from the low intertidal zone to depths of around 8 m and is extensively cultivated for food. Sporophytes can be up to 2 m long and are usually annual, with reduced abundance in summer related to high water temperatures (Saito 1975). It was intentionally introduced into the Northeast Atlantic for aquaculture in 1983 and spread widely in this region and into the Mediterranean Sea, most likely via ships. Shipping and boating were also the likely vectors for introductions and spreading in Argentina, Australasia, and the Northeast Pacific (reviews in Valentine and Johnson 2003, Thornber et al. 2004, Uwai et al. 2006). It now occurs as part of the understory in giant kelp forests in Tasmania (Valentine and Johnson 2003), New Zealand (Russell et al. 2008, Schiel and Thompson 2012), and Catalina Island in southern California (Miller and Engle 2009). Observations and removals of native canopies composed primarily of large fucoids and the kelp *Ecklonia radiata* by Valentine and Johnson (2003) showed that *U. pinnatifida* occurs primarily in disturbed areas in Tasmania, especially when the disturbance coincides with the growth season of the sporophytes. They found that invader densities reached over 19 plants m⁻², and declined as native species recolonized. In this case, however, the disturbance caused an increase in sedimentation, and the native species that recolonized differed from those present prior to the disturbance. Later experiments showed that a matrix formed by sedimentation around small, filamentous algae could inhibit the recovery of native species (Valentine and Johnson 2005). Edgar et al. (2004) found a similar invasion of *U. pinnatifida* when native canopies were removed. Russell et al. (2008) described dense stands of the invader above and below the distribution of giant kelp and large native fucoids at sites in New Zealand, but only patches of *U. pinnatifida* within the distribution of these natives. Experiments and observations by Thompson and Schiel (2012) and Schiel and Thompson (2012) show that this invasive kelp is inhibited by the reduced light under native canopies, and invades by keying into small disturbances in native algal stands in southern New Zealand. It is also adept at recruiting into coralline turfs, which mostly exclude fucoid recruitment, but is a poor competitor and does not displace native species. Overall, *U. pinnatifida* seems to act as a refuge species and there are no known effects of the invasive kelp on giant kelp.

*U. pinnatifida* was reported to be invasive and reduce the diversity of native species at a site in Argentina that lacked an established native canopy (Casas et al. 2004) but such displacement has not been observed in intertidal or subtidal areas with established native canopies (review and data in Schiel and Thompson 2012). There have been efforts to reduce or eradicate *U. pinnatifida* (example and review in Hewitt et al. 2005). These can be expensive and labor intensive, and eradication is probably impossible given the life history and dispersal characteristics of this and other algal invaders. One invasion was apparently prevented in the Chatham Islands, New Zealand, by eradicating *Undaria* attached to a wrecked vessel before the alga could spread (Wotton et al. 2004).

As discussed above, *S. muticum* has invaded kelp forests in abundance only when giant kelp has been substantially reduced by oceanographic conditions, and it has declined when giant kelp returned. This lack of long-term invasion success may be the result of the interaction of the life history features and light requirements of potential invaders in areas where native canopies can greatly reduce light. Optimal light for the growth of small *S. muticum* is 44 $\mu$M m$^{-2}$ sec$^{-1}$ (Hales and Fletcher 1989), while average irradiance at 8 m under a well-developed giant canopy can be around only 35 $\mu$M m$^{-2}$ sec$^{-1}$ (Gerard 1984), and less if reduction by understory canopies is included. The fronds of the fucoid are also annual, and are absent in summer (Ambrose and Nelson 1982) or winter (Britton-Simmons 2004), depending on location in the Northeast Pacific. Britton-Simmons (2004) found that at maximum development their effects on light varied, with a maximum reduction of only 20%. The available data therefore suggest that if disturbance to giant kelp results in sufficient light to allow *S. muticum* to invade, it is unable to prevent giant kelp recolonization when its fronds are absent, and may only slow recolonization when fronds are present. Engelen and Santos (2009) suggest that the key demographic trait allowing *S. muticum* to dominate tide pools and outcompete native species is the accumulation of perennial holdfasts and remnant fronds that regenerate during the growing season, a process unlikely to occur in kelp forests where suitable growing conditions occur episodically. The light requirements of *S. horneri* are unknown but it is primarily a subtidal species in its native range (review in Riosmena-Rodriguez et al. 2012) so may be better adapted to the low-light conditions common in giant kelp forests. Whether it will be able to sustain large populations that affect native species in this habitat is unknown.

Like *S. muticum*, *U. pinnatifida* is also an annual with sporophytes absent or reduced in abundance in summer to autumn, probably in response to increased water temperature (Saito 1975). It may have overlapping generations in some areas (review and data in Thornber et al. 2004, Schiel and Thompson 2012). Saturating irradiances, 60 $\mu$M m$^{-2}$ s$^{-1}$ for gametophytes (Kim and Nam 1997) and an average of 80 $\mu$M m$^{-2}$ sec$^{-1}$ for sporophytes (Dean and Hurd 2007), are also considerably above those that can be found in giant kelp forests. Morelissen et al. (2013) found very little gametophyte growth and no development at 8.5 $\mu$M m$^{-2}$ sec$^{-1}$, and sporophyte production at 28 $\mu$M m$^{-2}$ sec$^{-1}$ was high only at elevated concentrations of dissolved inorganic nitrogen.

There is also increasing evidence from the field that *U. pinnatifida* invasion is inhibited by light reduction from native canopies (Valentine and Johnson 2003, Schiel and Thompson 2012, Thompson and Schiel 2012), so it is unlikely to prosper under a giant kelp canopy.

These introduced species are likely to be transitory in giant kelp forests, behaving like the large, native brown alga *Desmarestia ligulata* in California kelp forests (Edwards 1998): sporophytes become abundant where native canopies are disturbed but then decline as native species recolonize. As discussed above, however, *U. pinnatifida* has been observed in giant kelp forests when giant kelp did not appear to be recently disturbed. This may be because disturbances were not well documented or the surface canopies of giant kelp and other large brown algae and / or understory kelp canopies were thin or patchy. Irradiance beneath a giant kelp canopy alone varies exponentially with canopy density (Gerard 1984). Understanding the effects of giant kelp canopies on native understory species has been hampered by lack of comparative information on the light regimes at sites where the studies were done (Chapter 7). This is equally true for invaders in the understory. Moreover, information on the light requirements of introduced species is not particularly useful to understanding their ecology if the light regimes in forests where they occur or might occur are unknown. Our knowledge of light within giant kelp forests comes primarily from a few forests in California with extensive, thick surface canopies. Forests in other regions with thinner and / or patchier canopies or different understory species may have light regimes more favorable to invaders.

Although there is considerable concern over the possible effects of *U. pinnatifida* and other introduced seaweeds in giant kelp forests, eradication is probably impossible, and the costs of control can be considerable. Thomsen et al. (2009) pointed out the paucity of evidence for the negative effects of introduced macroalgae, and argued the need for more and better designed field experiments. We agree that such experiments would help evaluate environmental concerns and the costs versus benefits of control. In giant kelp forests and other subtidal habitats with established native canopies, the usefulness of experiments would be greatly enhanced by information on disturbance and light regimes, densities of the introduced species, and characteristics of the habitats within a forest where introduced species do and do not occur. It is highly likely that the wave of species being transported over international borders will only increase and, because eradication has proved to be difficult or impossible in most cases, it seems clear that effective prevention should be a major focus (Floerl et al. 2005).

## ANIMALS

The literature on introductions of animals into giant kelp forests is sparse, with few species reported. It may be that invasions have been under-reported, that most invasive animals are native to harbors so kelp forests are not a suitable habitat, or that giant kelp forests have not been thoroughly surveyed. However, surveys for introduced invertebrates have been done at over 20 sites along the open coast of California (Maloney et al. 2007). Of the introduced species found, five occurred in kelp forests: the bryozoans *Watersipora* spp., *Bowerbankia gracilisi*, and *Bugula neritina*, the polychaete *Hydroides elegans*, and the amphipod *Elasmopus rapax* (K. Hammerstrom, pers. comm.). Four of the five occurred in giant kelp holdfasts and most were rare. A few other introduced species were considered possibly invasive and are awaiting better identification, and four species were labeled cryptogenic because their origins have yet to be determined. Of the known invaders, *Watesipora* spp. are the best studied. These bryozoans have been introduced to numerous bays and harbors around the world through shipping, but their native range is presently uncertain (Mackie et al. 2012). *Watersipora* spp. have a complex structure and might be expected to enhance habitat. Studies in Bodega Bay in northern California, where they are part of the fouling community and can cover >10% of the substratum (Stachowicz and Byrnes 2006), found they had mixed effects on associated species. They increased the species richness of some mobile invertebrates but not others, but had no significant effects on sessile species richness (Sellheim et al. 2010). These introduced bryozoans have recently been found in giant kelp forests in the Monterey area (seanet.stanford.edu) but their effects have not been investigated. There are as yet no reports of introduced fishes that regularly inhabit California giant kelp forests (Schroeter and Moyle 2006).

The sea urchin *Centrostephanus rodgersii* has received considerable attention because it has had very large impacts in Tasmania, and interacts strongly with other species (reviews and data in Ling 2008, Ling et al. 2009; see Chapter 8). Strictly speaking, *C rodgersii* is not an invasive species because it is native to southeastern Australia and its spread south into Tasmania is thought to have occurred by larval transport as a result of a poleward extension of the southward-flowing East Australian Current, a possible manifestation of ocean climate change (Ling 2008). The urchin has occupied large areas of kelp forest in eastern and southern Tasmania, and experiments by Ling (2008) indicate its grazing activity can eliminate 150 macroalgal and invertebrate taxa. As discussed previously, disturbances to native algal species, including those caused by sea urchins, could also increase *Undaria pinnatifida* abundance although this may be a "love–hate" relationship as the urchins may ultimately remove the invasive alga (Valentine and Johnson 2005). The success and persistence of *C. rodgersii* is affected by lobster predation which may, in turn, be moderated by regulation of the lobster fishery and establishment of marine protected areas. These are discussed in later chapters on predation, fishing, and marine protected areas. Castilla et al. (2005) summarized information on non-indigenous marine algae and invertebrates in Chile, most of which are found in the intertidal zone, bays, and aquaculture facilities. No species were characterized as associated with giant kelp forests.

At present, giant kelp forests appear to be little affected by introduced seaweeds or invertebrates, probably because the invasive algae do not compete well with giant kelp and other native canopy species for light, and invasive invertebrates are mostly bay and estuarine species. While we have tried to avoid Chicken Little syndrome ("the sky is falling") that characterizes many environmental predictions, there are science-based reasons for predicting that giant kelp forests could become highly modified as multiple stressors like runoff, ocean warming, more frequent El Niños, and perhaps more grazer introductions interact to affect giant kelp. It can only be hoped that no particularly noxious or aggressive species get into giant kelp forests and that the innate production and diversity of these forests will help buffer them from untoward effects.

# 11

# HUMAN USAGE OF GIANT KELP AND KELP FOREST ORGANISMS

There is no doubt that whatever may be the direct food value of algae they are useful because of their iodine content, which serves as a protection against goitre, whilst the bulk and water prevent constipation.

—Chapman and Chapman (1980)

Seaweed offers kelping hand for biofuels.

—Businessgreen (2012)

Giant kelp and its associated species have supported many human uses and activities, from food and chemicals to recreation and cultural enrichment. As discussed in Chapter 1, the impetus for a better understanding of the biology and ecology of giant kelp beginning in the early 1900s was the development of harvesting and processing facilities in California for the production of potash and, later, alginate. Although *Macrocystis* is now only a minor species in the extraction industry, its other uses are varied and some are important for human activities. In this chapter, we discuss the uses of giant kelp as well as forest-associated species, including those comprising important recreational and commercial fisheries.

## FOOD AND FERTILIZER

Seaweeds have been consumed by humans since they first inhabited coastlines, and giant kelp is no exception. Most current consumption of kelp is in Asia of species other than *Macrocystis*. Consumption is especially common in Japan and China where there is a long tradition of eating kelp, especially *Laminaria* and *Undaria* (Japanese: kombu, wakame; Chinese: haidai, quandai-cai; other names are used depending on the species and preparation). These are prepared and consumed in various ways, most commonly as part of soups, stews, and salads (review in Chapman and Chapman 1980). Human consumption is often based on cultural traditions, flavor, and texture, and not necessarily on nutritional value. Such traditions and tastes are not as common in the temperate areas of the world where giant kelp occurs and, although some *Macrocystis* is consumed, it is not discussed as human food in reviews by either Chapman and Chapman (1980) or McHugh (2003). In the Northeast Pacific, herring (*Clupea pallasi*) spawn on *Macrocystis* and other kelps, and the resulting "spawn on kelp" is eaten by humans (Turner 1995). This phenomenon has become the basis of a small fishery in California and other areas in the Northeast Pacific where giant kelp is attached to lines or floats in herring spawning areas and collected after spawning has occurred (van Tamelen and Woodby 2001, Waters et al. 2001). The majority of the harvest is shipped for sale to Japan. Kelp in various forms is also sold as a "health product," but the kind of kelp is often not specified. Giant kelp is currently harvested in New Zealand, where it is dried, flaked, and sold as a food seasoning called "Valēre Kelp Pepper" by NZ Kelp.

The primary nutritional value of kelps is in their mineral and vitamin content, but these are not exceptional in most cases. As an example, we compare a similar list of nutrients for kelp (*Laminaria* spp.) and a common vegetable (carrots; see Table 11.1). Complete comparative data could not be found for *Macrocystis*, but data for some giant kelp minerals in North (1987a) indicate that *Laminaria* and *Macrocystis* are likely to be similar. Generally, carrots and kelp contain similar nutrients, although there are exceptions (e.g., calcium, potassium, sodium, iodine, folate, vitamin A, carotenoids). However, iodine content is a particularly important difference because it is essential to prevent goiter, a significant health problem in areas with little iodine in the soil and without access to iodized salt (reviewed in Zimmerman 2009). Historically at least, this was a primary reason for kelp consumption in China, where iodine was scarce in the vast inland areas away from the coast (Cheng 1969). Kelp contains carbohydrates and proteins, but the carbohydrates are in forms such as alginate and mannitol that are not readily digested by humans (Grabitske and Slavin 2008), and digestion of proteins from kelp may not produce soluble nitrogen compounds (Chapman and Chapman 1980). Moreover, kelp can concentrate elements such as arsenic which, when eaten in large quantities as a herbal supplement, can cause poisoning (Amster et al. 2007). The benefits of kelp consumption as a source of energy and amino acids are therefore unclear. For those with a predilection for eating giant kelp, therefore, its consumption is a matter of taste, availability, suitable alternatives, and price.

There is also a long history of seaweed use as food for terrestrial animals. As for humans, the nutritional value and digestibility indicate the main benefits are in mineral content, particularly iodine, in areas where pasturelands are mineral deficient (review in Chapman and Chapman 1980). Feral or free-ranging cattle will graze on kelp, as seen historically for other kelp genera in Ireland and more recently in the uninhabited, subantarctic Auckland Islands, where cattle roamed freely for over 100 years until they were removed for ecological restoration. Giant kelp is currently used occasionally as an additive in sheep food because it is believed to improve their coats (R. Beattie, pers.

TABLE 11.1    Nutrients in raw carrots and the wet, edible part (blade)
of *Laminaria* spp.

| Nutrient | Units | Carrots (amount: 100 g$^{-1}$ raw) | Kelp (amount: 100 g$^{-1}$ wet) |
|---|---|---|---|
| Water | g | 88.29 | 90.3[a] |
| Protein | g | 0.93 | 1.68 |
| Total fat | g | 0.24 | 0.56 |
| Total fiber | g | 2.8 | 1.3 |
| Total sugars | g | 4.74 | 0.6 |
| **MINERALS** | | | |
| Calcium | mg | 33 | 168 |
| Iron | mg | 0.3 | 2.85 |
| Magnesium | mg | 12 | 1.21 |
| Phosphorus | mg | 35 | 42 |
| Potassium | mg | 320 | 89 |
| Sodium | mg | 69 | 233 |
| Zinc | mg | 0.24 | 1.23 |
| Copper | mg | 0.5 | 0.13 |
| Manganese | mg | 0.14 | 0.2 |
| Selenium | mg | 0.1 | 0.7 |
| Iodine | mg | Low[b] | 13–86[c] |
| **VITAMINS** | | | |
| Vitamin C | mg | 5.9 | 3.0 |
| Thiamin | mg | 0.07 | 0.05 |
| Riboblavin | mg | 0.06 | 0.15 |
| Niacin | mg | 0.98 | 0.47 |
| Pantothenic acid | mg | 0.27 | 0.62 |
| Vitamin B6 | mg | 0.14 | 0.00 |
| Total folate | µg | 19 | 180 |
| Choline | mg | 8.8 | 12.8 |
| Vitamin A | µg | 835 | 6 |
| β-carotene | µg | 8285 | 70 |
| α-carotene | µg | 3477 | 0 |
| Lycopene | µg | 1 | 0 |
| Vitamin E | mg | 0.66 | 0.87 |
| Vitamin K | µg | 13.2 | 0.25 |

SOURCE: Data from USDA 2012 except

a. for *Macrocystis* from Rassweiler et al. (2008);

b. from Zimmerman (2009);

c. from Chapman (1970).

NOTES: Fatty acids low for both (not shown). Amino acid composition and amounts similar and low for both (not shown).

FIGURE 11.1

Kelp harvesting in New Zealand. Roger and Nicki Beattie with drying crop of giant kelp, used for food seasoning ("Kelp Pepper," Valēre), soil conditioner ("Zelp"), and as food for New Zealand abalone (*Haliotis iris*) in culturing blister pearls ("Blue Pearls," Eyris pearls, http://www.bluepearls.com/).

SOURCE: Photo by D. Schiel.

comm.; figure 11.1). Although once harvested in California as a food additive and seasoning (Neushul 1987), *Macrocystis* is now a relatively small part of this use. It is dried, processed into flakes or powder, and sold as a food supplement (e.g., "Zelp" by NZ Kelp in New Zealand, "Seaweed Flour" by SurAlga in Argentina). It may also be marketed as a nutritional additive of food for aquarium fish (e.g., "Omega One Kelp Flakes" by Omega Sea Ltd. in the United States), and has been evaluated as a nutritional additive for farmed rainbow trout, *Oncorhynchus mykiss* (e.g., Mansilla and Ávila 2011). It may also be used simply as a binding agent in pelletized food (McHugh 2003).

*Macrocystis* and other seaweeds are the natural foods of marine grazers so one would expect they would be good food for the aquaculture of these species. With the decline of giant kelp as a source of alginate and with decreases in wild abalone stocks, the largest use of giant kelp is now as food for farmed abalone. For the 5 years from 2007 to 2011, an average of 3728 metric tons were harvested annually in California (figure 11.2), and around 96% of it was used to feed farmed abalone (primarily the red abalone *Haliotis*

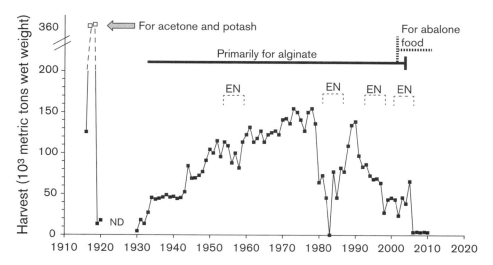

FIGURE 11.2

Giant kelp harvest in California. Data to 1999 from Bedford (2001) and from 1999 to 2010 from R. Flores-Miller (pers. comm.). No data available from 1921 to 1930.

EN = El Niño.

NOTES: Major uses are indicated at the top of graph. Note that the harvest from 2006 onward was around 3000 tonnes.

*rufescens* (R. Flores Miller, pers. comm.). Vásquez (2008) reported that over 4500 wet metric tons were collected for abalone food in Chile in 2005–2006. Farmed abalone in Chile are primarily *H. rufescens* introduced from California but also *H. discus hannai* introduced from Japan (Flores-Aguilar et al. 2007). Around 400 metric tons of *Macrocystis* are used annually in southern New Zealand as food for endemic abalone (called paua, *H. iris*), which are raised for food and "blister pearl" production (e.g., Schiel 1997). Abalone are farmed in many other places in the world (review in Cook and Gordon 2010) but because giant kelp is not abundant, collection of wild seaweed is restricted, or the abalone grow better on alternative food, other algae or manufactured feed are used.

The fact that seaweeds have long been used as fertilizer is exemplified by the word "kelp," which is derived from Middle English as a term for the ash derived from burning primarily large brown algae and used, in part, to fertilize fields. As discussed in Chapter 1, the initial impetus for harvesting large quantities of giant kelp in California was to produce potash for fertilizer. Giant kelp and other seaweeds are still marketed in several areas of the world for fertilizer, but this industry no longer exists in California. As McHugh (2003) pointed out, however, kelps are relatively low in phosphorous and their trace elements are low relative to plant requirements. A more important benefit may be as a soil conditioner. However, the efficacy of several liquid fertilizers, including 15 derived from seaweeds, was rigorously examined by Edmeades (2002) in a meta-analysis of nearly 800 trials involving a range of test crops. His analyses showed there were "no practical effects when applied as recommended" and he concluded that this is not surprising since the products do not contain high enough concentrations of nutrients, or organic and growth compounds. These results are, no doubt, a rather large disappointment to the industry, but there are many true believers in the merits of kelp fertilizer and so a small industry is likely to persist.

## CHEMICALS

As indicated above, the large-scale production of chemicals from giant kelp was stimulated by the need for potash fertilizer in the western United States, and the need for potash and acetone as components of the smokeless gunpowder, cordite, used by the British during World War I. Hercules Powder Company secured a contract to supply cordite and developed a large harvesting operation and processing plant near San Diego, California, in 1916 (review in Neushul 1987). Hercules also produced other chemicals including various organic solvents, tars, and alginate. The resource demand of this and other kelp harvesting companies resulted in harvests exceeding 350,000 metric tons (figure 11.1) and concerns that demand would eventually exceed the productivity of the resource (Neushul 1987). The high demand, however, was largely the result of a war economy. After World War I ended and other chemical sources became available at lower cost, the potash and gunpowder industry rapidly declined and was replaced by a smaller (in terms of giant kelp harvest) alginate industry.

The alginate industry expanded as a result of creative chemistry and product marketing (Neushul 1987). Algin is a complex polysaccharide (see Chapter 1, History of Research) that can be processed into various types of alginate, such as sodium, calcium, and propylene glycol alginate, which vary in viscosity, gel strength, and other properties depending on their intended use. The many uses include as a suspending agent in food, in paper sizing, and as molding material for dental impressions (Chapman and Chapman 1980). Algin is abundant in giant kelp, ranging from 13% to 24% dry weight (Chapman and Chapman 1980, Draget et al. 2005). At the peak of the alginate industry, large harvesters (figure 11.3) cut over 150,000 wet metric tons annually in forests from Pacific Baja California, Mexico, to Monterey in central California (figure 11.2), and transported it to San Diego, California, for processing. Assuming that most of this was harvested in California, it amounted to about 5% of the total giant kelp production (see Biofuels below). With increased use of dried kelp from other areas of the world, advances in processing that can produce a variety of products in one processing plant, and rising costs of production in California (including the need for large amounts of water as well as waste disposal), large-scale giant kelp harvesting and algin processing in California ended in 2006. Most of the current supply of algin comes from Europe and Asia, extracted from *Laminaria* spp. harvested in many areas and cultivated in China, *Lessonia* spp. harvested in Chile, and the fucoid *Ascophyllum nodosum* harvested in Europe. Dried plants from the countries of harvest are now commonly shipped to

FIGURE 11.3

The giant kelp-harvesting vessel "Kelstar," built in 1977 and operated by
Kelco Co. in San Diego, California. The vessel was 52 m long and could hold
545 metric tons. Cutting blades were located at the bottom of the conveyors,
and the entire cutting apparatus could be raised out of the water while cruis-
ing to a harvesting site. Cutting was regulated by law to no more than 1.2 m
below the surface.
SOURCE: Photo courtesy of Ron J. McPeak, UCSB Library Digital Collection.

algin production facilities elsewhere (Bixler and Porse 2011). Sources might change
again, however, as new uses for alginate are discovered, such as in the anodes of high-
capacity batteries (Kovalenko et al. 2011). Bixler and Porse (2011) suggested the cessation of alginate production in California was also a result of concern that harvesters
disturbed sea otters (*Enhydra lutris*). This disturbance was minor, however, and the
cost of sea otter observers on harvesters to monitor it was considered to be insignificant
by Kelco, the kelp harvesting company (D. C. Barilotti, pers. comm.). Bixler and Porse
(2011) also stated that the cessation of *Macrocystis* harvesting in California has increased
the amount of giant kelp wrack on beaches, causing an "environmental problem." The
environmental problem was not stated, and we are unaware of hard evidence of any
such increases. If there were increases, they may well be beneficial to the animals that
consume beach wrack and aid in recycling it into nearshore waters (see Chapter 6, Giant
kelp "beaches").

## BIOFUEL AND CARBON SEQUESTRATION

Sharp increases in the price of gasoline in the 1970s heightened interest in energy
self-sufficiency and stimulated research on algae as a source of fuel (reviews in Bird
and Benson 1987). Interest has recently been rekindled for these same reasons as well

as being a way to inhibit climate change by reducing $CO_2$ in the atmosphere and ocean. The conversion of algae into biofuel results in recycling of $CO_2$ rather than releasing more of it from fossil fuel (e.g., Chung et al. 2011) or sequestering it for long periods of time such as in wood or peat. Seaweeds have their advantages and disadvantages compared to producing biofuels from terrestrial plants. Seaweeds generally lack lignin so their carbohydrates are more amenable to digestion and conversion into fuel. Conversion has been enhanced by the recent genetic engineering of a microbe that directly degrades and metabolizes alginate, and produces bioethanol with a yield of 0.281 weight of ethanol per dry weight of macroalgae. This is about 80% of the theoretical maximum yield from algal sugars (Wargacki et al. 2012). Other advantages are that seaweeds would not remove land and other resources for food crops (Stokstad 2012) and would not cause $CO_2$ release associated with clearing new land (Fargione et al. 2008).

The high productivity of giant kelp and relative ease of harvesting its surface canopy make it a prime target for conversion to fuel (e.g., Roesijadi et al. 2010, Stokstad 2012). A critical examination of its potential, however, indicates several disadvantages. Assuming a conversion rate of 0.281 weight of ethanol-to-dry weight of kelp and given that *Macrocystis* is around 90% water (Table 11.1), giant kelp would yield about 0.356 L of ethanol per dry kg of biomass (ethanol = 0.789 kg $L^{-1}$). Assuming further that the entire maximum of around 150,000 wet metric tons that was harvested annually for alginates in California was used to produce ethanol, the amount of ethanol produced would be $4.7 \times 10^6$ L or $1.2 \times 10^6$ gallons. Although variable from year to year, kelp forests in California cover some 10,400 ha (review in Bedford 2001), and Reed et al. (2008) determined the annual productivity of *Macrocystis 'pyrifera'* in natural stands to be around 2.3 kg dry weight per square meter of kelp forest. If all stands were similarly productive (actually, they are not), the entire production of giant kelp in California, if harvested and converted to ethanol, would yield around $85 \times 10^6$ L or $23 \times 10^6$ gallons. This may seem like a considerable amount, but Californians combusted $18 \times 10^9$ gallons of motor fuel in 2009 (USDT 2011). Using all the giant kelp productivity in the state would therefore replace only 0.13% of the fossil motor fuel demand. This accounting does not subtract the energy needed to harvest, process, and transport the ethanol, or that compared to 100% gasoline from fossil fuel, a mixture of 85% ethanol and 15% gasoline reduces the distance travelled per unit fuel consumed by 25–30% (USDE 2012).

The calculations above indicate that harvesting natural stands of giant kelp cannot provide important amounts of fuel, although this might be accomplished by growing it in monocultures on offshore farms. Offshore farming was investigated in California from the late 1970s to mid-1980s (reviews in Neushul 1987, North 1987b). The research included studies of potential deep-water farm structures (North 1987b), harvestable production (Gerard and North 1984), and, because of low nutrients in offshore surface waters, nutrients available in artificially upwelled water (Kuwabara 1982). The work of Gerard and North (1984) is noteworthy as it was done in the largest container ever used to grow giant kelp in the ocean. The container was moored in a cove at Catalina

Island in southern California. It consisted of a hemispherical bag, 15 m in diameter and 12 m deep, made from polyvinyl chloride (PVC) and held open and floating at the surface by a support structure. Viewed underwater from the outside it looked like the nose of a blimp, and was called "The Hemidome." It could hold around 50 adult plants, and nutrients could be added to the water pumped into it from the surrounding water. Unfortunately, after the first set of experiments in the spring of 1982, the 1982–1983 El Niño arrived in the region. Elevated temperatures stimulated bacterial growth that caused the plants to degenerate, strong winds caused the bag to split, and experiments were discontinued (North 1998).

The effects of plant density and artificial fertilization were also examined on an experimental nearshore farm established on a sand bottom by anchoring plant hold-fasts in mesh bags filled with gravel. This farm produced a harvestable biomass of only around 1 kg (dry weight) per square meter annually (Neushul and Harger 1985). Assuming that the harvestable biomass from the canopy is around 50% of total biomass produced (Gerard and North 1984), the total biomass production of experimental farms was similar to that of natural stands.

Work on offshore farm structures culminated in the placement of a 0.25 acre (0.1 ha) module that would be part of an array of such modules (North 1987b, 1998). The quarter acre module (QAM; figure 11.4) included a central buoy with diesel pumps to bring up nutrient-rich water from a depth of 450 m, and a surrounding curtain to reduce currents and retain the nutrients in the artificially upwelled water. The QAM was installed 8 km off the southern California mainland over 550 m of water in 1978 and populated with 100 adult giant kelp transplants. However, as with many good ideas, this system did not accomplish its intended purpose and, in fact, ended in instructive failure. The motion of the module caused plants to entangle with the structure, and the curtain and plants were soon removed by winter storms. A new cohort of plants, most likely recruiting from spores produced by the original transplants, were lost due to abrasion on barnacles that settled on the structure, entanglement, grazing by fish attracted to it, and large numbers of sea urchins settling on it and grazing on plant remnants. Finally, the project was abandoned when the structure of the QAM was lost during a storm in 1979. These studies not only demonstrated that giant kelp could potentially be grown in the open ocean using artificially upwelled water, but also highlighted that numerous engineering and biological problems will need to be solved before it can be done successfully, especially on a large scale. Without knowing the solutions to these problems as well as how giant kelp would be harvested and how fuel would be produced and delivered to the user, costs cannot be accurately determined. So far, none of the trials has been particularly encouraging.

With the exception of pilot studies in Chile (Buschmann, Varela, et al. 2008) there are no current projects (as of 2013) focused on large-scale *Macrocystis* farming, but farming of other kelps for biofuel is being investigated elsewhere (e.g., Seaweed Energy Solutions 2012). The information for *Macrocystis* highlights the engineering, biological,

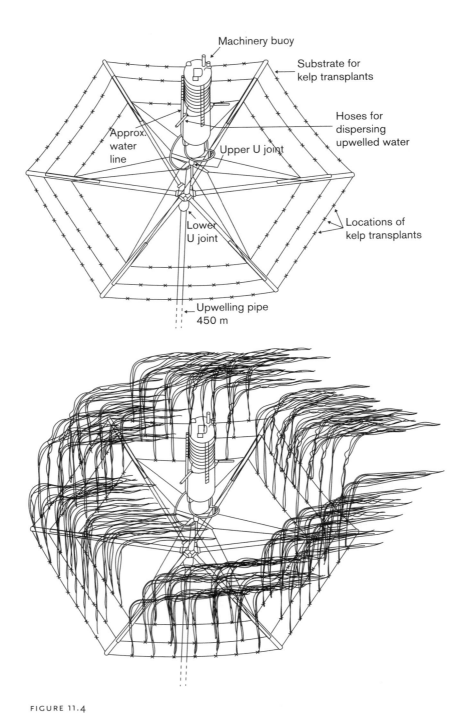

**FIGURE 11.4**

The quarter acre module for growing giant kelp at deep, offshore sites. The module was 32 m in diameter at its widest point. The lower diagram is an artist's conception of what the module would have looked like when fully populated with giant kelp.

SOURCE: From North (1987b).

and ecological difficulties likely to be encountered in large-scale cultivation of seaweeds in offshore farms. Moreover, the fuel production estimates discussed above show that producing enough fuel to replace the yearly fossil motor fuel consumption in California alone would require farms in the order of 10,000 km². Roesijadi et al. (2010) estimated that around $10^4$ km² of seaweed growth was needed to replace 1% of the U. S. gasoline demand. They used higher productivity values and lower water content than in our calculations, and apparently assumed that all the production could be harvested. Their calculated area, therefore, is most likely an underestimate. Nevertheless, both their estimates and ours involve very large areas. Farms at this scale could cause large changes in water quality, affect fisheries, and have conflicts with shipping (Hruby 1978) and other activities. These challenges and conflicts remain largely uninvestigated and are nowhere near being resolved.

Analyses of using giant kelp as biofuel indicate that farming the land is comparatively easy. It highlights the scaling and consequent environmental and engineering problems of replacing fossil fuel with kelp or other seaweed biofuel and, therefore, the importance of energy efficiency and conservation. It also enhances our appreciation of the vast amounts of plant biomass accumulated over geological time that was naturally processed into fossil fuel. It may be, however, that biofuel from giant kelp harvested or grown on a smaller scale, combined with biofuels from other sources such as microalgae, as well as solar, wind, geothermal, and nuclear energy, will make an important contribution to fuel self-sufficiency and the reduction of greenhouse gasses. Time will tell.

Sequestering of carbon dioxide in kelp and other seaweeds has also been discussed. Atmospheric carbon dioxide could potentially be reduced by increasing the biomass of algae that might serve as a reservoir for $CO_2$ which would otherwise be taken up in the atmosphere and ocean waters. The general argument is that carbon in the biomass of seaweeds could also be sequestered in a place where it could not return to the atmosphere for a very long time. Relative to terrestrial plants, however, giant kelp has a high biomass turnover, low litter accumulation, and lacks structural carbohydrates like lignin that are difficult to metabolize or decompose (see Chapter 6). Chung et al. (2011) concluded that carbon sequestration by seaweeds was potentially important only if the stored carbon were used to make biofuel that would replace fossil fuel. Wilmers et al. (2012) argued that an increase in standing stock of natural kelp stands through the removal of sea urchins by sea otters would increase the reservoir of carbon that would otherwise accumulate in the atmosphere and ocean. Some of this increase would be exported as drift into the deep sea, in their opinion resulting in long-term carbon sequestration. Their calculations, however, do not consider potentially large seasonal changes in kelp biomass, or large variations in biomass due to El Niños or other episodic disturbances. Moreover, the amount of kelp that ends up in the deep sea at a depth where carbon would be sequestered from the atmosphere for long periods is unknown.

## AQUACULTURE

Aquaculture of giant kelp for fuel, other chemicals, and animal feed has been explored and continues to be considered, but has yet to become a regular source for commercial use. This is partly because harvests from the large natural stands in California and elsewhere have been sufficient to meet the needs of various users, presumably at a cost less than that of farming kelp. Whether this will continue is unknown. Certainly, kelp aquaculture is big business in parts of Asia, particularly China, Japan, and Korea, where inlets and bays often have large cultivated crops of kelp. These are mostly *Saccharina (=Laminaria) japonica* and *Undaria pinnatifida*, which are grown for human consumption (McHugh 2003) and, more recently, as a source of algin (Bixler and Porse 2011). Such uses of coastal waters, often with attendant problems of excessive nutrients, are uncommon in the rest of the world, both because of the relative unimportance of seaweed in diets and because of user conflicts and regulatory issues. Giant kelp aquaculture is being actively pursued in Chile where demand for it as food for abalone may eventually exceed possible harvest from natural stands (Vásquez 2008). In Chile, experimental giant kelp farms have been developed on floating lines anchored to the bottom and planted with small sporophytes grown in the laboratory (Gutierrez et al. 2006) or with holdfast fragments from *Macrocystis 'integrifolia'* (Westermeier et al. 2013). An integrated aquaculture system is being investigated for growing *Macrocystis* that uses nutrients from salmon or other marine animal farms to produce useful giant kelp biomass (Buschmann, Hernández-González, et al. 2008). Hybridization experiments with plants from different geographic regions in Chile indicate selecting particular hybrids for cultivation can potentially increase yields (Westermeier et al. 2011).

The goal of giant kelp aquaculture is to produce a sustainable source of large sporophytes from which biomass can be extracted, most easily with a harvester at the sea surface. The questions are where to place the farm, which involves consideration of regulatory constraints and use conflicts, how to construct it, how to populate it, and what are the optimal depths and plant densities. The answers, of course, will vary depending on production requirements and where the farming is to be done. Results from experimental farming (e.g., Druehl 1979, Neushul and Harger 1985, North 1987b) show the importance of grazers and ocean climate to farm success, particularly variation in temperature and nutrients. Because *Macrocystis* has floats and can occupy the entire water column, a critical consideration is water motion because it can cause damage to farm structures and to plants as they entangle with the structures. Planting of experimental farms has been done using all phases in the life history of giant kelp (figure 2.1). "Seeding" with very small stages is easier than transplanting larger sporophytes but must be done when conditions are suitable for reproduction and growth at the depth where planting is done.

Because of its fast growth, large biomass, and other perceived production benefits, *Macrocystis* has been grown outside its natural range. In the late 1980s to early 1990s, a commercial venture attempted to grow California red abalone, *Haliotis rufescens*, and giant kelp in an onshore integrated system in Hawaii. The giant kelp was grown in a very large outdoor tank. Because warm, nutrient-poor tropical waters are unsuitable for the growth of both of these species, the venture took advantage of cold, nutrient-rich seawater pumped to land from deep water on the Island of Hawai'i as part of what is now the Natural Energy Laboratory of Hawaii Authority. The venture failed, at least partially because the high light and nutrient levels required for kelp growth caused plankton blooms in the tank, which then inhibited giant kelp growth (M. Foster, pers. obs.) This again highlights the difficulties of managing isolated marine systems. Giant kelp was also introduced to the coast of Brittany, France, in the early 1970s as a potential source of alginate (Boalch 1981). This generated criticism by marine scientists concerned about the potential spread of giant kelp as well as species associated with it to areas outside their natural range, and consequent impacts to native marine communities (Druehl 1973). The British Admiralty was rumored to be concerned that invading *Macrocystis* would impede ship traffic, and power companies argued that drift plants might clog the cooling water intakes of coastal power plants. The project was terminated in response to these concerns, plants in the sea were removed, and there have been no reports of escapees. Giant kelp from Baja California, Mexico, was introduced to China in 1978 and again in 1982. Gametophytes were imported and cultured, and the resulting sporophytes were transplanted to the field for growth on subsurface lines (Tianjing et al. 1984, North et al. 1988). Farming developed using subsurface and surface farm structures, and plants were harvested for abalone food, algin, and the production of kelp pickles. Cultivation has recently ceased due to problems with plant fitness (possibly inbreeding depression; Raimondi et al. 2004), but may be resumed in the future (J. Chen, pers. com.).

There have been efforts to increase the productivity and size of natural forests, in part to improve or sustain potential harvesting. For example, giant kelp productivity declines in southern California when nutrients are low. To examine this relationship experimentally and determine if adding nutrients might increase harvestable yield, North et al. (1981) fertilized natural beds with ammonium sulfate and phosphate using a helicopter. The results were inconclusive, but suggestive that such fertilization might be economically justifiable at times when nutrients are very low. There have been numerous efforts to increase forest size along the southern California mainland by eliminating sea urchins (*Strongylocentrotus franciscanus* and *S. purpuratus*). Urchins were vacuumed from the sea floor, macerated, and pumped back into the sea in kelp forests in southern California in the 1960s. In other areas, quicklime was dumped over urchin grounds with the aim of killing vast numbers of them. Giant kelp forests increased in some of these areas eventually, but the overall increase around this time was more likely the result of improvements in water quality rather than the removal of urchins (Foster and Schiel 2010). Transplanting juveniles or holdfast fragments of *M. 'integrifolia'* is being considered in Chile as a way to enhance natural beds and help mitigate the effects of overharvesting (Westermeier et al. 2012).

FIGURE 11.5

Sport diver swimming in a kelp forest (upper)
and visitors viewing the giant kelp forest tank
in the Monterey Bay Aquarium (lower).
SOURCE: Upper photo is courtesy of John Heine;
lower photo is courtesy of Ed Bierman.

## RECREATIONAL USE

The diversity, beauty, and bounty of kelp forests attract recreational users worldwide. In California, for example, over 12 million people enjoy some form of marine recreation yearly (Pendleton and Rooke 2006) and although there is little hard data on the amount of use associated with kelp forests, it is no doubt very large. Sport divers, kayakers, and boaters are common in and around kelp forests during good weather (figure 11.5). Sport diving in particular has become a major industry, with diving facilities increasingly available even in remote locations (e.g., Scuba Advisor 2012). Many divers now enjoy just taking pictures or observing wildlife, while others spear fish or catch lobsters, abalone, and other benthic animals for food. Hook-and-line fishers catch a wide variety of species in and around kelp forests. Other users simply enjoy observing birds and seals from shore or in tour boats; sea otters are a special attraction in central California. Altogether, the recreational and tourism aspects of kelp forest usage represent major industries and enjoyment, and are potentially in conflict with other commercial activities. The dynamic tension between such uses presents considerable challenges to management of the coastal zone and will undoubtedly assume increasing importance in the decades to come.

## SPORT FISHING

Recreational or sport fishing can be extensive around giant kelp forests. In California, anything from a single person on a rock or pier to large party boats with dozens of fishers testing their luck can be seen on most days near the edges of kelp forests from central California southward. Potentially, dozens or even hundreds of species of fish can be caught on lines. Divers also spear many of these species as well as gathering invertebrates. Species associated with giant kelp along the west coast of North America are primarily rockfish (*Sebastes* spp.), surfperch (Embioticidae), cabezon (*Scorpaenichthys marmoratus*), greenling (*Hexagrammos* spp.), lingcod (*Ophiodon elongates*), sheephead (*Semicossyphus pulcher*), spiny lobster (*Panulirus interruptus*), and abalone (*Haliotis* spp.). There is also extensive spearfishing in Chile. In both California and Chile, the effects of recreational fishers on key species can be extensive (Dayton et al. 1998, Godoy et al. 2010).

Recreational (i.e., noncommercial) fisheries are notoriously difficult to regulate and manage. Unlike commercial fisheries, which usually have total catch limits for individual species as a key part of management, recreational fisheries have catch limits for individual fishers, usually as a daily "bag" limit per species. As the number of fishers increase, these catch limits must be adjusted, often with sparse or inadequate information of what has actually been landed. Commercial fisheries are often seen as the major culprit in the demise of fish populations and in potential trophic effects cascading through various ecosystems (e.g., Jackson et al. 2001), but it is now widely recognized that recreational fisheries can contribute to fisheries declines. Cooke and Cowx (2004, 2006), for example, examined common issues among recreational and commercial fisheries and found that fisheries-induced selection, trophic changes, habitat degradation, and fishing effort were remarkably similar among the two fishery sectors. They argued that management of recreational fishing should have the same urgency as commercial fisheries. Although it is often difficult to judge the scale and scope of recreational fisheries because of the diffuse nature of the fishery and under-reporting, recreational landings of inshore species targeted by this sector was around 23% of landings nationwide in the United States for "populations of concern"; for example, 87% of the landings of bocaccio (*Sebastes paucispinus*) on the Pacific coast were from recreational fishing (Coleman et al. 2004). The extent of the recreational fishery in California was demonstrated by Schroeder and Love (2002) for 19 nearshore fish species that included 12 rockfish species and California sheephead (figure 11.6). They found that landings overall declined from the late 1980s onward for these species, that recreational fishers caught 87% of total landings during the 1980s, and that this decreased to 60% in the 1990s. The declining catch in the 1990s saw a relative increase in commercial landings at least partially due to development of the commercial live finfish market.

In response to overfishing and the decline of many stocks, regulatory agencies such as the California Department of Fish and Wildlife brought in many regulations to limit fish landings. These include marine reserves, areas closed to fishing, size limits, bag limits, depth limits for particular species, closed seasons, and prohibition on landing some species altogether (CDFW 2014a). The decline in the fishery of rockfish has been so great over the past two decades (Mason 2004) that several species are no

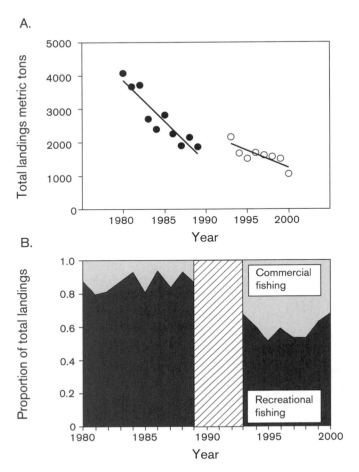

FIGURE 11.6

Annual landings of fish species in the nearshore fishery off California.
No recreational data were collected in 1990–1992. (A) Total landings,
summing both recreational and commercial catches. Straight lines for
each data set were calculated using the least squares method. (B) Proportion of total landings caught in each year by recreational or commercial fishers. Fish included were 13 species of *Sebastes*, 2 *Hexagrammos*,
*S. pulcher, Cebidichthys violaceus, Scorpaena guttata*, and *S. marmoratus*.
SOURCE: From Schroeder and Love (2002).

longer allowed to be taken by sport fishers. These include yellow rockfish (*S. ruber-rimus*), bronze-spotted rockfish (*S. gilli*), Canary rockfish (*S. pinniger*) and cowcod (*S. levis*) (CDFW 2014a). The onus, of course, is on fishers to be able to tell the difference between the numerous species of rockfish and ensure that they do not land prohibited species.

Of all the fished species associated with giant kelp forests, abalone are almost entirely reliant on kelp for food and their populations have been severely depleted over time. In California, there has been such severe depletion of abalone populations of all species that the commercial fishery was closed. Abalone "report cards" are required for sport fishing of abalone in California. These entail recording details of date and area, and a tag must be attached to all individuals landed. Report cards of individual catches must be reported to CDFW to ensure that no one exceeds the limit of 10 *H. rufescens* annually from areas open to fishing. This allows a fairly accurate assessment of landings and an adjustment of the bag limit, if necessary. Since data were kept from 2002 onward, the recreational landings have averaged 256,000 individuals annually (CDFW 2014b).

Chile presents an interesting contrast to California. In a paper entitled "Spearfishing to depletion," Godoy et al. (2010) described the severe effects of recreational and artisanal spearfishing on three temperate nearshore species. *Graus nigra* (vieja negra), *S. darwini* (Chilean sheephead or pejerro), and *Medialuna ancietae* (acha) comprise >98% of spearfish landings. Since the mid-1980s, landings of these species have declined from a peak of ca. 900 metric tons to <100 metric tons by the mid-2000s, and their sizes are considerably smaller. One species (*M. ancietae*) has been so depleted that it is reportedly now practically unknown to a younger generation of spearfishers. The other two species show high fidelity to particular reefs and have proven to be especially vulnerable to spearfishing. Godoy et al. (2010) concluded that catches of reef fishes have shifted from large carnivorous species to small omnivorous and herbivorous species, with potential consequences on subtidal reef community structure.

It is difficult, if not impossible, to find reliable recreational landings data for most kelp forest species worldwide. It is nevertheless clear that this sector of fishing can have impacts and requires better data and good regulation, as does the commercial sector. The lobbying power of the recreational fishing sector is strong and widely dispersed, which adds special challenges to effective management. The management of recreational fishing, however, must continue to be more fully integrated into overall management plans of kelp forests that include all types of extractions (see Chapter 12).

## COMMERCIAL FISHING

The high productivity and structure provided by giant kelp supports numerous species, many of which are or have been harvested from prehistoric times (Lotze et al. 2011) to the present. The commercial catch in or very near kelp forests, however, is difficult to assess accurately because most species also occur on rocky reefs too deep to support kelp. Moreover, although giant kelp stands generally dominate shallow rocky reefs in central and northern California, kelp forests are generally more discontinuous in other regions so landings are more difficult to associate directly with giant kelp habitat. This is further complicated by a paucity of catch records or area-specific data in many regions. Nevertheless, it is clear that kelp forests constitute an important habitat for commercial fish and shellfish species. Commercial fishers in California currently fish in and around kelp forests primarily for lobster, sea urchins, and a variety of fish (see above), the latter increasingly for the live fish market (Leet et al. 2001). The topic of inshore fisheries with all of its ramifications, regulations, and management could fill a volume. There is also considerable concern about historical overfishing, the consequent reduction in abundances and sizes of key species, and effects of their decline on kelp forest ecosystems (e.g., Tegner and Dayton 2000). Dayton et al. (1998) highlighted many of the declines in fish abundances and the fact that there are sliding baselines for gauging what a natural marine system would be like without excess fishing and multiple other anthropogenic stressors. Here, we briefly discuss some of the major features of commercial fisheries associated with giant kelp in California and elsewhere. Management relating to overfishing and kelp is discussed in Chapter 12.

## CALIFORNIA KELP-ASSOCIATED FISHERIES

Mason (2004) analyzed commercial landings from California waters over a period of 74 years from 1928. Historically, rockfish (genus *Sebastes*) constituted most of the kelp-associated catch along California from the mid-1970s to the late 1980s, peaking at around 12,000 metric tons annually. By 2000, however, rockfish catches had declined to 12% of their 1980s average and by 2003 were only a small proportion of the commercial fishery. California sheephead (*Semicossyphus pulcher*), another inshore species that is common within giant kelp forests, has been fished commercially since the early 1900s, with several peaks of commercial landings (figure 11.7). One was around 160 metric tons in the late 1920s, another at 120 metric tons in the mid-1940s, and another at around 160 metric tons again during the 1990s, after which landings declined (Alonzo et al. 2004). This fishing has reduced the average sizes of sheephead and affected their life history traits and demographics (Caselle et al. 2011). The live fish trade has greatly expanded and, by the late 1990s, the live fishery for California sheephead accounted for over 70% of the total landings of this species (Stephens 2001). Together, the landings of this suite of nearshore species have declined considerably since the 1980s (figure 11.6) and there are many management measures in place, including closures, seasonal fishing, and depth restrictions (see above).

Invertebrate fisheries have a long history in California, especially those for spiny lobsters and abalone. The California spiny lobster *Panulirus interruptus* has been fished since the early 1900s. From around 1920–1940, annual catches averaged about 182 metric tons (CDFW 2014b; figure 11.7). This more than doubled in the early 1950s, declined to the early 1970s, and then increased to the late 1990s. The 2013 landing of this species was 351 metric tons. A new fishery management plan is being developed for spiny lobsters, which will evaluate integration of marine protected areas and different fishing techniques into the management of the lobster fishery (CDFG 2011).

The fishery for sea urchins escalated massively after 1970 (figure 11.7; Kato 1972). Both the red urchin *Strongylocentrotus franciscanus* and purple urchins *S. purpuratus* are

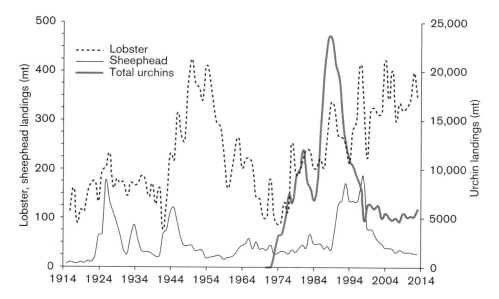

FIGURE 11.7

Landings of California spiny lobsters (*Panulirus interruptus*), California sheephead (*Semicossyphus pulcher*), and sea urchins (combined totals for purple urchins, *Strongylocentrotus purpuratus*, and red urchins, *S. franciscanus*).

SOURCE: Data from the California Department of Fish and Wildlife 2014 (data available on their website).

fished commercially, but red urchins constitute around 99% of the catch. The roe from urchins is processed and sold mostly in Japan (Kramer and Nordin 1979, Reynolds and Wilen 2000). Because urchins are harvested by divers, landings can be quite variable among years due to sea conditions; they tend to be lower during El Niño periods when seas are rough for long periods of time. This fishery peaked at around 24,000 metric tons in the late 1980s and is currently around 6000 metric tons. This fishery was seen as somewhat of a replacement for the declining fishery for abalone in California, and many of the same fishers were involved in both of these fisheries. The removal of urchins in this fishery has been viewed by some as a substitute for the predation effects of sea otters, thereby allowing kelp forests to persist (see Chapter 12). Although this conclusion is contentious, there is no doubt that vast numbers of urchins are removed annually in and around kelp forests, particularly in southern California.

The fishery for abalone in California is a classic example of serial depletion. Because abalone tend to be so visible on reefs and are collected by hand, fishers get very good at finding them until populations are at very low levels and it takes considerable effort to find the remaining few. One species after another was fished until populations of all five of them were in severe decline and no longer commercially viable (figure 11.8). Red abalone, *Haliotis rufescens*, was fished commercially for most of the 20th century, peaking at around 1900 metric tons in 1935. The fishery peaked again after World War

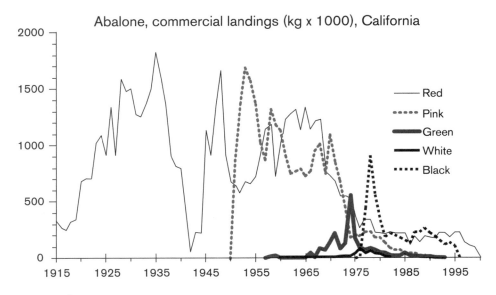

FIGURE 11.8

Commercial landings of five species of abalone that were sequentially fished in southern California. SOURCE: Data from California Department of Fish and Wildlife, and Foster and Schiel's (2010) analysis.

II and went into decline from the mid-1960s. From the mid-1970s onward, landings hovered around 200 metric tons until the decline in the mid-1990s. Pink abalone (*H. corrugata*) was fished from the early 1950s and was depleted by the early 1990s. The other species of abalone were boom-and-bust fishing, with sharp peaks and precipitous declines. Up to 43% of the total catch of white abalone (*H. sorenseni*) was taken in a single year (Karpov et al. 2000, Hobday et al. 2001).

It is clear that overfishing was the major cause of the decline of abalone in areas not reoccupied by sea otters. Minimum size limits, which are meant to allow species to grow to reproductive size and release gametes, thereby sustaining populations, were clearly a failure. In 1997, all commercial and recreational fishing of abalone south of San Francisco was banned (Karpov et al. 2000, Rogers-Bennett et al. 2002). A combination of illegal fishing, disease, habitat loss, recruitment failure, and climate variability were contributing factors to the decline of abalone species, as well as predation by sea otters in central California (Hobday et al. 2001, Tegner et al. 2001). For example, black abalone, *H. cracherodii*, suffered massive mortalities from disease from the mid-1980s onward (Lafferty and Kuris 1993, Neuman et al. 2010). This was caused by a pathogen that led to a wasting disease called "withering syndrome." Raimondi et al. (2002) speculated that the spread of the disease was related to increased temperatures associated with El Niño events. The timing of the initiation of mortalities coincided with the strong El Niño in 1982–1983 and also occurred as an isolated outbreak in Diablo Cove, north of Point Conception, following warm-water discharge from a power plant (cf. Schiel et al. 2004).

Around 95% of black abalone south of Monterey in central California died (Neuman et al. 2010). The spread of withering syndrome is now known to be enhanced by periods of ocean warming. Neuman et al. (2010) identified this disease as a continuing threat to black abalone. Black abalone was listed as endangered under the U.S. Endangered Species Act by the U.S. National Marine Fisheries Service in 2009. Another species, the white abalone *H. sorenseni*, underwent massive declines in numbers. Davis et al. (1996) reported that this deep-water abalone, normally found from around 25 to 65 m depth, was found in high densities at the Channel Islands in the early 1970s but had virtually disappeared by the early 1990s, so much so that some were concerned about the possibility of extinction (Tegner et al. 1996, Davis et al. 1998). More recent surveys have shown that the total population is larger than previously reported because more habitat was surveyed, although white abalone are still orders of magnitude less abundant than indicated in historical estimates (Butler et al. 2006). In 2001, this species became the first invertebrate to be listed as endangered under the U.S. Endangered Species Act. Abalone, particularly *H. rufescens*, continue to be fished commercially along the Pacific Coast of Baja California, Mexico, and, as discussed above, there is still a recreational fishery allowed in some areas of the California coastline.

## GIANT KELP-ASSOCIATED FISHERIES ELSEWHERE

Commercial fishing associated with giant kelp forests in other regions of the world varies, depending on the extent of kelp forests and their accessibility. Chilean kelp forests are inhabited by many species of commercial and recreational importance, including the Chilean "abalone," *Concholepas concholepas* ("loco," a carnivorous muricid), the sheephead, *Semicossyphus darwini* ("pejeperro"), and the sea chubs *Graus nigra* ("vieja negra") and *Medialuna ancietae* ("acha"). These and other species are commonly taken by divers, either by hand or by spear fishing (Castilla and Fernandez 1998, Godoy et al. 2010; see "Recreational Use" above). *Concholepas*, however, is one of the most important commercial species associated with kelp in Chile. This fishery was a main impetus for inclusion of Management and Exploitation Areas for Benthic Resources (MEABRs) in the Fishery and Aquaculture Law of 1991, which regulated benthic and pelagic coastal resources by artisanal fishers. This history of this fishery has been described by Castilla and Gelcich (2008). This fishery was sustained from 1960 to 1974 at between 3000 and 6000 metric tons, but was severely overfished at up to around 25,000 metric tons in 1980. Since the management plan came into being, which included territorial user rights for fisheries (TURFS), the fishery fluctuated between 2000 and 5000 metric tons and seems to be sustainable at this higher level. Castilla and Gelcich (2008) pointed out that loco has been the most important species to be harvested from MEABRs, accounting for around 30–60% of fish landings, and that the numerous fishers who participate in this fishery attach important non-economic values to it, such as pride, accountability, and cultural values.

Similar cultural and non-economic values are attached to New Zealand abalone, *Haliotis iris*, which is fished at around 1200 metric tons annually under a commercial quota management system. This species, however, is not particularly common in giant kelp forests but instead occupies shallower inshore areas (Schiel et al. 1995). There is also an extensive and large fishery for greenlip (*H. laevigata*) and blacklip (*H. rubra*) abalone in Australia, but, again, these are not particularly associated with giant kelp forests. The rock lobster *Jasus edwardsii* has constituted important fisheries in New Zealand and Australia for much of the 20th century. This species is not particularly associated with giant kelp forests in New Zealand, but there is an important kelp-associated fishery in Tasmania and South Australia (Barrett et al. 2009; see Chapter 12). Hamon et al. (2009) reviewed the history of this species and its management in Tasmania. Tasmanian lobsters have been fished for over 200 years and first came under regulated management in 1889. The fishery eventually increased to >2000 metric tons by the 1980s and declined to a historically low level of around 1400 metric tons in 1994. An individual transferable quota system was introduced in the late 1990s and catches have stabilized at around 1500 metric tons. There are still some concerns about this fishery with the regression of kelp stands in eastern Tasmania and an extensive loss of habitat (see Chapter 12).

## OTHER USES

It is noteworthy that the inspirational or even spiritual aspects of giant kelp forests have received increasing attention in modern times. This is not only by indigenous people, such as the New Zealand Maori who speak of the "mauri" or life force of the sea, and "kaitiakitanga" or humans as guardians who must assist in protecting the environment and its life forces. These aspects are seen in most areas of the world, as exemplified in the poem "Enigmas" by Chilean poet Pablo Neruda:

> You've asked me what the lobster is weaving
> there with his golden feet?
> I reply, the ocean knows this.
> You say, what is the ascidia waiting for in its transparent bell?
> What is it waiting for?
> I tell you it is waiting for time, like you.
> You ask me whom the *Macrocystis* alga hugs in its arms?
> Study, study it, at a certain hour, in a certain sea I know.

For most people the charisma of giant kelp pales in comparison to whales and sea otters, but the plants and forests themselves have achieved a certain cultural significance. Historically, the abundant human food available in giant and other kelp forests may have influenced the migration routes of the first maritime peoples from Asia to the west coast of the Americas in the late Pleistocene (Erlandson et al. 2007). Giant kelp and its forests continue to inspire poets and artists, stimulate public interest and

"The Macrocystis pyrifera of the great Pacific kelp forest is the world's fastest-growing plant. You may know it by its common name—the giant bladder kelp."

FIGURE 11.9

*Macrocystis* cartoon and logos. Cartoon is drawn by Booth for *New Yorker* magazine, 1984. The Western Society of Naturalists© is one of the oldest scientific societies on the west coast of the United States. Members are primarily marine scientists (www.westsocnat.com). The Monterey Bay Aquarium is a large public aquarium in Monterey, California. Both logos are based on the growing end of a frond.

aquarium displays (figure 11.5), and enhance trips to the beach. The most common art is made from portions of fronds pressed onto thick paper and mounted to hang like pictures. Stipes are woven and dried to make decorative baskets and other objects. Artistic renditions are used as logos and *Macrocystis* has even become the stuff of cartoons (figure 11.9). Giant kelp forests were recognized for their "beauty and complexity" in a set of stamps issued by the United States Postal Service in 2008 (figure 11.10), and the Kelp'thar Forest is visited by millions as a featured habitat in Vashj'ir, an underwater zone in the online game World of Warcraft (www.wowwiki.com/Kelp'thar_Forest).

Public interest and aquarium attendance have been increased by exhibiting live plants and kelp forests, exhibits pioneered by the Monterey Bay Aquarium with a 30 m tall, $1.25 \times 10^6$ L display tank that includes a surge machine (figure 11.5). Live plants are now displayed in numerous other aquaria in the United States, as well as the National Museum of Marine Biology and Aquarium in Taiwan where adult plants must be imported from overseas. Given the difficulties of obtaining and maintaining live plants at locations remote from natural sources, a few companies now produce plastic giant kelp that can be ordered to size (e.g., several companies have elaborate displays on their

# KELP FOREST

**FIGURE 11.10**

U. S. Postal Service© kelp forest stamp set. The stamps are various parts of the drawing that can be peeled off and used for postage, each indicated by "USA 44." Note the densely populated and diverse community depicted above and below the sea surface.

internet sites). Various parts of the plants can also be obtained separately and "snapped" together. Anticipating that *Macrocystis* will be part of an exhibit including animals, some of this artificial kelp is marketed as "seal and penguin resistant" (Living Color 2012). Giant kelp wrack becomes the stuff of play at the beach, serving as whips and impromptu decorations for sand sculptures. For those wishing for longer-range interactions, floats from fresh wrack can be used as ammunition for a giant kelp blowgun, a short length of PVC pipe with about 2 cm internal diameter. Eye protection is recommended for live targets.

Importantly, kelp forests are great laboratories for education about the wonders and workings of nature. Giant kelp forests are used to train and inspire undergraduate and postgraduate students in focused, field-oriented university courses, usually entitled "Kelp Forest Ecology" or "Subtidal Ecology" (review in Pearse et al. 2013). The primary objectives of courses in California are to introduce students to the natural history and ecology of giant kelp forests using field observations, sampling, and, in some cases, individual or group research projects. Students must be certified divers and the fieldwork, like all in situ research, requires surmounting various logistical and regulatory hurdles. The resulting knowledge and experience is well worth the effort; some students have become kelp forest biologists and ecologists as a result, and many others benefit simply by experiencing an extraordinarily rich and diverse habitat. Furthermore, when courses are offered consistently over many years, the data obtained from sampling by students can be an important contribution to understanding kelp forest dynamics.

## RESTORATION OF KELP FORESTS

Attempts at restoring former forests and creating new ones began in the late 1950s, stimulated by the severe reduction in size of the two of the largest stands of giant kelp in California, the Palos Verdes forest near Los Angeles and the Point Loma forest off San Diego. These declines were associated with increased ocean sewage discharge and coastal development related to rapidly growing urban populations, and an intense El Niño (Foster and Schiel 2010). The largest restoration project was the Kelp Habitat Improvement Project under the direction of Wheeler North (review in North 1976). Kelco, the largest kelp harvesting company at the time, also endeavored to increase *Macrocystis* abundance, particularly in the Point Loma area close to their processing plant. Later restoration activities were undertaken by the California Department of Fish and Game (e.g., Wilson and McPeak 1983), including efforts to restore giant kelp to areas north of Santa Barbara that were deforested by extreme storms during the 1982–1983 El Niño (Schiel and Foster 1992). More recent restoration efforts in southern California have been supported by the U.S. National Oceanic and Atmospheric Administration (NOAA) as part of a settlement related to environmental damage due to the discharge of DDT by Montrose Chemical Corporation (NOAA 2012).

The first consideration in any restoration effort is identifying the cause of the decline (Schiel and Foster 1992). If, as has been the case in many areas, the cause is deterioration of water quality, such as increased turbidity or increased sedimentation, then reintroducing giant kelp will not be successful unless the physical and chemical environment is restored so that it is conducive to the growth and reproduction of *Macrocystis* and associated species. If these general improvements are made, however, it is likely that kelp will recover naturally, particularly if there is a natural source of spores nearby. For example, even though transplanting kelp and removing sea urchins may have helped, the kelp forests at Palos Verdes and Point Loma recovered rapidly when sewage effluent quality improved and outfalls were moved further offshore (Foster and Schiel 2010). Unfortunately, kelp forests around large urban areas such as San Diego and Los Angeles have ongoing legacy effects, and poor water quality and accumulated sediments from coastal development may still be impeding the recovery of local kelp forests to their historical levels.

Even in local areas where sea urchin grazing has reduced the size of forests, the ultimate cause may be a decline in habitat quality that reduces kelp abundance, inducing indirect effects by altering the foraging behavior of sea urchins. The common sea urchins in California, *Strongylocentrotus franciscanus* and *S. purpuratus*, shift from passive feeding on captured drift algae to active foraging on attached plants when kelp abundance declines (e.g., Harrold and Reed 1985). Sea urchin removal alone may not be successful, or necessary: if the quality of habitat is restored and giant kelp is likely to recolonize naturally from nearby sources, it is conceivable that kelp restoration could be accelerated by inducing a shift in urchin feeding from active back to passive by adding drift algae or other food. This occurs naturally (Duggins 1981). In any case, attempts at restoring giant kelp forests have been fraught with difficulties that highlight the complexities of hysteresis of natural ecosystems that have been compromised or eliminated.

If one assumes that sea urchin grazing alone is reducing kelp forest size and that this is an un-natural, indirect result of a reduction in sea urchin predators such as some fish and lobsters from fishing (e.g., Jackson et al. 2001; discussed in Chapter 9), then marine protected areas or other forms of fishery regulation can be an effective restoration method if they result in an increase in these predators. The clearest example where this is the case is in Tasmanian giant kelp forests where changes in ocean climate caused a range extension of the sea urchin *Centrostephanus rodgersii* which, combined with other factors, resulted in severe deforestation. These sea urchins are eaten by the lobster *Jasus edwardsii*, and deforestation is inhibited in some areas where the lobster is protected from fishing (Ling et al. 2009; see Chapter 12).

Several methods of giant kelp seeding and transplanting have been used in attempts to enhance recolonization, most of which are reviewed by North (1976). As for farming discussed above, the giant kelp life history offers several choices. Spores can be provided by attaching mesh bags of fertile sporophylls to the bottom (Hernández-Carmona et al. 2000), and microscopic stages can be grown in the laboratory and dispersed over the bottom. The success of approaches using microscopic stages, however, is dependent on having suitable benthic conditions for attachment, reproduction, and/or development of the life history stage used. Creating suitable benthic conditions may include removing understory algae if they are inhibiting giant kelp recruitment (Hernández-Carmona et al. 2000). As discussed in Chapter 3, such conditions may occur only sporadically and may be difficult to predict. An alternative to the equivalent of seeding is to grow small sporophytes (<1 cm tall) on artificial substrata in the laboratory and then place the substrata in the field (Vásquez et al. 2014), or transplant the much larger juvenile and adult sporophytes that are not as influenced by bottom conditions. Transplantation is much more labor intensive because the plants must be collected, transported to the restoration site, and individually attached to the bottom. Juveniles are much easier to transplant than adults, and can be attached to the "stumps" (holdfasts plus short sections of the stipes) of understory kelps after upper portions have been removed (Hernández-Carmona et al. 2000). The scale of transplantation is important because small transplanted stands may be consumed by grazing fishes (North 1976). The survival of transplants in areas with abundant sea urchins can be enhanced by surrounding the transplants with strips of plastic that wave over the bottom. The whiplash and "touch" of the strips apparently cause urchins to move away (Vásquez and McPeak 1998). Timing is also important as storms may remove plants before they securely attach. So far, most, if not all, of these transplanting efforts have failed. As previously stated, none are likely to be effective unless the causes of the original decline are eliminated. Furthermore, transplanting large areas requires enormous effort, plants are relatively short-lived, and unless growing conditions improve a transplanted population is not sustainable.

## ARTIFICIAL REEFS, CREATION OF NEW KELP FORESTS, AND "MITIGATION"

Artificial reefs placed on a sand bottom are used to produce new kelp forests in California. Initial efforts at producing artificial reefs looked more like rubbish disposal than anything that could be construed as natural habitat. Materials such as old streetcars and tires were dumped into the ocean primarily to provide fish habitat and enhance fishing opportunities, and a few of these small reefs were colonized by giant kelp (Lewis and McKee 1989). Since the early 1980s, there has been demand, at least in California, to "mitigate" anthropogenic impacts along the coastline by providing new habitat as compensation for damage. In particular, this has occurred with respect to power plants, which draw a large amount of seawater into their cooling intakes, through heat exchangers and back into the ocean (see Chapter 10). The major focal areas of this have been at the San Onofre Nuclear Generating Station (SONGS) near San Clemente in southern California and the Diablo Canyon Power Plant (DCPP) near Avila Beach in central California. Various forms of mitigation have occurred or been proposed, but the main efforts have been in the creation of artificial reefs to provide new kelp forest habitat (Carter, Jessee, et al. 1985, Ambrose 1994, Reed et al. 2004, Reed et al. 2006—"An experimental investigation"). These have met with mixed success over the years because of differences in overall size, depths, and configuration, but reef designs are now far more effective than they were originally in being suitable for their intended purpose of supporting giant kelp forests and associated communities.

The first of these "mitigation" artificial reefs was Pendleton Artificial Reef (PAR) in 1980, which comprised 1.4 ha of quarry rock deposited as eight modules on a sand bottom at around 11 m depth and situated 5–6 km from the nearest kelp forest (Carter, Carpenter, et al. 1985). Giant kelp, the understory kelp *Pterygophora californica*, and abalone (*Haliotis rufescens*) were transplanted to the reef in efforts to establish a kelp forest rapidly. Unfortunately, the results were not those intended. Herbivorous fishes (halfmoon *Medialuna californiensis* and opaleye *Girella nigricans*) were attracted to the isolated modules and ate the kelp. Abalone did not survive, most likely due to lack of food and consumption by predators (Carter, Jessee, et al. 1985). Since then, giant kelp has occasionally colonized PAR, but the modules have not developed a giant kelp community similar to nearby natural reefs. It was concluded that this initial attempt at

FIGURE 11.11

Satellite images of the surface canopy of giant kelp on Wheeler North Reef, off southern California.
Left: canopy on experimental modules (dark spots) in 2003 from SPOT satellite (10 m resolution).
Right: canopy on the completed reef in 2010 from color-corrected Landsat image (30 m resolution).
AR = experimental artificial reef modules. SM = San Mateo kelp forest, a natural forest adjacent to the artificial reef. "Wheeler North" = completed Wheeler North Reef. "San Mateo" = San Mateo kelp forest.
NOTES: Note the variation in the extent of the San Mateo kelp canopy between years. The distance between AR and SM in left image is approximately 3 km.
SOURCE: Left image courtesy of Google Earth from the United States Geological Survey. Right image courtesy of D. Reed.

production of a kelp community did not accomplish its aims because of isolation of the modules from other reefs, their placement in an area where light is not always favorable to kelp growth, and the high relief of the modules (Carter, Jessee, et al. 1985, Ambrose 1994).

Subsequent installations of artificial reefs have proven to be more successful. Based on PAR and other assessments of conditions likely to produce kelp forests on artificial reefs that resembled natural forests, a new design was developed for the construction of what is now called Wheeler North Reef (WNR) as mitigation for "loss of habitat" related to the discharge from SONGS. This reef, a series of modules spaced along the coastline of southern California north of San Diego, was installed as a large-scale experiment. WNR was placed on a mostly sand bottom 500 m northwest of a natural kelp forest at depth of 13–16 m. Construction occurred in two phases. In 1999, Phase I placed 56 modules, each with an area of 1600 m², in seven experimental groups to evaluate the effects of substratum type (rock versus concrete), substratum size (boulder vs. rubble), and bottom cover (Reed et al. 2004, Reed et al. 2006—"An experimental investigation", Reed et al. 2006—"Quantitative assessment"; figure 11.11). The vertical relief of all modules was <1 m, to obviate the problems experienced at PAR. The modules continue to be monitored since placement. *Macrocystis* colonized via spores from the natural forest within 6 months, and within 2 years produced a dense surface canopy on each module that has persisted (figure 11.11). The reef was completed (Phase II) in 2008 by placing rock around the experimental modules to create a reef area of 61 ha. Giant kelp colonized the new rock (figure 11.11) and monitoring has continued to evaluate how well WNR meets established performance standards used to determine whether it successfully mitigates for the kelp forest habitat destroyed by the discharge from the power plant. Results so far have been encouraging (UCSB 2012). This project exemplifies the value of clear objectives and rigorous experiments and assessments to the success of artificial reefs. The scale, planning, construction, and evaluation of such a reef, however, do not come cheaply. The California Coastal Commission, the state agency overseeing the project, estimated the total cost of construction and ongoing monitoring to be in the order of US $50 million (UCSB 2012, California Coastal Commission reports), which is being paid by the owners of the power plant.

Giant kelp forests, like their terrestrial analogues, are clearly "forests of many uses" that enhance human experience and commerce, especially when they occur near heavily populated coasts, and their resources are economically important. Use can lead to abuse, however, as has occurred when forests are used as disposal sites for sewage and runoff (discussed in Chapter 10) or when resources are extracted without regulations that ensure sustainability (Chapter 12). No doubt, future uses will require continued evaluation, particularly in areas most affected by changes in ocean climate (Chapter 13) where currently benign activities may become cumulative multiple stressors. One hope is that the greatest benefit of marine protected areas (Chapter 12), which include protection of water and benthic quality, will be to preserve some giant kelp forests in as natural a state as possible.

# 12

# MARINE PROTECTED AREAS
# AND FISHERIES EFFECTS

We have been gambling with the future by establishing a poor balance between
short-term profit and long-term risks. The absence of meaningful, fully
protected reserves has produced a situation in which there are virtually no
areas north of the Antarctic in the world's oceans that have exploitable
resources where scientists can study natural marine systems.

—Dayton et al. (2000)

Although numerous reviews and meta-analyses have built the case for
increased use of PAs (protected areas) . . ., few have dealt with failures of PAs
or with the general effectiveness of PAs at halting global biodiversity loss.
Evaluation of the performance of PAs is critical since failure of PAs to protect
biodiversity could erode public and political support for conservation.

—Mora and Sale (2011)

If marine reserves and other MPAs are to provide significant conservation
benefits to species, they must be scaled up. . .. By aggregating the benefits of
multiple MPAs, the network can have larger impacts. More importantly, a
number of theoretical models suggest that networks can have emergent
benefits that make the network more than the sum of its individual parts.

—Gaines, White, et al. (2010)

There are many conservation, ethical, and societal reasons to implement marine protected areas (MPAs) (e.g., Airame et al. 2003), but MPAs and overfishing seem linked
like the Gemini twins, with perhaps fishing in the role of Castor (the mortal who per-
ished before being revived) and MPAs in the role of Pollux (the immortal who came
to the rescue); if they are not actually bound together, they are at least within each
other's gravitational pull. Although there are many human activities that have impacts
on marine areas (e.g., Sousa 2001, Schiel 2009, for reviews), fishing is the chief means
of extraction of organisms with potentially large impacts on the functioning of marine
ecosystems (e.g., Baum and Worm 2009). Of the thousands of publications on MPAs,

most mention the cessation of fishing as a principal avenue to restoration of function (e.g., Davis 1989, Dugan and Davis 1993, Gell and Roberts 2003). Various types of MPAs have been established worldwide for a wide variety of reasons, including prevention of overfishing, protection of species and habitats, preservation of special areas, as insurance policies for large-scale impacts, as nurseries for exploited species, and to restore trophic linkages and ecosystem functioning (Lauck et al. 1998, Babcock et al. 1999, Carr 2000, Syms and Carr 2001, Halpern and Warner 2002, Allison et al. 2003, Gell and Roberts 2003, Lubchenco et al. 2003, Lester et al. 2009, Gaines, White, et al. 2010, Gaines, Lester, et al. 2010). The formation of MPAs has been a long-fought battle by scientists and environmentalists (Ballantine and Langlois 2008) that now has considerable credence with a wide range of stakeholders, who have increasingly recognized the stresses on coastal ecosystems and the need for some form of protection. In many ways, however, MPAs are simple management tools, affording partial or full protection from exploitation of species and habitats within their borders. There is a diverse and rich literature discussing their positive effects (Babcock et al. 1999, Murray et al. 1999, Halpern and Warner 2002, Shears and Babcock 2002, Lubchenco et al. 2003, Shears and Babcock 2003, Lester et al. 2009, Gaines, White, et al. 2010, Gaines, Lester, et al. 2010, Leleu et al. 2012), but there are also other tools in spatial and other forms of management that require consideration (e.g., Mora and Sale 2011, Thrush and Dayton 2010).

It is interesting to note that two of the world's first MPAs were established for the purpose of scientific study (Ballantine and Gordon 1979, Fernández and Castilla 2005). One of these, in northeastern New Zealand, had extensive urchin barren grounds at the time of establishment in 1975 (Choat and Schiel 1982) and it has become one of the world's focal points for trophic changes following decades of protection (Babcock et al. 2010). The prescience of a few pioneers of MPAs (e.g., Bill Ballantine [Ballantine and Langois 2008] and Juan Carlos Castilla [Castilla and Duran 1985]; see also Murray et al. 1999, Davis 2005) blossomed as an idea whose time had come, with increasing pressures on marine ecosystems worldwide and the apparent failure of traditional management, especially of many fisheries. Early discussions about MPAs along coastlines where giant kelp are prominent focused particularly on urchin barren grounds, overfishing, and the loss of function. In sometimes highly charged circumstances, the benefits of MPAs to fishing, the so-called "spillover effect" of enhanced fishing outside reserve boundaries (e.g., McClanahan and Mangi 2000, Halpern 2003, Halpern and Warner 2003, Lester et al. 2009, Halpern et al. 2010), were argued as a benefit of MPAs, but not to universal approval (Agardy et al. 2003). Although there is always a dynamic tension between stakeholders in the marine environment, perhaps especially relating to commercial and recreational fishers and environmentalists (Klein et al. 2008), there has been a gradual societal shift toward the recognition of potential benefits of MPAs and the idea that in the sea, as on the land, there is great merit in setting aside some areas largely free from extraction and other forms of exploitation.

The focus now has shifted largely from the question of "are marine reserves worthwhile?" to "how can they best be implemented to gain maximal benefit?" This involves questions of where to put them, how big they should be, how many are needed, and how they should be spaced and networked along a coastline (Gaines et al. 2003, Gaines, White, et al. 2010). In this chapter, MPAs are discussed with particular reference to *Macrocystis* forests. Perforce, much of this discussion relates to California, where the major MPAs with giant kelp occur and where the preponderance of MPA research relating to *Macrocystis* forests has been done. There have been several compendia of papers to which readers may refer for a fuller treatment of MPA issues (Dayton et al. 2000, Lubchenco et al. 2003, Babcock et al. 2010, Gaines, Lester, et al. 2010). Below we present an overview of what MPAs are, their design and implementation, fishing and other effects, and management beyond MPAs. Much of the theory and practice have been based on giant kelp forests, particularly along the west coast of North America.

## WHAT IS A MARINE PROTECTED AREA?

The definitions, types, and purposes of MPAs tend to vary by country and application, and come in several varieties. The International Union for Conservation of Nature (IUCN), an organization aimed at finding pragmatic solutions to the world's environmental and development challenges, defines a protected area as "A clearly defined geographical space, recognised, dedicated and managed, through legal or other effective means, to achieve the long-term conservation of nature with associated ecosystem services and cultural values" (http://www.iucn.org/). Under this, they define six management categories, ranging from full protection to protected areas with sustainable use of resources (Day et al. 2012). The IUCN defines MPAs as "Any area of the intertidal or subtidal terrain, together with its overlying water and associated flora, fauna, historical and cultural features, which has been reserved by law or other effective means to protect part or all of the enclosed environment." The Convention on Biological Diversity extends MPAs as "Marine and Coastal Protected Areas" to include both coastal areas and areas that cross the land–sea interface such as estuaries and marine salt marshes. These constitute "an area within or adjacent to the marine environment, together with its overlying waters and associated flora, fauna, and historical and cultural features, which has been reserved by legislation or other effective means, including custom, with the effect that its marine and/or coastal biodiversity enjoys a higher level of protection than its surroundings."

There are considerable issues worldwide in what does or does not constitute an MPA. Day et al. (2012), for the IUCN, outline the critical issues relating to the definition and functions of MPAs. Among others, conservation must be the main priority and they must have precise boundaries and legally defined zones. Day et al. pointed out that other forms of management may overlap with the intentions of MPAs but have quite different objectives. For example, the ecosystem-based approach advocated by the Convention on Biological Diversity has a principle that land and water management should contribute to conservation. According to the IUCN, the critical criterion is whether nature conservation is the primary objective. Therefore, fisheries management areas, community areas set aside for sustainable harvest of marine products, and areas set aside for other purposes, such as military use or pipeline protection, are not MPAs. Furthermore, it is expected that MPAs should be managed in perpetuity and not as short-term or temporary management strategies.

Because they are enacted on a country-by-country, and sometimes state-by-state, basis, MPA legislation is tailored to particular circumstances. One of the earliest pieces of legislation designating MPAs was the Marine Reserves Act 1971 of New Zealand (http://www.legislation.govt.nz/act/public/1971/0015/latest/DLM397838.html). This was enacted "for the purpose of preserving, as marine reserves for the scientific study of marine life, areas of New Zealand that contain underwater scenery, natural features, or marine life, of such distinctive quality, or so typical, or beautiful, or unique, that their continued preservation is in the national interest," and the public were guaranteed to have right of entry. It is interesting to note the key element of "scientific study," a feature which infuses much of subsequent legislation, either directly or implicitly through management and monitoring (e.g., see Dayton quote at start of this Chapter). Even in a conservation-oriented country like New Zealand, MPAs of various types protect around 7% of New Zealand's territorial waters, with 99% of this area being within two large offshore, deep-water marine reserves, and a total of only 0.31% of the exclusive economic zone is being managed under MPA legislation (Department of Conservation and Ministry of Fisheries 2005, Bess and Rallapudi 2007).

Legislation for establishment of MPAs in the United States was enacted in 2000 by United States Executive Order 13158 (http://www.gsa.gov/portal/content/101568), designating them as "any area of the marine environment that has been reserved by federal, state, tribal, territorial, or local laws or regulations to provide lasting protection for part or all of the natural and cultural resources therein." Their designated purpose is to "help protect the significant natural and cultural resources within the marine environment for the benefit of present and future generations by strengthening and expanding the Nation's system of marine protected areas," and included science-based identification and prioritization of natural and cultural resources for additional protection as well as ecological assessments of linkages between MPAs. In California, where the greatest number of MPAs encompassing giant kelp occurs, the setup and administration of protected areas come under the ambit of the California Department of Fish and Wildlife (CDFW; until January 1, 2013, it was called the California Department of Fish and Game). The California State Legislature passed the Marine Life Protection Act in 1999 that required CDFW to "increase its coherence and effectiveness at protecting the state's marine life, habitats, and ecosystems" (California Department of Fish and Wildlife 2013). This included three types of MPA designation (state marine reserve, state marine conservation area, and state marine park) as well as other types of management areas. MPAs are developed regionally in the north, north-central, central, and southern parts of the state. There is a very extensive network of hundreds of protected areas of various types covering numerous habitats and ecosystems.

Similarly, MPAs in Australia were enacted to be "parts of the ocean that are managed primarily for the conservation of their ecosystems, habitats and the marine life they support" and to include a network of Australia's different marine habitats and marine ecosystems (http://www.environment.gov.au/marinereserves/overview.html). Australia purports to have the largest marine reserve network in the world, with 60 covering 3.1 million square kilometers. Fernández and Castilla (2005) reviewed the progress and impediments to establishing MPAs in Chile. Unlike most other areas of the world, they reported that 11 of 27 natural sanctuaries were sponsored and administered by private organizations, with these beginning as early as 1941 in a concession of 0.013 km² to the University of Valparaiso for research. Management and exploitation areas, rather than those with full protection, represent the great majority (around 97%) of reserves in Chile.

## DESIGN AND IMPLEMENTATION

A great deal of research has been done on the optimal design, size, and spacing of marine reserves. Because they occur in largely open coastlines, their effectiveness depends to a great extent on flows and connectivity through hydrographic processes, species movement patterns, larval drift and retention, and so, to a great extent, on their spacing and the condition of the surrounding habitats (e.g., Allison et al. 1998). Dayton et al. (2000) pointed out that the main objectives of coastal zone management are conservation, the maintenance of biodiversity and intrinsic ecosystem services, and maintenance of sustainable fisheries. These and other objectives of MPAs factor critically in their design. The primary purpose of MPAs may be to conserve, but the implicit objectives are to rehabilitate areas and return them to full function. This is challenging enough when considering sessile and locally mobile species, but becomes even more difficult when considering highly mobile fishes. There are also economic and social dimensions in the design and implementation of MPAs, particularly where fishing and long-term usage have been well established (Castilla 1999).

It was recognized early that the effectiveness of marine reserves as a conservation tool would have limitations because oceanographic processes that influence dispersal and population replenishment and human impacts such as oil spills often occur over much larger scales than reserves can accommodate (Allison et al. 1998). Kinlan and Gaines (2003) assessed the differences in long-distance dispersal (LDD) of several key taxa in California coastal communities and found that LDD ranged from meters to hundreds of kilometers, with fish and some invertebrates clustering at the high end and macroalgae at the lower end of dispersal ranges. Kinlan et al. (2005) further pointed out that few organisms could be simply classified as short or long dispersers and that dispersal must be defined only with respect to the explicit definitions of the scale and process of interest. Similarly, Shanks et al. (2003) examined dispersal distance and time spent dispersing for 25 taxa for which data were available, including *Macrocystis* and several macroalgae, tunicates, bryozoans, molluscs, and crustaceans. Like Kinlan and Gaines, they found that dispersal ranged from meters to thousands of kilometers. Based on their analyses, they recommended that a reserve between 4 and 6 km in diameter should be large enough to retain larvae of short-distance dispersers and that reserves should be spaced 1–20 km apart so that they are close enough to capture propagules of distant dispersers from adjacent reserves.

Carr and Reed (1993), in considering giant kelp forests, pointed out that with respect to fish and fisheries, the necessary size of a reserve will depend on the rate at which reserve populations and those in the surrounding area supply new recruits. They distinguished four modes of dispersal relevant to reserve design (figure 12.1). These included self-seeding closed populations, single sources feeding many areas, many areas feeding larvae into a large larval pool, and limited dispersal through stepping stones of nearby areas. Gaines et al. (2003) provided some of the theoretical underpinnings for explicitly accounting for larval dispersal in marine reserve design. They found, perhaps unsurprisingly in retrospect, that advection plays a dominant role in determining the effectiveness of reserve configuration, but the important consequences of this are that where currents are strong, several reserves are far more effective than single reserves of the same total size and that reserves can outperform traditional, effort-based strategies in terms of fisheries yields.

Others have explored the implications of reserves as fisheries management tools, which was one of the original reasons for MPAs in California. Nowlis (2000) gives a good perspective on this topic. He pointed out that closed-area management is an idea that stretches back many decades (Beverton and Holt 1957) and that studies have repeatedly shown that the number and average sizes of fished species and the total number of species increase within closed areas (e.g., Roberts and Polunin 1991, Dugan and Davis 1993, Rowley 1994, Bohnsack 1996). Because marine reserves entail short-term fisheries losses and are touted as a tool for rehabilitating populations, he examined these effects with three models. His results indicated that if reserves are properly designed to minimize spillover effects of adults and allow abundant larval transport, they can maximize fisheries harvests, at least for the groups he examined. Roberts (2000) pointed out that, up to that point, the location of marine reserves depended more on social criteria and opportunity than on scientific study. He advocated the urgency of establishing reserves on the basis of what is known rather than delay them until long-term studies are done. He concluded that a network of interacting reserves should be established to hedge bets against uncertainty and variability in ecological processes and in management, such as compliance, that can affect reserve performance.

A network of reserves was clearly needed to achieve the aims and objectives of protecting marine ecosystems. Lubchenco and colleagues (e.g., Lubchenco et al. 2003)

## A. Closed population

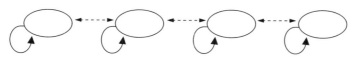

## B. Single source

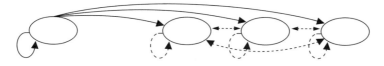

## C. Multiple source

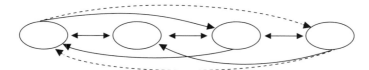

Larval pool

## D. Limited distance

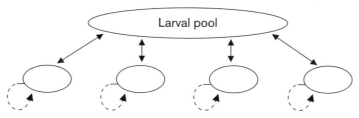

FIGURE 12.1

Schematic depiction of models of larval replenishment: (A) closed population, (B) single source, (C) multiple source, and (D) limited distance (see text
for details). Ellipses represent isolated adult populations. Solid lines indicate
high recruitment rates within and / or between isolated adult populations.
Broken lines indicate low recruitment rates.

SOURCE: From Carr and Reed (1993)

were not only strong advocates of MPAs but also did important underpinning work
on design, benefits, and implementation of reserves along the western United States
coast. Lubchenco et al. (2003) highlighted several features that came to dominate design
principles. These included the span of the network from end to end, the size, shape,
and number of individual reserve units, and their placement, which together determine
important features such as connectivity. To date, however, single isolated reserves still
comprise most of those globally (Lester et al. 2009).

An interesting contrast to most previous studies on marine reserve design was provided by Parnell et al. (2006), who argued that single reserves are important. They examined the issue of locating of a marine reserve and the requirements for size. They presented a case study of the giant kelp forest at Point Loma, San Diego (one of the longest-studied kelp beds in the world), in which they examined fine-scale habitats and the species associated with them, mobility of adults, and the physical forcing affecting the dynamics of the habitats. They argued there is an inherent "continuum of certainty" in marine reserve design, with habitat quality and distribution within a reserve being the most certain factors with respect to a broad range of species, and ecological links to nearby or remote habitat via dispersal being less certain. They argue for the importance of fine-scale approaches to defining habitats and their value for exploited species. Their argument is essentially that reserves must be large enough to protect a self-sustaining assemblage of exploited species, which they calculated to be around 42% of the kelp forest at Point Loma.

The diversity of approaches, research, and even arguments within the scientific community highlight the importance of the issues and the various perspectives on solving widespread marine problems. Layered on this, of course, is the spectrum of societal involvement, from support to opposition, of sequestering portions of the marine environment in perpetuity, and also the various governmental agencies that can be involved. Airame et al. (2003) discussed the marine reserve design at the Channel Islands, off southern California. This reserve had goals relating to ecosystem biodiversity, sustainable fisheries, economic viability, natural and cultural heritage, and education; 11 federal, state, and local agencies had some jurisdiction in the planning region. Key to working through issues was having a working group across agencies. Their analyses included ecological criteria but, as in many other studies, they found that data were often sparse in key areas and that a habitat-based approach could be effective in identifying reserves based on individual species' needs. The result of this process was that in 2003 the California Department of Fish and Game (= Wildlife) established a network of 10 no-take marine reserves and two conservation reserves with limited fishing allowed, covering around 12% of state-controlled waters within the Channel Islands National Marine Sanctuary. With expansion into federally controlled waters in 2007, the network covers 488 km² around five islands (Hamilton et al. 2010).

Klein et al. (2008) argued that a balance must be reached between conservation and socioeconomic viability in reserve design. They discussed "consumptive" and "nonconsumptive" interests of stakeholders, including divers, fishers, scientists, conservationists, and managers, in the design network for an area of central California. The scientific advisory committee recommended the inclusion of habitats across five depth zones from the intertidal zone to >200 m depth (CDFG 2005). The stakeholder group adopted some of the scientific recommendations and added areas of biodiversity and special significance to include in protected areas. They found that the expertise of stakeholders, such as in providing data on fishing grounds (although these are often not included in protected areas), was crucial to the design process, and concluded that their inclusive approach would aid in the robustness of MPAs. In a more general sense, Peterson et al. (2005) advocated the merits of argument, or at least an eagerness for debate, among stakeholders rather than the power of consensus to achieve conservation goals. However, anyone who has participated in multi-stakeholder meetings will know that, whether or not a "strategic consensus around a common vision" is reached, there is always what is often euphemistically called "lively debate."

Many developing countries present particular challenges in marine conservation, not least because of cultural values and the great degree of dependency of people on local production of marine resources. Castilla and colleagues in Chile have discussed the numerous issues involved, the benefits from even very localized protection of species and related functions, and the issues relating to enforcement (e.g., Castilla 1999, 2000, Fernández and Castilla 2005). This seems to be a case mostly of get what you can where you can.

## MPA EFFECTS ON FISH, FISHERIES, AND COMMUNITIES

Many studies highlight the benefits of MPAs to fish and fisheries (references in the reviews cited in the introduction to this chapter). In the great majority of fully protected, no-take reserves, the sizes and abundances of formerly exploited species increase and there are often increases in diversity and other community metrics (e.g., Gaines, White, et al. 2010, Gaines, Lester, et al. 2010). These responses are both intuitive and encouraging, showing that real responses and benefits can accrue in relatively short time periods. However, it is also known that responses among even nominally similar areas can be quite variable, and the flow-on effects to lower trophic levels are far from clear in most places. Furthermore, the occurrence and magnitude of spillover effects, where species in protected areas enhance adjacent areas through the provision of settlers or strays that wander outside reserve borders, are generally unclear. Within the context of giant kelp forests and MPAs, the evidence for effects comes mostly from California and Tasmania. It is these areas where there has been a conjunction of heavy exploitation of top predators, formation of urchin barren grounds, and established MPAs. A great part of the impetus for establishing reserves was not only the formation of urchins barren grounds, which at first were suggested to possibly result from the elimination of sea otters from hunting (North and Pearse 1970), but also the coincident overfishing of many fish stocks (e.g., Dayton et al. 1998, Tegner and Dayton 2000). The serial depletion of fisheries and large fish (e.g., fishing down the food web, Pauly et al. 1998; loss of ecosystem services, Worm et al. 2006) and the accumulation of detrimental effects (e.g., Dayton et al. 2000, Thrush and Dayton 2010) have heightened the awareness that something is seriously wrong and needs fixing.

There is a long history of fisheries in and around the giant kelp forests of California. Assessing the effects of protection from fishing, although a criterion of the benefits of

MPA status, can be fraught with difficulties. Fish assemblages are highly associated with habitat and specific seascape features, and therefore assessments inside and outside of reserves must account for significant variations in habitat features. Furthermore, until relatively recently there were not enough "replicate" reserves to compare within regional networks. Therefore, comparisons of MPA with non-MPA areas could be confounded by both varying habitats among reserve "treatments" and pseudo-replication at the treatment level. These caveats must be kept in mind in an evaluation of the evidence for MPA effects as they are crucial to understanding both the average responses of particular taxa but also their variation across the wider seascape. It has become fashionable to call such variation "context dependency" (e.g., Shears et al. 2008), but this type of site-to-site variation has been known since the early days of ecology (often called a site × treatment effect in experiments). Fortunately, there are now networks of reserves in many countries that will allow assessments of effects and variation, and therefore are starting to provide a much more robust understanding of how, where, and to what extent MPAs yield benefits to fish, fisheries, and the wider kelp forest ecosystem. Gaines, Lester, et al. (2010) called this the "evolving science of marine reserves," which will undoubtedly provide many insights into the effectiveness of reserve networks for meeting conservation, fishery, and commercial objectives in the years to come.

## FISHERIES EFFECTS

Fish assemblages in nearshore waters of California are known to be highly associated with habitat features at several spatial scales. For example, Anderson and Yoklavich (2007) found this to be the case at a scale of kilometers, at which habitat strata such as substratum type, depth, relief, and habitat patchiness were a strong predictor of fish assemblage composition. At meters to hundreds of meters, individual species were associated with specific habitat features, and at the scale of a meter or less species could show microhabitat specificity. Yoklavich et al. (2002) examined a deep-water (20–100 m) fish assemblage in an area protected from fishing for 4 years in central California using a submersible. They identified 70,000 fish from 82 taxa, 36 species of which were rockfishes (*Sebastes* spp.) that comprised 93% of fishes within the Big Creek Marine Ecological Reserve. They found there were four distinct assemblages, including associated fine sediments in deep water, bedrock with an uneven surface in deep water, sand waves, and shell hash in shallow water, and boulders and biogenic habitats on rock in shallow water. However, they found there were no differences in fish densities inside and outside the reserve and no consistent differences in fish size. They attributed this lack of significance to a combination of not enough time for the reserve to have had an effect on long-lived fish species, the fact that the somewhat remote reserve might not have been fished heavily anyway, and / or that the reserve might not have adequately protected some of the highly mobile species. Similarly, Schroeder and Love (2002) compared fished and unfished areas in southern California for the density and sizes of *Sebastes* spp., particularly with reference to the recreational fishery. One of their study areas, in Santa Monica Bay, had been closed to commercial fishing since the 1950s but was open to some forms of recreational fishing. Another site around an oil platform was essentially closed to all fishing, and a third site was open to all fishing. The density of all rockfish was greatest at the site open to all forms of fishing, although fishes were generally small and most abundant at the site where fishing did not occur. Cowcod (*Sebastes levis*) abundance was 32 times greater, and bocaccio (*S. paucispinis*) were 408 times greater, in the closed area than in the area where recreational fishing was allowed.

Hamilton et al. (2014) examined the dynamics of sheephead (*Semicossyphus pulcher*) populations around the giant kelp forests of San Nicolas Island, southern California, from the 1990s through a period including a decade of management intervention. They used gut contents and stable isotope analyses to show that there was a niche expansion after the period of recovery, related to a large shift in size structure (figure 12.2). They found that in the early 1990s a rapid expansion of an unregulated commercial live-fish fishery to supply the Asian market led to a decrease in sizes and in size at maturity. Fisheries management regulations in the late 1990s addressed this and commercial landings declined. They also found that large sheephead, which were mostly males, ate more sea urchins, whereas smaller fish ate more barnacles, bivalves, and shrimp. They concluded that these changes were due to changes in sizes and not in prey availability. It was interesting to note that the density of sea urchins, the main agent of maintaining urchin barrens putatively due to a reduction in top predators, did not vary significantly among the periods, and average densities were similar or higher in the 2000s than during the 1980s to early 1990s. The community effects detected in this study involved increases in some sessile invertebrates (mostly the sea cucumber *Pachythyone rubra* and tube-dwelling polychaetes), but not other macroinvertebrates, giant kelp, or foliose red algae. In their careful analyses, the authors note that the community effects were difficult to disentangle from environmental forcing, such as El Niño and the Pacific Decadal Oscillation, and the foraging effects of other urchin predators, especially sea otters, which were introduced to San Nicolas Island in the late 1980s.

The serial depletion of the abalone fishery over the course of a century was a major focal point of concern about the status of the coastal marine environment and the need for increased protection (Tegner 1993, Dayton et al. 2000, Dayton 2003). The very low abundances of abalone in kelp forests was due to a combination of overfishing and the return of sea otters to many areas of the coastline, particularly in central California (e.g., Dayton et al. 1998). Also, a combination of disease, illegal fishing, loss of habitat, failure in recruitment, and variable climatic conditions has contributed to the decline of abalone (Hobday et al. 2001, Tegner et al. 2001). There are now tight regulations on removing these iconic species and all fishing of abalone south of San Francisco has been banned since 1997 (Haaker et al. 2001, Rogers-Bennett et al. 2002; discussed in Chapter 11).

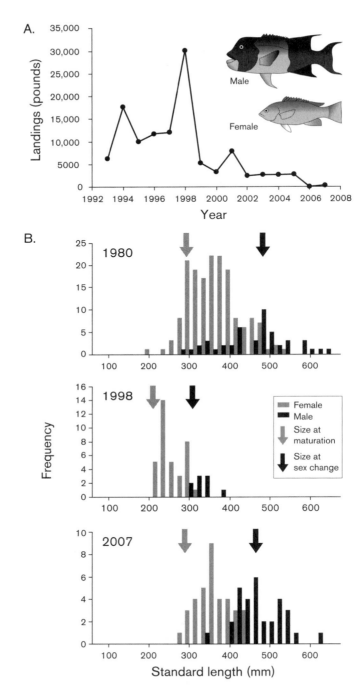

FIGURE 12.2

Exploitation and recovery of California sheephead, *S. pulcher*, at San Nicolas Island off southern California. (A) Commercial landings from 1993 to 2007 from the study area, illustrating rapid exploitation and a subsequent decline in landings following management intervention (catch restrictions). (B) Shifts in size structure of males and females, and changes in the timing of maturation and sex change of sheephead across three decades.

SOURCE: (A) Sheephead images by Larry G. Allen. (B) From Hamilton et al. (2014).

These impacts on somewhat iconic species associated with kelp forests were a major thread in the urge for protection of coastal species and habitats (e.g., Dayton et. al. 2000). How do they fare with the implementation of MPAs? The jury is still out on this, but the mounting evidence is encouraging for the revival of many species. Micheli et al. (2008) examined the persistence and possible recovery of red (*Haliotis rufescens*) and black (*H. cracherodii*) abalone around Monterey in central California, an area inhabited by otters. In the Hopkins Marine Life Refuge, they compared their surveys with those of John Pearse and colleagues from 1972 to 1981 (Lowry and Pearse 1973, Cooper et al. 1977, Hines and Pearse, 1982) and repeated in 1990 and 1992 by graduate students (unpublished theses). Neither the density nor the size structure of red abalone changed significantly over 32 years, although they were quite variable between surveys. Similarly, there were no difference in black abalone among years, but there was significant site-to-site variation. They concluded there was high mortality and that there was no return to a level that could support a fishery because of the presence of sea otters. Rogers-Bennett et al. (2002) provided mostly site-by-site descriptions and it was not possible to derive anything definitive about the role of MPAs in abalone densities or sizes. Current evidence from other areas of the world provides some support for the benefits to abalone in reserves where there are no voracious predators like sea otters. For example, a reserve at Kaipiti Island in central New Zealand has seen dramatic increases in number and sizes of the native abalone *H. iris* since 1992 when the reserve was established (C. Battershill, pers. comm.).

Sea urchins should be a group that responds to MPA status, as the trophic effects involving them were a major impetus in establishing MPAs. However, the mechanisms affecting their abundances and sizes can work in different directions and over different time scales. A release from fishing pressure (where it exists) should potentially lead to an increase in both numbers and sizes, but indirect effects from potentially increasing numbers and sizes of predatory fishes and lobsters could reduce urchin numbers and sizes. Shears et al. (2012) used abundance and size data for two species of sea urchins at the Channel Islands, off southern California, to assess this problem. They found that the unfished purple urchin, *Strongylocentrotus purpuratus*, had consistently low densities since 1982 within the reserve sites clustered at the end of Anacapa Island and that there were variable but generally higher densities at the numerous non-reserve sites across the oceanographic regions of the Channel Islands. The fished species, *S. franciscanus*, displayed similar patterns of abundances between the reserve and non-reserve sites, but within the reserve the urchins were much bigger and therefore had much greater reproductive output. They argued that the low densities of *S. purpuratus* at the reserve sites were "consistent with the top-down density-mediated model," whereas the larger red urchins within the reserve sites were from cessation of fishing pressure. They acknowledged, however, that the overall level of inference from their study was limited because of the limited number of reserve sites, which were near each other in one area (see comments in previous chapter on lobster effects), and that their study demonstrated the difficulties in generalizing about the effects of MPAs on sea urchin densities. This is particularly so when there is a paucity of quality field- and lab-based experiments to test key hypotheses, and a reliance mainly on correlative data across highly variable sites and oceanographic regimes.

Lobsters appear to respond well to reserves, despite the potential for increased predation on them if predatory fishes increase in sizes and numbers. Kay et al. (2012) tested hypotheses relating to abundance, trap yield, and spillover effects of the lobster *Panulirus interruptus* at and around three marine reserve sites with giant kelp at the Channel Islands, off southern California, over 6 years. Visual underwater surveys showed that there was a fourfold to sixfold increase in lobster abundances within reserves 6 years after reserve status, relative to abundance before the reserve was established and at sites outside reserves (figure 12.3). These results mirrored those from setting lobster traps in the same areas; traps set within reserve sites had a fourfold to eightfold increase in trap yields. No "spillover" effects were found, with numbers at adjacent sites and those several kilometers away being far fewer than those within the reserve sites. As would be expected, they also found that lobsters within reserve sites had morality rates similar to estimates of natural mortality, whereas those outside reserves were higher and reflected mortality due to fishing. Although Kay et al. (2012) found significant variation over time among locations, there was clearly a positive effect of reserve status on lobster numbers.

The story is also complicated for multispecies assemblages. Guenther et al. (2012) examined sea urchin and macroalgal abundance data with reference to fishery data for landings of the lobster *P. interruptus* using 8 years of catch data from eight rocky subtidal reefs with giant kelp forests near Santa Barbara, southern California. All sites were subjected to fishing. They found that sea urchins were abundant at all sites but there was no relationship between lobster fishing landings and sea urchin abundances overall. They also found that macroalgal biomass varied independently of urchin density across sites and years, and that there was no relationship between lobster fishing and macroalgal biomass. They concluded there was "little evidence that intensity of lobster fishing, as measured by fishing effort, induced a trophic cascade leading to low macroalgal biomass," a similar conclusion to that of Foster and Schiel (2010).

Further afield in Australia, Edgar and Barrett (1999) examined the responses of a suite of species to four no-take reserves, using associated reference sites for comparison on the eastern and southeastern coasts of Tasmania. They used multiple sites within reserves and reference areas, enabling location-by-location analyses. They found that at the largest reserve (Maria Island), the number of fish species increased by 5% within the reserve compared to −23% in the reference area. The number of mobile invertebrates increased by 25% and algal richness by 11% compared to drops of 7–5% outside the reserve. None of these taxon groups changed in the 5 years of reserve status at the two smaller reserves. Some of the biggest changes at Maria Island were in rock lobsters,

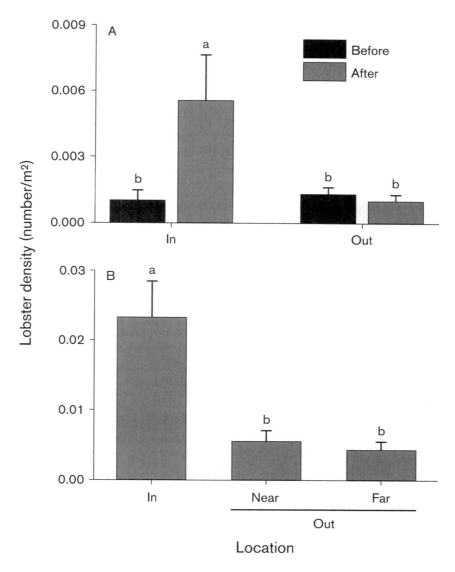

FIGURE 12.3

The results of visual underwater surveys (mean + SE) of the abundances of the California spiny lobster *P. interruptus* within reserve (n = 3) and non-reserve sites at the Channel Islands, southern California. (A) The numbers per m² "Before" = 1997–2003 and "After" = 2007–2010 reserve status. This includes all lobsters observed. (B) Abundances "In" = inside reserve sites, "Near" = adjacent to reserve sites, and "Far" = 2–6 km outside of reserve sites in "after" period of 2008; these surveys included only legal-sized lobsters (>82.5 mm carapace length). Letters represent results of Tukey's post hoc test (designation "a" significantly greater than those with "b," which are similar to each other).

from an average of 0.8 per 50 m transect in 1992 to 2.8 per transect in 1997, compared to an increase from 1.3 to 1.5 per transect outside the reserve. The number of abalone inside and outside reserves did not change, but there were more large ones within the reserve. These studies were extended to 10 years, using a BACI design (before–after control impact). Barrett et al. (2009) examined four reserves and their references sites, analyzing them separately and collectively. They found increases in the destructive urchin *Centrostephanus rodgersii* after 7 years outside of reserves, driven mostly by large recruitment in two reference sites. Another sea urchin, *Heliocidaris erythrogramma*, declined by around 30% after 7 years inside the reserves but not in the fished sites. The abundances of the abalone *H. rubra* fluctuated greatly in both reserve and reference sites over the years, and declined significantly at Maria Island. There was a decline in both cryptic juveniles and intermediate-sized (sub-legal) abalone, which was not seen at the reference site. At another site (Tinderbox), abalone increased at fished sites but not in the reserve. Similarly, the abundances of lobsters, *Jasus edwardsii*, increased greatly in the Maria Island reserve, almost entirely due to the increase in numbers of large ("legal" sized) individuals. They estimated that total biomass of lobsters within the reserve increased by an order of magnitude over the 10 years, but remained relatively stable in the reference area. Similarly at Tinderbox, lobster numbers increased twofold and biomass ninefold relative to the reference site. Other metrics, like invertebrate and algal taxa richness, did not vary greatly between reserves and reference sites. Community analyses showed there was no obvious divergence between reserve and reference sites. They concluded that one implication of their study is the complex nature of trophic changes because in some habitats, abalone may not increase in MPAs when predators are also protected and increasing in abundance. If the goal, therefore, is to conserve abalone, then other protection measures may be needed.

Babcock et al. (2010) reviewed decadal trends of direct and indirect effects of MPAs in several countries, including those with giant kelp forests. They found that direct effects on target species take an average of around 5 years to be manifested and indirect effects on other taxa on average of 13 years. In 78% of the cases examined, they found that populations of exploited species increased in reserves. These included lobsters and predatory fish, although almost all effects had quite large variation. In half of the cases they examined, indirect effects occurred that involved increases in basal groups such as macroalgae. They noted that not all effects were due to reserve status, as there was evidence of considerable variation in recruitment among key species, locations, and years.

As discussed briefly above for lobsters in California, the issue of spillover effects from reserves into nearby areas is hard to tie down. It makes intuitive sense that if populations of targeted species increase in reserves, then some of these might find their way outside of reserves and also that the increased reproductive output of the larger individuals accumulating in reserves would contribute to areas outside. There are some documented cases of spillover effects, but many effects, particularly involving beneficial recruitment effects of key species within reserves spilling over to outside areas, remain elusive so far. In one tropical example, Abesamis and Russ (2010) found evidence for density-dependent export of a planktivorous fish after around 15 years in a coastal marine reserve in the Philippines. The effect on numbers and sizes of fish extended, however, to no more than 0.5 km from the reserve. They and others have highlighted how technically and logistically difficult it is to demonstrate export from reserves to nearby areas (e.g., Hilborn et al. 2004, Sale et al. 2005).

## GAPS AND THE FUTURE OF MPAS

In reading through the voluminous literature on MPAs, it is clear that their benefits are substantial and yet many are unrealized, at least to this point. There is no doubt there has been a fair degree of uncritical advocacy involved in the debate about MPAs, their beneficial effects, and their utility as management tools (cf. Sale et al. 2005). There are several thoughtful perspectives on gaps in our understanding and how future MPAs might be developed. Steve Gaines and colleagues refer to this evolving science of marine reserves with respect to new developments and emerging research frontiers (Gaines, Lester, et al. 2010). In a review of progress to date, they note that the protocols for analyzing responses for multiple interacting reserves are not well developed. They advocate that a framework is needed for implementing adaptive management so that optimal networks can be developed that collectively maximize a range of objectives, including conservation and viable fisheries outside the reserves. They view it as essential to have an integration of marine reserve and fisheries science, an argument also made by Gell and Roberts (2003). Reserves are most likely to be effective in their multiple objectives if they are part of an ecosystems-based management approach that includes consideration of areas outside reserves. Given the multiple users and uses of coastal resources, this may require some form of bioeconomic and socioeconomic modeling. Sale et al. (2005) identified gaps in the ecological science of no-take reserves. These include understanding the distance and direction that larvae disperse, knowing more about the movement patterns of juvenile and adult fishery species, knowing more about indirect effects such as those relating to fisheries extractions, a better knowledge of water masses along coastlines, and better designed reserves that can rigorously test effects on nearby areas where fisheries and other activities are allowed. Of course, this can read as a compendium of work across the marine science domain but, nevertheless, illustrates the great unknowns in such highly variable systems.

We end this chapter noting that MPAs are powerful tools with enormous societal resonance, but they are not a panacea to coastal problems. Schiel (2013) noted that even in New Zealand, which has an advanced and largely sustainable commercial fisheries adaptively managed through a transferable quota system, and a long-standing and advanced marine reserve system, at least 93% of coastal waters have no formal protection. With increasing anthropogenic stressors along the coastline, this implies either that MPA networks will need to be expanded considerably or that other management tools will be required. One of the biggest threats to coastal ecosystems worldwide is the terrestrial runoff and sediments, which are essentially land-use and catchment-based issues. As pointed out relatively early in the MPA debate (e.g., Allison et al. 1998) and highlighted in recent reviews (Mora and Sale 2011), no reserve is likely to be effective if wide-ranging and diffuse impacts are affecting whole coastlines. This may well have happened in some heavily urbanized areas of California, the legacies of which are still being felt (Foster and Schiel 2010). Furthermore, the problems are often too great and expansive to be solved by MPAs without some other forms of management.

# PART IV

# GLOBAL CHANGE AND THE FUTURE

# 13

# GLOBAL CHANGE

Warming of the climate system is unequivocal, and since the 1950s, many of
the observed changes are unprecedented over decades to millennia.
The atmosphere and ocean have warmed, the amounts of snow and
ice have diminished, sea level has risen, and the concentrations of
greenhouse gases have increased.

—IPCC (2013)

Whereas forecasts of changes in species' geographic ranges typically predict
severe declines . . ., paleoecological studies suggest resilience to past climatic
warming . . . Superficially, it seems that either forecasts of future response
are overestimating impacts . . . or that history is somehow an unreliable
guide to the future.

—Moritz and Agudo (2013)

The preponderance of evidence is that the earth's climate is changing at an unprecedented rate, and many predictions are dire about the consequent changes on ecosystems. In regard to the ocean, the major physical variables of change are likely to be temperature, wave forces, sea level, and pH from increasing atmospheric $CO_2$ (Brierley and Kingsford 2009, IPCC 2013). Climate change must be viewed, however, within the context of numerous other stressors from increasing human populations, urbanization, intensified land use, coastal runoff of sediments, nutrients, and contaminants, species extractions, and the spread of nonindigenous species (e.g., Schiel 2009, Smale et al. 2013; see Chapter 10). These in combination may have numerous nonlinear and perhaps unpredictable effects on species distributions, community structure, disease vectoring, food and interaction webs, and ecosystem functions and services (Mooney et al. 2009, Harley et al. 2012, Smale et al. 2013). Here we use "global change" as a more inclusive term to indicate climate change within the broader context of other stressors and impacts.

It seems clear there are no "free lunches" with respect to the environment, and that almost all activities have flow-on effects to other parts of the broader ecosystem. Given this connectivity and interaction among ecosystem components, isolating the effects of one of even a few variables acting on kelp forest dynamics is challenging. We know that impacts have already occurred and that we are not proceeding from a baseline of pristine conditions (cf. Dayton et al. 1998). Nevertheless, the impacts of extreme physical conditions, especially at the limits of kelp distributions, give clues about changes that might occur in giant kelp forests over the coming decades. As long-term studies continue in more places, we will get a sharper view of changes, their environmental correlates, and causes. Furthermore, a wide range of laboratory-based studies is beginning to show the sensitivities of many key species, for example to changes in water temperature and pH levels, although it is currently far from clear how these might translate into natural communities. The scenarios for species responses are usually designated as tolerate, move, adapt, or die (e.g., Harley et al. 2012). If this book and similar analyses stand the test of time, future readers will be able to see how accurately present-day scientists predicted effects and changes, or if we were alarmist or wide of the mark in how extreme the changes turned out to be. Certainly, there has been an exponential increase in the literature on climate change over the past few years, and this seems likely to continue as science further engages in trying to understand the mechanisms, processes, and consequences of global change. Here we present an overview of climate changes and discuss potential effects on giant kelp and associated organisms.

## CLIMATE CHANGE VARIABLES AND PREDICTED TRENDS

The Intergovernmental Panel on Climate Change (IPCC) has gone through periodic updates of climate model refinements and predictions. The bad news, with a high degree of certainty, is that a wide range of climate-related variables are undergoing large and rapid change. The IPCC reports that the averaged globally combined land and ocean surface temperature has increased by 0.78°C during the period 2003–2012 compared to 1850–1900 (figure 13.1). There has been substantial interannual and inter-decadal variability due, for example, to strong ENSO and Pacific Decadal Oscillation periods. The IPCC (2013) reports with "high confidence" that the ocean accounts for >90% of the energy accumulated between 1971 and 2010, with the upper 75 m having warmed by 0.11°C per decade, and with 60% of this energy stored in the upper ocean (0–700 m depth). In conjunction with soot fallout and loss of ice albedo (Hansen and Nazarenko 2004), this has led to melting sea ice, thermal expansion, and increases in mean sea level by around 0.19 m over the past century (figure 13.2). Carbon dioxide concentrations have risen by around 40% since pre-industrial times, and the ocean has absorbed around 30% of this, which has caused increased acidification (IPCC 2013; figure 13.3). The rise in $CO_2$ concentration in the atmosphere is the greatest in several hundred-thousand years (Hoegh-Guldberg et al. 2007) and the adsorption of $CO_2$ has led to a decrease of around 0.2 pH units per decade over the past 30 years (Hoegh-Guldberg and Bruno 2010). There is also accumulating evidence that changes to the earth's wind fields are leading to increased wave action in many parts of the world's oceans (Young

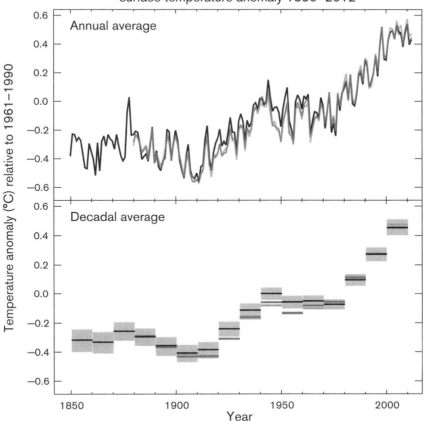

Observed globally averaged combined land and ocean
surface temperature anomaly 1850–2012

FIGURE 13.1

Observed data of global mean temperatures (combined land and ocean surface temperature anomalies, relative to the mean from 1901 to 2012) from 1850 to 2012. Top panel: annual means; bottom panel: decadal means, including error estimates (shown in shaded blocks). Different shaded lines represent different data sets.

SOURCE: From figure SPM.1 in IPCC (2013).

et al. 2011), which potentially will have direct impacts on kelp forests. The most recent climate models show all of these trends continuing through this century. The IPCC climate models for future scenarios are based on Representative Concentration Pathways (RCPs) carried out under the World Climate Research Programme. Based on these models, the most likely projections are for continued increases in surface temperature, loss of sea ice, increases in sea level, and increasing acidification in the ocean surface. Furthermore, IPCC (2013) concludes that most aspects of climate change are likely to last for many centuries.

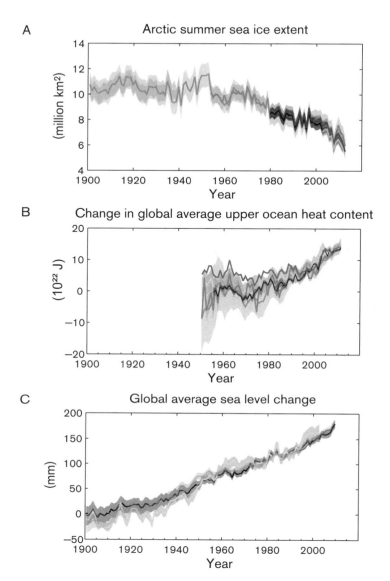

FIGURE 13.2

Other observed indicators of changes in the global climate (annual averages
of different data sets in darker lines and error estimates in shaded areas).
(A) Arctic summer sea ice extent; (B) global mean heat content of the upper
ocean (0–700 m) (aligned to 2006–2010, relative to mean for 1970); (C)
global mean sea level.

SOURCE: From figure SPM.3 in IPCC (2013).

A

## Atmospheric CO₂

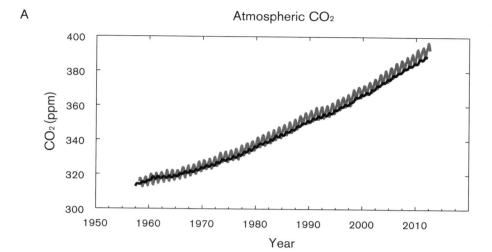

B

## Surface ocean CO₂ and pH

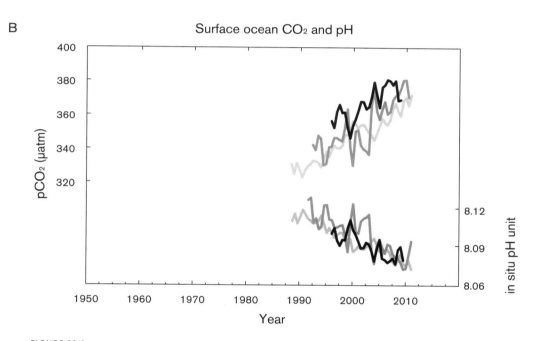

FIGURE 13.3

(A) Atmospheric $CO_2$ levels since 1958 from Mauna Loa (gray line) and the South Pole (black line).
(B) Ocean surface $CO_2$ (as partial pressure of $CO_2$, top lines) and pH (a measure of sea water acidity, bottom lines); lines indicate different data sets from the Atlantic and Pacific Oceans.
SOURCE: From figure SPM.4 in IPCC (2013).

Taken together, these changes and their interactions provide a dramatically altered environmental envelope in which ocean ecosystems are likely to operate. The potential changes to kelp forest ecosystems are far less certain and largely surmised, due not only to the, as yet, unpredicted and complex ways that climate variables may act on species and their interactions, but also in how they may act in a more direct way on giant kelp itself with its complex requirements and physiological–ecological responses across its life stages (see Chapters 2–4). Helmuth et al. (2006) raised several questions that must be addressed in forecasting the impact of climate change on the distribution and abundances of organisms. Among these are, how close are individuals living to the limits of their physiological tolerances within their range? Essentially, are they at the limit of their "fundamental niche"? Other questions involve knowing where distributions are determined by biotic interactions that may be affected by climate indirectly, whereby boundaries and limits determine the "realized niche" space, understanding whether species can adapt and evolve, and how patterns of fertilization success and propagule dispersal are affected. Because of the complexities involved in answering these questions, it is often the case that community changes do not necessarily correlate well with any single factor. With reference to predictions of future responses, Harley et al. (2012) termed these "gaps in the ecophysiological framework."

These complexities are exemplified by a reconstruction of giant kelp distribution and biomass in southern California since the last glacial maximum, around 20,000 years ago. Graham et al. (2010) made the assumption, which they recognized as untestable, that the responses of *Macrocystis* to environmental forcing have stayed relatively constant over this long time period. Their model showed that the area cover of rocky reef and the biomass of giant kelp peaked around 13,500 years ago, but then declined by 40–70% to present levels. Peaks in kelp abundance were the result of increases in habitable forest area and a transition to more productive ocean conditions. Production was surmised from the known relationship between temperature and nutrients in that region. A transition to a sand-dominated system in some areas around 4000–6000 years ago was associated with the demise in kelp distribution and abundance. This reconstruction is based on known, present-day responses of kelp, but, of course, these constitute only one of a suite of potential responses of giant kelp to environmental forcing. How instructive this might be for the rapid environmental changes that are occurring now cannot be determined at this stage.

The complexity of processes relating to global change encompasses most of the topics of this book. These span broad spatial changes in ocean conditions to local-scale effects within coastlines and kelp forests, temporal changes over decades to short-term effects within seasons, and potentially cumulative effects that compound impacts (Pandolfi et al. 2011, Harley et al. 2012, Maxwell et al. 2013). Here, however, we discuss potential changes to giant kelp forests under the rubric of climate-related stressors, and refer back to relevant chapters for more thorough treatment of the effects of these factors on giant kelp.

Brierley and Kingsford (2009) argued that temperature has the most pervasive climate-related influence on biological function. It seems clear that temperature is a major factor in die-off events of major habitat formers such as corals (e.g., McClanahan et al. 2004, Donner et al. 2005) and kelp (Wernberg et al. 2010), although there are potentially numerous contributing factors such as increases in solar radiation and decreases in nutrients (Hoegh-Guldberg et al. 2007). An increase in extreme thermal events (e.g., Denny et al. 2009) contributes to the frequency of these die-offs. For kelps, diebacks have tended to occur at the limits of biogeographic distributions where physical tolerances may be exceeded (e.g., Wernberg et al. 2010). For example, Helmuth et al. (2006) reported that the poleward contraction of a cold-water kelp, *Alaria esculenta*, from its southern limit in Britain has reflected a poleward shift in the summer SST isotherm of 16°C, which can be fatal to mature plants of this species (Widdowson 1970). The occasional change in the southern, warm-water limit of *Macrocystis* has been well described along the west coast of North America. It recedes from its southern limit off Baja California, Mexico, during El Niño years, most likely because of a combination of high temperature, low nutrients, and extreme storm events (see Chapter 5, Edwards and Estes 2006). Similarly, canopy dieback during summer in southern California is correlated with both higher seawater temperatures and nutrient depletion (Jackson 1977, 1987) but may also be related to natural frond senescence (Rodriguez et al. 2013). Because numerous factors act in concert in natural situations, there can be some lability of kelp responses even during extreme events, whereby some local populations of giant kelp survive (e.g., Foster and Schiel 1993, Ladah and Zertuche-González 2004). These refugia may act as source populations for recovery.

One of the longest studies involving giant kelp in which temperature was manipulated involved a small cove (18 m maximum depth, 2 km shoreline, 15 ha area) in central California that was heated by an average of 3.5°C above ambient seawater temperatures for 10 years (Schiel et al. 2004, Steinbeck et al. 2005). The heating was the result of a once-through cooling unit (see Chapter 10) of the Diablo Canyon nuclear power plant, which draws seawater from outside the cove, uses it in a cooling system to recondense freshwater steam, and discharges it into the cove at a maximum rate of $9.5 \times 10^9$ L day$^{-1}$, creating a thermal plume that spreads throughout the cove with the warmest water near the sea surface. As this cove was near the biogeographic boundary of Point Conception between central and southern California, it was hypothesized that cold-water species might decline and warm-water species might dominate, as predicted by other research (Sagarin et al. 1999). Although 150 of the 172 subtidal species recorded changed significantly in abundance with elevated temperatures relative to controls, there was no consistent pattern of change in abundance related to geographic affinities. Most species were cosmopolitan, with broad geographic distributions, and most of these underwent significant changes in abundance. Subtidal understory kelps, especially *Pterygophora californica* and *Laminaria setchelli*, declined by 82%, from an average of 7 m⁻² to 1 m⁻² (figure 13.4). The cold-water bull kelp *Nereocystis luetkeana*, which had dominated the surface canopy in the cove, was almost entirely replaced by *Macrocystis*. This study showed that a rise of around 3.5°C in temperature can trigger wide-ranging changes in communities, but also that changes involve complex biotic interactions. Although control areas underwent large changes over 20 years, they were remarkably similar to initial conditions by the end of the study. The changes to thermally affected areas, however, were great. Not only was there replacement of the large kelps, but the understory kelps also had no major recruitment over the 10 years of temperature increase; sea urchins increased for 3 years and then decreased as foliose understory (mostly red) algae came into dominance. These fluctuating changes in understory algae and invertebrates in the thermally elevated cove have continued since Schiel et al. (2004) completed their study (J. Steinbeck, pers. comm.). It is interesting to note that although the once-dominant kelp *Nereocystis* recruited annually into the cove, the plants did not grow to reach the canopy, presumably because of thermal stress in the water column. Although it is unwarranted to ascribe a single cause to these persistent changes, it nevertheless is instructive that the major factor that initiated change was the elevation in temperature within the cove. Furthermore, because this was set up as a before–after control-impact study, there was thorough sampling for 10 years in both the cove and outside control areas before the heating treatment affected the cove, thereby providing robust statistical analyses of changes, rather than just correlations between disparate time periods.

One other study is particularly instructive for discussion of temperature effects on giant kelp communities. The studies of Craig Johnson and colleagues in Tasmania, discussed in a previous chapter (see Chapter 8), presented clear mechanistic ways in which environmental effects on a major herbivore contributed to the significant demise of kelp forests. In this case, the onshore movement of a warm water current brought larvae of the sea urchin *Centrostephanus rodgersii* to the coast of Tasmania. Because urchin larvae had suitable temperatures and development time in this current, they were able to survive the transport from the Australia mainland and establish large populations along coastal Tasmania (Johnson et al. 2011). Their arrival and then population expansion coincided with both a long-term decline in giant kelp and a large increase in fishing of their major predator, the lobster *Jasus edwardsii*, especially of the large individuals that are capable of eating mature urchins (Ling et al. 2009). It is unlikely anyone could have predicted these types of extensive changes to giant kelp communities because they were the result of complex interactions between kelp dynamics, the behavior of a current, larval life history, and transport of a key grazer that had not previously been important in Tasmanian waters, buildup of urchin populations over time resulting in intense grazing effects, and size-selective overfishing of a key predator. As in the example of the warming of a cove above, temperature played a key role in triggering ecological changes, but the resultant cascades through the kelp communities were dictated by a wide range of other factors.

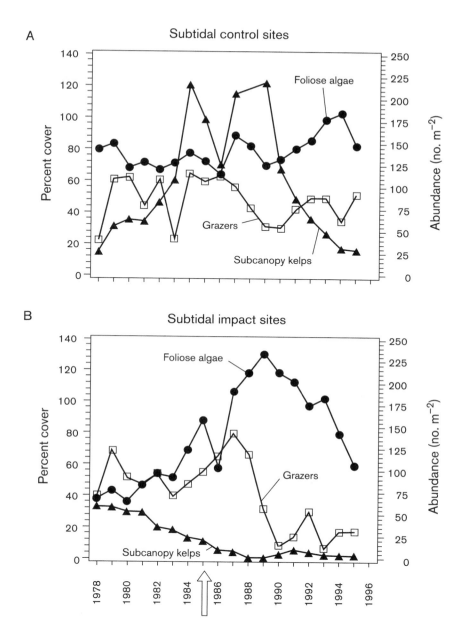

FIGURE 13.4

Changes in cover of kelps, foliose algae, and invertebrate grazers at Diablo Cove (central California) at sites under the influence of heating from the discharge of heated water of a coastal power plant, and of nearby control sites under ambient conditions. Annual mean percent cover of foliose algae (left axes) and numbers of subcanopy kelps and grazers (right axes) at (A) subtidal control sites and (B) subtidal impact sites. Note that percent cover can total >100% because of layering of algae; the sample area was 7 m². Arrow indicates start of thermal increase of seawater temperature by an average of 3.5°C, which lasted from 1985 onward. SOURCE: From Schiel et al. (2004).

It is clear that even with the potentially quick recovery of kelp after disturbances, short-term changes can have long-term consequences. As has been argued for intertidal assemblages (Schiel 2011), these may be due to sporadic recruitment, changes to benthic substrata such as occupation by corallines and foliose algae, alteration of canopy structure and therefore of the light environment to benthic areas but also, as seen in subtidal kelp assemblages, the continued presence of stressors such as sediments (Connell 2005). Collectively, these impacts may alter the resilience of kelp communities and their competitive and facilitative interactions, as well as make them more vulnerable to a changing climate (Wernberg et al. 2010).

There is also evidence that some kelp-dominated areas are becoming "tropicalized." Verges et al. (2014) showed that regions with continuous tropical-to-temperate coastlines influenced by Western Boundary Currents, such as Japan, the eastern United States, eastern Australia, and southeastern South Africa, are vulnerable to kelp declines and replacement by barren communities or even corals. They concluded that these phase shifts are largely due to warm-water intrusion and range-shifting tropical herbivorous fishes that severely graze temperate macrophytes, including kelps. These types of effects have not yet been seen in giant kelp forests, but this study offers instructive scenarios on how temperature-mediated complex interactions can drastically alter warm-temperate communities.

## NUTRIENTS AND WAVE FORCES

A fundamental relationship in the coastal zone around giant kelp forests is the negative correlation between seawater temperature and nutrients (see Chapter 3). This is particularly well described along southern California where seasonal upwelling brings a rapid drop in seawater temperature and a corresponding increase in upwelled nitrogen (Jackson 1977, McPhee-Shaw et al. 2007, Lucas et al. 2011). The combination of these two factors can have a great effect on giant kelp, and it is the frequency and magnitude of high SST and low nutrients with climate change that are likely to produce the greatest effects, particularly if SSTs get beyond around 26°C (Hernández-Carmona et al. 2001).

As has been shown in strong El Niño periods, *Macrocystis* is vulnerable to extended periods of low nutrients because of its limited nutrient storage capacity of around 2 weeks (Gerard 1982b). As described in Chapter 3, water column stratification can lead to warm, nutrient-depleted surface water (Lucas et al. 2011) and deterioration of plants (Gerard 1982a). The reversal of these conditions during upwelling episodes with low SST and great pulses of nitrogen sweeping through kelp forests is a principal driver of giant kelp forest growth and the persistence of giant kelp at the equatorward limits of its range in Mexico (Hernández-Carmona et al. 2001) and Chile (review in Vásquez et al. 2006). Any alteration of the periodicity, intensity, or chemical characteristics of upwelling may therefore interact with larger-scale increases in temperature and decreases in nutrients to further truncate the biogeographic limits of *Macrocystis* under future climate change scenarios, especially as they impinge on the physiological tolerances of giant kelp.

Extensions of the biogeographic distribution of *Macrocystis* in the opposite direction toward the poles could conceivably occur in the northern hemisphere. This would depend not only on the extent of warmer SSTs but also to a considerable extent on the ability of giant kelp to compete with a diverse range of resident northern kelps in the lower light and truncated seasons of the north (Chapter 3). The southern hemisphere poleward distribution of *Macrocystis* is unlikely to change in future scenarios. The current southern distribution is set by the absence of shallow rocky reefs south of the subantarctic islands (see Chapter 5), and considerable warming of the Antarctic continent as well as considerable dispersal would need to occur for giant kelp to occupy it.

Exposure to waves has large effects on the local-scale distribution and dynamics of *Macrocystis* (see Chapter 3). In very exposed situations, such as the subantarctic islands and many outer coast sites, *Macrocystis* is limited to the lee side or protected inlets. Reed et al. (2011) showed that winter storms remove most of the canopy of giant kelp annually in central California, essentially overwhelming many other processes affecting primary production in forests. El Niño storms may provide the best mimics of a changing wave climate, whereby increasing temperatures, decreasing nutrients, and increased storm waves combine to remove large tracts of giant kelp forests, with recovery taking up to several years (Edwards 2004). The wind fields and wave climate are changing across the globe. Young et al. (2011) showed that there has been a significant increase in wave height, particularly of the largest waves (95th and 99th percentiles), since 1985. If this trend persists, then exposed outer coasts will continue to be challenging environments for developing kelp forests, regardless of what other changes occur in temperature increases and nutrient provision.

## OCEAN ACIDIFICATION (OA)

Acidification of the world's oceans from increasing amounts of atmospheric $CO_2$ is one of the more insidious problems of climate change. The past decade or so has seen a growing literature on the effects of OA on marine organisms and an increasing awareness of potential effects on species and ecosystems (reviews in The Royal Society 2005, Doney et al. 2009, Hurd et al. 2009, Hofmann et al. 2010, Harley et al. 2012, Dupont and Pörtner 2013). Doney et al. (2009) pointed out that OA is not prone to the uncertainties of climate change forecasts but is a predictable consequence of rising atmospheric $CO_2$. They state that "absorption of anthropogenic $CO_2$, reduced pH, and lower calcium carbonate ($CaCO_3$) saturation in surface waters, where the bulk of oceanic production occurs, are well verified from models, hydrographic surveys, and time series data."

The processes affected by acidification include calcification of plankton, macroalgae and invertebrates, carbon and nutrient assimilation, primary production, and acid–base balance in the oceans (Blackford 2010). Much of the literature relating to algal beds is derived from laboratory-based studies or those around thermal vents involving natural gradients in pH levels. There are myriad problems, however, in translating these types of studies into the far more complex processes in natural ecosystems. Calcification responses in nature are complicated, for example, by interactions with changing temperatures or nutrients. Doney et al. (2009) also identify that most studies relating to acidification have been short term, usually no more than a few weeks, and that chronic exposure to increased $CO_2$ could have complex effects on growth and reproduction. Other problems are technological and methodological in working out the best ways to build lab systems that can realistically alter pH levels without introducing artifacts in seawater conditions. Hurd et al. (2009) pointed out that elucidating acidification effects on algae is not simple because when pH is altered, so too is the carbon speciation in seawater, which can affect photosynthesis and calcification rates. Other problems arise because pH levels in seawater are not constant but change in response to photosynthesis and respiration, which release $CO_2$ and lowers pH (Hurd et al. 2009). Hofmann et al. (2010) argued that the core issues involve understanding the extent to which organisms can tolerate future acidification and the acclimatization capacity of populations, neither one of which is well known. Because of the extensive potential effects of acidification on calcifying organisms throughout marine ecosystems, and the roles of these organisms in primary production, benthic cover, competitive effects, water chemistry, and food web dynamics, the topic of OA is vast. We confine ourselves here largely to consideration of potential effects in giant kelp forests.

EFFECTS OF OA ON ALGAE

Studies on giant kelp have so far shown there will be few, if any, direct negative effects of acidification. Roleda, Morris, et al. (2012) found in a lab experiment that a lowered seawater pH of 7.60 (but achieved with the addition of HCl) led to a 6–9% reduction in meiospore germination, but increased dissolved inorganic carbon had the opposite effect. They emphasized the need for appropriate manipulation of seawater carbonate chemistry when testing OA on photosynthetic organisms. They also found that gametophytes were slightly larger under conditions of lowered pH. They concluded that metabolically active cells may compensate for seawater acidification. Also in southern New Zealand, Hepburn et al. (2011) used isotopic analysis of $\delta^{13}C$ across varying pH levels at three depth strata to 12 m to separate noncalcifying macroalgae into functional groups. They combined these field studies with "pH drift experiments" in the lab, whereby changes in pH are measured during incubations. They found that all canopy-forming algae, including *Macrocystis*, appeared to have active uptake of inorganic carbon, but this was affected by low light. Noncalcifying red algae relied on diffusive uptake of

$CO_2$ and were more common in low-light habitats. They concluded that increased $CO_2$ would negatively affect only coralline species. Similarly, Harley et al. (2012) pointed out that some red and most brown and green algae use bicarbonate ($HCO_3^-$) by converting it intracellularly with $CO_2$ concentrating mechanisms. They concluded that even with variation in carbon use strategies across all taxa, noncalcifying seaweeds will probably have a positive response to increased $CO_2$. The responses of understory corallines are not straightforward. For example, Kamenos et al. (2013) found that *Lithothamnion glaciale* increased its calcification rate at low pH (7.7) during the daytime, apparently compensating for OA-induced dissolution at night, but this was not supported by a change in photosynthesis. When pH was changed rapidly, there were changes in the calcite skeleton making the fronds structurally weak, which may compromise their competitive abilities and the communities they support.

At this stage, there are reasonably consistent conclusions about how macroalgae are likely to respond to increasing levels of $CO_2$. Calcifying organisms generally exhibit larger negative effects than noncalcifying organisms, although with some exceptions (Kroeker et al. 2010). As in other summaries, this could change the competitive dynamics of benthic calcifying and noncalcifying seaweeds (Kroeker et al. 2013). The result of this type of modification to assemblage structure (although not involving giant kelp) was shown along a series of volcanic vents, which acidified the seawater to a pH gradient ranging from 8.20 to 6.07. Porzio et al. (2011) found that the vast majority of 100 macroalgal species occurred at almost all the sites along this gradient but their abundances varied; calcifying turf species had disproportionate decreases in coverage down to pH 7.8. At the extreme pH level of 6.7, however, where carbonate saturation levels were <1, calcareous species were absent and there was a 72% decline in species richness compared to the less acidified sites. Because these were resident communities, and volcanic bents can persist for millennia, Porzio et al. highlighted the worth of further studies on tolerant species, which may have undergone genetic changes and adaptations to a high $CO_2$ environment.

## EFFECTS OF CLIMATE CHANGE ON NON-ALGAL SPECIES

A review of the climate change literature highlights both the potential for massive changes in algal-dominated systems through direct effects on organisms and the importance of interactive effects across the wide spectrum of physical, biological, chemical, and ecological changes. There are numerous examples of changes in the distributions of marine fishes in response to increases in seawater temperature (e.g., Perry et al. 2005, Hsieh et al. 2009, Verges et al. 2014). Temperature effects were examined, for example, in southern California by Holbrook et al. (1997), who assessed changes in fish assemblages at two sites between a cold-water period from 1960 to 1975 and a warm-water period from 1975 to 1995 when average SST increased by 1.0–1.5°C. They found that fish species richness declined by 15–25% and abundances by >90%

between these two periods. They attributed most of this decline to poor recruitment, which coincided with a decrease in macro-zooplankton biomass in the California Current. Other studies have shown that recruitment success of species at higher trophic levels can depend greatly on synchronization of larvae and pulses of planktonic production. The so-called match–mismatch of marine pelagic communities can be affected by climate warming where responses differ over the seasonal cycle, with a consequent mismatch between trophic levels and different functional groups (Edwards and Richardson 2004).

Of more direct influence on kelp forests are the potential effects of climate change on sea urchins. Dupont et al. (2013) showed that acclimation time affected the responses of the sea urchin *Strongylocentrotus droebachiensis* to increased seawater $CO_2$. This urchin species had impaired egg production and increased larval mortality when exposed to high partial pressures of $CO_2$ (1200 µatm) compared to 400 µatm $pCO_2$. However, these effects disappeared when urchins had acclimated over 16 months. In a review that included corals, polychaetes, molluscs, and echinoderms, Byrne (2011) found that warming of 4–6°C did not impair fertilization success. Similarly, the great majority of the species she reviewed showed that fertilization was robust to pH levels of 7.4–7.6 ($pCO_2 \geq 1000$ ppm). Although there is concern about the effects of OA on the development of calcifying larval stages because of reduced aragonite and calcite saturation in seawater, there is only limited knowledge about these effects. Some echinoderms can increase their metabolic rates and ability to calcify in response to increasing seawater acidity. However, this may come at the cost of muscle wastage (Wood et al. 2008). Abalone larvae may be quite sensitive to climate change, with embryo development being impaired in conditions of increasing temperature and decreasing pH (Byrne 2011). For many taxa, there is also evidence for phenological shifts induced by ocean warming that create mismatches between larval production and their food (Schofield et al. 2010).

Global sea level rise over the next 100 years is likely to be around 0.5 m due primarily to the melting of the Greenland and Antarctic ice sheets and thermal expansion, although this is considerably across the globe (IPCC 2013). This would undoubtedly have great impacts on the coastal environment, but, other than potentially providing more hard substrata for giant kelp, it seems unlikely it will have much of an effect on its distribution. However, as in the distant past (Graham et al. 2010), associated changes in habitats, such as sand inundation, could affect patch structure and abundances.

---

The overwhelming conclusion of studies on climate change is that "the impact of humans on the biotic systems of the earth is dramatic and is accelerating" (Mooney et al. 2009), that responses of species are quite variable, and that the "proximate causes of species decline relative to resilience remain largely obscure" (Moritz and Agudo 2013).

It seems likely that in many places, we will be dealing with new mixtures of species, as alien species continue to establish and spread across the globe, perhaps acclimatizing better than native species (Occhipinti-Ambrogi 2007), species shift in response to changing conditions, and ecological relationships and dynamics change as new assemblages are formed. Many processes are, of themselves, quite complex. For example, Roleda, Boyd, et al. (2012) offer a reminder that before OA is discussed there must be "calcifier chemistry lessons." They highlight that many published OA studies have overlooked the fundamental issue that most calcifying organisms do not rely on carbonate from seawater to calcify, but can use bicarbonate ($HCO_3^-$) or metabolically produced $CO_2$. Mollusc shell carbonate comes from three sources, and calcifying seaweeds also vary in their substrate for calcification. It seems we are still some way from understanding both fundamental processes across taxa and how these might combine to act across communities.

Given the projections in changes to SST over the 21st century, it may well be the case that the distribution of *Macrocystis* contracts along its present equatorward limits (figure 13.5). SST changes are projected to be in the range of 1–2°C, even with the conservative RCP4.5 models, and above the global average in much of the area occupied by giant kelp (IPCC 2013). Given that giant kelp has limited nutrient storage capacity, and tends to decline at temperatures greater than around 23°C where nutrients are usually below 1 µmol $L^{-1}$ (Zimmerman and Robertson 1985), there may be considerable shrinking of its range away from the equator. This has been seen, for example, during El Niño years in the Mexico and southern California range of *Macrocystis*, which has contracted up to several hundred kilometers during these periods (e.g., Edwards and Estes 2006). Even if upwelling remains the same in the future or perhaps increases (Sydeman et al. 2014), the higher temperatures between events may prevent recovery of temperature- and nutrient-stressed populations. Other areas of potential contraction are in the Peru–northern Chile region, the northern range of giant kelp along Argentina, the warm-water area along South Africa, and along the coast of Tasmania where warm-water events are increasing. Any expansion of the range of giant kelp toward the poles is likely to be light limited, because of shorter seasons and the low angle of incidence. Jackson's (1987) modeling (see Chapter 3) seems fairly convincing that long periods of low or no light in far northern latitudes probably limits giant kelp to where it is now, with competition from other species only making things more difficult for *Macrocystis* to expand its range in that direction.

More localized, but potentially global, threats to giant kelp forests are likely to be cumulative impacts across multiple changes and stressors, which can occur even where there is some form of management protection (Maxwell et al. 2013). Our best guess is that *Macrocystis* will continue to survive and thrive along most of its present midlatitude distribution in the face of these threats because of its broad physiological tolerances, fast growth, and massive reproductive output. On the other hand, the ecological

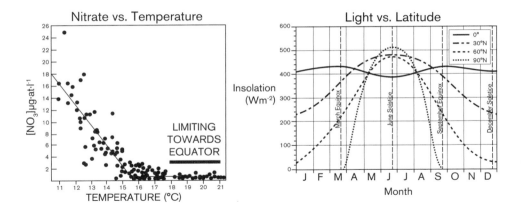

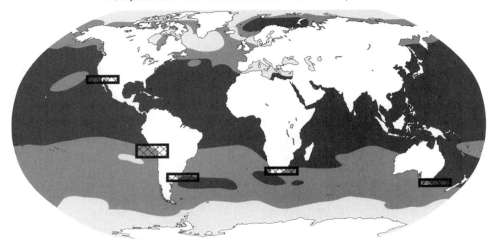

FIGURE 13.5

Potential contraction of giant kelp distribution based on its temperature, nutrient, and light requirements. Giant kelp does not thrive at temperatures beyond ca. 23°C (upper left), where in most areas toward the equator nutrients (especially nitrate) are well below 1 µg atm L$^{-1}$ (see also figure 3.2). In the poleward direction, light becomes limiting because of its seasonal duration and angle of incidence (upper right). Distributional changes in giant kelp are most likely to occur near its current warm-water limits. Lower panel depicts projected changes in sea surface temperature for 2016–2035 relative to 1986–2005 under IPCC RCP4.5 models; darker shading is an increase of 0.75–1°C, middle shading is 0.25–0.5°C, lightest shading is 0–0.25°C. The likely areas of giant kelp loss are indicated by boxes with xx.

SOURCES: Upper left from Zimmerman and Kremer 1984; upper right from Pidwirny (2006); lower panel from figure 11.20 in IPCC (2013), Kirtman et al. (2013).

surprises and lessons from Tasmania are that overfishing, combined with changes in currents, warming seawater, and the population explosion of a formerly rare sea urchin, can have devastating consequences. Time will tell whether lessons are learned, stressors under our control can be reduced, new and more effective types of management can be enacted, and the degree to which these will make a difference in the face of global climate change.

# 14

# GIANT KELP FORESTS: CONCLUSIONS AND FINAL THOUGHTS

> If nothing else, asking what species do in ecosystems and trying to cast replies in a structured way, forces increased integration of ecosystem and population biology, I believe to the benefit of the subject as a whole.
>
> —Lawton (1994)

> We want to emphasize that our usage applies not to those models but only to the real world . . ..Rather than the physicists' classical nature of stability, the concept of persistence within stochastically defined bounds is, in our opinion, more applicable to real ecological systems.
>
> —Connell and Sousa (1983)

Giant kelp forests have achieved an iconic status among the world's marine ecosystems and within the scientific literature. With ever-increasing numbers of nature tourists, scuba divers, recreational fishers, and wide-reaching nature documentaries, both the natural beauty of giant kelp forests and the key role they play in the provision of "services" are much better appreciated globally. Within the scientific literature, giant kelp forests have been a focal point for discussions about regime shifts and alternate states of coastal ecosystems, and they have fed into a growing nexus of ecological theory and management models related to spatial planning (e.g., Foley et al. 2010). In many ways, they have achieved the status of "rain forests" of temperate seas in being the most widely recognized inhabitant of temperate reefs and known for the high diversity they support.

We are therefore happy to report that giant kelp forests are largely thriving and have remained so despite the numerous stressors on coastal ecosystems throughout the past few centuries. To paraphrase the American humorist Mark Twain (1897), reports of their demise are greatly exaggerated. This is both a cause for optimism, because giant kelp populations worldwide themselves seem to be mostly intact, and for caution in that there are clearly losses or severe reductions in many species associated with kelp forests as well as many impacts and sources of degradation that may well increase in the future to the detriment of giant kelp forests and the species and functions they support. To say that *Macrocystis* forests are thriving is somewhat at odds with widely cited views that they are in a highly compromised state because of the loss of key predators and increases in intensive grazing that have shifted kelp forests into an alternate state of degraded structure and collapsed functioning (Jackson et al. 2001, Steneck et al. 2002). Based on our reading of the worldwide literature, however, our summary of the state of giant kelp forests is not a naive Panglossian view (Voltaire 1759) involving "the best of all possible worlds," but a recognition that, so far as we can determine, giant kelp still occupies the great majority of its postglacial period distribution. There are modern exceptions, of course, such as portions of the southern California mainland, where urban expansion coincided with greatly increased sewage discharge, sedimentation, and altered dynamics involving sea urchins and reduced numbers of predators (e.g., Foster and Schiel 2010). Instead of the narrative from one or a few sites that formerly dominated the literature, however, we now possess detailed knowledge about many giant kelp forests across both hemispheres. What is revealed is a rich panoply of forest types, regional species pools, and a wide range of ecological interactions and functional relationships across the globe. Even though dominant trophic groups include predacious fishes, lobsters, and sea urchins in most kelp forests worldwide, there is no set of simple structures or ecological interactions that universally apply to kelp forests, and no set of simple environmental conditions that span the diversity of kelp forest communities across hemispheres. It is the combination of biotic and abiotic conditions within and across sites and regions and their effects on giant kelp communities that have provided a more comprehensive knowledge of functioning and a greater understanding of what *Macrocystis* does in the wider ecosystems it occupies. With respect to the opening quote of this chapter, understanding the population biology of *Macrocystis* and its relationship to the abiotic environment has provided the springboard for our wider and more comprehensive appreciation of giant kelp forests. We also look to the future, as we begin to experience changing global and regional climates, recognizing that the present state is far from static and that there is a critical interplay between the life history and ecology of giant kelp and the wider environment in which it lives.

Our conclusions are based on an extensive scientific literature, which has burgeoned since our last treatise on *Macrocystis* (figure 14.1). Foster and Schiel (1985) cited 289 references that contained most of what was known about giant kelp forests at the time. The emphasis was on the kelp forests along the west coast of North America where most studies had been done. This present book, however, cites over 1000 references and contains a considerable expansion of knowledge on several fronts. Perhaps most notable of these is the worldwide increase in kelp forest studies and the cross-disciplinary nature of many studies that have achieved a great advancement in understanding the interplay between the complete life history of giant kelp, especially the early life history stages, and the environment. This has involved clever experimentation, detailed monitoring of structure and temporal changes, and a far greater knowledge of ocean processes and population dynamics of *Macrocystis*, dispersal of spores, settlement, delayed develop-

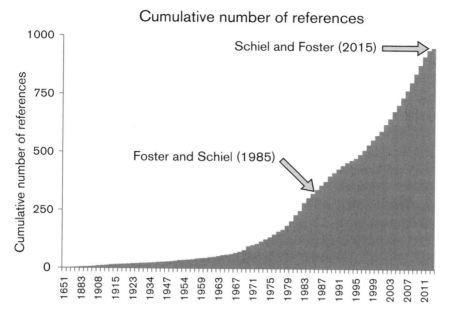

**Cumulative number of references**

FIGURE 14.1

Cumulative number of references cited in Foster and Schiel (1985) and this book.

ment and metapopulation processes, the spatial scaling of effects from boundary layer conditions to basin-wide climate impacts on giant kelp forests, and temporal scaling from wave periods to decadal oscillations.

Taken together, these have highlighted that about the only thing common among the giant kelp forests of the world is giant kelp itself. This has taken on new scope since the reunification of giant kelp into the single species *Macrocystis pyrifera* (Demes et al. 2009) which has resulted in a considerable expansion of the biogeographic range of the species (refer to figure 1.5) and changed our views of what constitutes a "giant kelp forest." Here we present an overview of what we have gleaned in our journey through the kelp forest literature with respect to the biology, community ecology, and management of giant kelp forests. We end with speculation about the future of kelp forest studies. In particular, we consider sustainability and the wider value of giant kelp forests in the context of "natural capital," which includes not just the "goods and services" to mankind but also intrinsic values and perhaps deep ethical considerations of ensuring the future of giant kelp forests for their own sake rather than ours.

## BIOLOGY OF GIANT KELP

A much greater understanding of the full life history of giant kelp has provided significant new insights into why it is so successful and so widely distributed. Concerns about degradation and altered functioning have largely focused on extractions of trophically important species and ecological processes that remove giant kelp, especially the large sporophytes (see Chapter 9). Population dynamics rely, however, on inputs and the processes that allow giant kelp to establish, maintain, and replenish populations. Given the obvious fact that *Macrocystis* is the sine qua non of giant kelp forest communities, the increased emphasis on the inter-relationships between its biology, life history characteristics, and the environment has yielded significant advancements in understanding processes that were elusive when there was a primary focus on how other members of the community, particularly grazers, predators and to a lesser extent competitors affected giant kelp.

There are several other kelp species that extend through the water column and form underwater forests with sea-surface canopies, but none of them has individuals that attain the multiple fronds, large biomass, or areal cover achieved by *Macrocystis*. At densities as low as one plant per 10–15 m² of reef, giant kelp can form closed canopies over extensive areas of reef, compared to around one plant per square meter for *Nereocystis*, *Ecklonia maxima*, or *Eualaria (=Alaria) fistulosa* (see Chapter 5). This potential for giant size alone means that recruitment at what would be considered marginal densities for other kelps can sustain a giant kelp forest. As discussed in Chapter 4, the capacity for rapid growth to the sea surface allows plants to take advantage of a high-light environment unavailable to benthic species without surface canopies. The ability of giant kelp to translocate materials to parts of the plant below in a low-light environment, especially the reproductive sporophylls at the base of plants, allows high productivity, high biomass, and repeated reproduction. *Macrocystis* is tolerant of a wide range of temperatures from just above freezing to around 23°C, although this upper range is often compromised by the low nutrients usually associated with higher temperatures.

In its population inputs, *Macrocystis* has a competitive advantage over most other macroalgae. Its microscopic stages can develop in very low light and, once above the substratum, larger sporophytes rapidly grow above turfing and foliose benthic species. It can mature in less than a year and its continual spore production, made possible because of it very high ratio of vegetative tissue to sporophyll biomass, probably saturates the forest floor with spores. Multiyear "seed banks" may not occur for *Macrocystis*, and it appears that the vast majority of recruitment results from newly settled spores (Reed et al. 1997), even though there can be persistence of microscopic stages on the substratum at least across multiple seasons (e.g., Carney et al. 2013).

In many ways, studies of giant kelp confirm its status as a mega-weed. Its prodigious reproductive output of up to trillions of spores per plant allows a quick response to disturbances that create light gaps and ensures that the local environment of kelp forests has the capacity for replenishment so long as adults are present nearby. Localized dispersal of spores within and around kelp forests is aided by tidal and local currents and turbulent mixing (Graham 2003, Gaylord et al. 2006a, 2006b). However, spores can disperse and settle in densities sufficient for fertilization and development up to several kilometers (Reed et al. 2004), and long-distance dispersal can occur through detached reproductive adults, which can form extensive floating rafts that persist and travel for many months (see Chapter 6).

The presence and persistence of *Macrocystis* are influenced by several constraints. The great biomass of adults in the water column makes plants highly susceptible to removal by waves, which results in the great seasonality seen in production dynamics along coastlines with severe winter storms, such as central California. Water motion generally confines *Macrocystis* populations to sites without continual extreme waves except in some places with exceptionally clear water where there are deep-water refugia. *Macrocystis* also has little storage capacity for nutrients, on the order of 2 weeks, and requires a continuous supply for populations to persist. This limited storage capacity can affect plants during periods of high temperature and low nutrients, such as in El Niño years in southern California and Mexico, when plants deteriorate (see Chapter 5). Conversely, the inability of *Macrocystis* to penetrate higher latitude reefs may be a function of low light, prolonged periods of darkness, and perhaps continuous cold temperatures and competition with other kelps.

On local scales, *Macrocystis*, like all kelps, is vulnerable to extensive grazing by sea urchins, even though the bulk of the plant in the water column is well separated from benthic grazers. Urchin-dominated areas can persist for many years before urchin mortality from storms or disease, poor urchin recruitment, predation, and / or successful episodes of kelp recruitment re-establish kelp dominance. Because of its unique combination of characteristics and disturbance-oriented life history, *Macrocystis* rarely fails to re-establish within a few years, even in sites severely impacted by disturbances, except on isolated patch reefs (Reed et al. 2006—"A metapopulation perspective"). Therefore, in answer to Lawton's (1994) question of "what do species do?," this one does a great deal in terms of establishment, growth, persistence, dominance, and habitat provision. Furthermore, it completes its life history with no need of facilitation by other species, and only the basic requirements of hard reef, light, and nutrients. As a biogenic habitat-former, it has no equal in temperate waters.

## COMMUNITIES

Delineating giant kelp communities is more a regional than a global exercise. The taxonomic revision to the single species *Macrocystis pyrifera* brought a much greater biogeographic cover of the species that included *M. integrifolia* along extensive regions of North America, and this species plus *M. angustifolia* and *M. laevis* in the southern hemisphere. The diversity of communities associated with these widely separated giant kelp forests is illustrated in Chapter 5. An experienced diver, for example, would note the poverty of understory kelps and water column fishes in southern hemisphere forests compared to those along California, where there is a lush and diverse group of stipitate kelps and a great abundance of rock fishes in the water column. Given that our notion of what constitutes a giant kelp "community" is now so expansive, it has also become necessary to extend our concepts of critical functions, food web dynamics, and structuring processes. Our conceptual frameworks may relate to a great extent on both hemispheric and continental differences. The possibility of coevolution shaping giant kelp communities and functions is far greater in the northern hemisphere where *Macrocystis* most likely evolved, but it is unlikely in the southern hemisphere where it appears to have spread relatively recently into existing communities (see Chapter 1 and figure 4.10). In this sense, therefore, *Macrocystis* has acted more as a highly successful invasive or immigrant species in the southern hemisphere and has gone through a process of acclimation in the communities it has invaded. Nevertheless, there are strong interactions involving *Macrocystis* in all communities, but these are most likely region- or area-specific.

From what we now know, it is enlightening to realize not only the vast differences in the composition of species associated with giant kelp but also the great differences in how trophically important groups such as sea urchins can have greatly different influences among regions. Several prominent studies argue for models of tight trophic control and cascading processes in the Northeast Pacific region (Estes et al. 1998, Jackson et al. 2001, Steneck et al. 2002) but, in contrast, there is little evidence for anything other than localized patch effects of sea urchins in giant kelp forests in Chile, New Zealand, and many other southern hemisphere regions (with the exception of Tasmania; below and Chapter 8). The strong trophic linkages in the north relate mostly to the voracious predator of sea urchins, the sea otter *Enhydra lutris*, which had been removed by hunting from much of the west coast of North America in historical times. The destabilization of evolved processes through hunting may well have had significant effects over millennia along the coast and offshore islands of California involving other herbivores such as abalone (Erlandson et al. 2005). Questions remain, however, about the extent, strength, and temporal occurrence of such trophic interactions and the degree to which they exerted a controlling influence on giant kelp worldwide.

There is somewhat of a dichotomy of perspectives in the ecological literature, broadly involving alternate stable states on the one hand or, as Connell and Sousa (1983) wrote, "persistence of communities within stochastically defined bounds." Several authors have reviewed this theme, pointing out the confusion of terminology involving stable states, regime shifts, and phase shifts, and the exceptions and limited site narratives used in many examples (Bruno et al. 2009, Dudgeon et al. 2010). There are a lot of theories involved that are attractive to many ecologists and which feed into system models. With respect to giant kelp forests and their structure, however, much of this seems un-natural, mainly because *Macrocystis* still occupies the great majority of its historical range. The dichotomy of perspectives is illustrated by two recent studies, one in Canada and one in Australia. Watson and Estes (2011) examined sites among parts of Vancouver Island that encompassed the re-habitation of sea otters over a period of 23 years. Through detailed surveys and following the development of plots over 20 sites, they found that sea urchins dominated many sites and large algae were sparse where otters were not present. Urchin density declined over a period of several years after otters returned and algae became more abundant. They called this a "phase shift," with the phase states being either algal- or urchin-dominated. However, their data showed not just the dominance of urchins or algae but also a mosaic of both. They also found that urchin dominance was depth restricted in many sites, with algae in shallower and deeper water. They concluded that the phase states "are predictable in space and time based on the presence or absence of sea otters," but also that there was "considerable variation in the composition of species and demography of populations within these phase states," apparently due to the "vagaries of foraging sea otters." Given the great variation in what they found, another way to view this is that the "phase states" represent end points of a continuum of algal bed structure, and that most states existed simultaneously across their sites, similar to what Bruno et al. (2009) and others (e.g., Selig and Bruno 2010) have argued for coral or algal dominance and phase states on tropical reefs. It seems clear from Watson and Estes (2011) and many other studies (see Chapter 9) that sea urchins along the northeastern Pacific can persist in high densities on open reef for long periods of time only when otters are absent or in low densities, but the corollary is that almost any state can exist when otters are present, and not just an alternate state of algal dominance. Similarly, in areas of southern California where sea otters do not occur, sea urchins can be locally abundant but there is a wide range of forest types among sites, in which urchins play only a minor role in kelp forest dynamics at most sites (e.g., Foster and Schiel 1988, Reed et al. 2011). The trophic dynamics–phase shift argument seems therefore to have limited generality, even in the regional context of multiple sites. Furthermore, it seems to obfuscate rather than clarify underlying processes of patch dynamics and community structure.

In Australia, Johnson and colleagues (e.g., Johnson et al. 2003, 2011) have argued for a "regime shift" affecting the kelp beds and forests of eastern Tasmania. In their case, however, the "catastrophic phase shift" (Ling et al. 2009) was the result of complex oceanographic and ecological changes (see Chapter 9). An invasion of the sea urchin *Centrostephanus rodgersii* from the Australian mainland occurred when larvae survived and were transported in the warm-water East Australian Current, which had moved onshore. The growth of *Centrostephanus* populations coincided with a natural regression of giant kelp over several decades, probably due to ocean warming, and severe overfishing of a once-abundant predator, the lobster *Jasus edwardsii*. In this case, therefore, a major underlying cause of change was an oceanographic shift but the effects were exacerbated by overfishing and ocean warming. The emphasis of Johnson and colleagues has been in restoring lobster populations, not least because of their importance to the local economy, retention of large lobsters that are capable of eating the long-spined urchins, and consequent restoration of kelp beds to support production (e.g., Johnson et al. 2005, 2011). This example presents a far more comprehensive perspective of kelp forest dynamics in seeking to understand the ways in which particular species have large effects on community structure, how these relate to environmental changes, the ways in which a regime shift in the environment can affect change, and, importantly, if hysteresis—the reversal of unwanted effects—is possible.

These are more than subtle differences in perspectives about alternate states. One relies on a straightforward trophic model to flip between phases, and the other deals with the complexities of kelp forest dynamics and their variation. Questions of scale, within and between site variability in structure and processes must be addressed if we are to understand the functioning of kelp forests across the many communities that exist worldwide. Recognizing complexity and incorporating it into models seems more fruitful than focusing on simple trophic models that emphasize only the end points in a continuum.

## INTERACTIONS

Numerous interactions, of course, affect kelp forest structure within sites. Across the large literature on kelp forests, the preemption of light by giant kelp canopies is the most important interaction identified. Canopy shading affects recruitment and growth of understory kelps and benthic algae within virtually all kelp forests, and this in turn affects benthic invertebrates. In the kelp forests of the northeastern Pacific, the detrital fall from kelp forests fuels food webs up to many kilometers outside of kelp forests. Grazing interactions, particularly by sea urchins, can set upper and lower boundaries of kelp, and numerous invertebrates can affect fronds. There are many examples of how these interactions can be important at the scale of patches within forests but are generally less important at larger scales. Spatial heterogeneity of reef structure, waves, nutrient provision, etc interact to constrict the co-occurrence of conditions necessary for most of these smaller-scale processes to have a great influence at larger spatial scales. Again, fundamentally, giant kelp forests occur if spores are available from a mixture of adult plants, rock is present, and the physicochemical conditions are suitable.

In trying to understand the status of kelp forest communities, there is no doubt that our collective reference points are what Paul Dayton has termed "sliding baselines" (Dayton et al. 1998). Historical overfishing of predators and grazers such as lobsters, fish, and abalone has removed potentially important trophic linkages in kelp forest communities. Unfortunately, the ecological roles of many of these species are not well known. Although key roles have been speculated for them in buffering against outbreaks of sea urchins, Dayton et al. pointed out how little the rest of the community seems to change when large fishes are extracted. We have seen massive populations of abalone reduced to small numbers after a few years of fishing, and lamented that their loss seems to have left few vestiges in the kelp forest community (Schiel 2006). How can we know what full and functioning communities were like, given these alterations and compounded with numerous other stressors? A goal of coastal management and marine protected areas (MPAs) is to restore these species and find out.

## MANAGEMENT

MPAs and marine spatial planning are initiatives to restore the structure and functioning of coastal ecosystems (see Chapter 12). The concept of MPAs and what they are meant to do have evolved over the years, from setting aside areas for protection and study to augmenting coastal fisheries to providing connected networks to safeguard coastal ecosystems from severe localized impacts. None of these goals has been fully realized yet, but it is clear and no surprise from a few decades of MPAs that target species such as some fishes, lobsters, and invertebrates have responded positively in abundances and sizes (Babcock et al. 2010). There is also evidence that sea urchins decline in numbers on open reef when predators return. "Spillover" effects in augmenting fisheries are more equivocal, but fishers are known to cluster near the borders of protected areas. MPAs have therefore been successful on several fronts, not least of which is increasing public awareness of the beauty and worth of coastal ecosystems.

At the same time, there is awareness of the numerous "goods and services" provided by coastal ecosystems and their vulnerability to numerous stressors, including alteration to coastal topography, increasing sediments, nutrient loading, contaminants, and a wide variety of diffuse land-use intensification impacts that affect algal beds worldwide. These sorts of stressors are not ameliorated by MPAs and yet may pose some of the greatest threats to coastal ecosystems. There is an increasing awareness that MPAs are necessary but not sufficient to protect the marine environment from degradation (e.g., Mora and Sale 2011) and that even in countries with well-developed networks of MPAs like New Zealand there is still over 90% of the marine environment not in protected areas and subject to a wide range of usages. The numerous societal benefits provided by the nearshore marine environment include food provision, recreational activities, and ecosystem benefits such as denitrification, refugia and nurseries for juveniles of many species, and habitat to support biodiversity. However, nearshore areas are also receiving waters for virtually all catchment-based land runoff, often resulting in occlusion of the light environment from sediments and phytoplankton blooms from excess nutrients, impaired recruitment of benthic species, altered structure and functional dynamics and occasional hypoxia, and "dead zones."

Townsend et al. (2011) argue for an "ecosystems principles approach" whereby ecology is incorporated into goods and services assessments in marine coastal ecosystems. They list 18 "key general ecosystem principles" relevant to service provision finding, for example, that services primarily depended on connectivity and that space and resource occupancy by native species can decrease invasion risk. They argue there is a need for integration of cross-disciplinary ecology and social science to achieve greater understanding of how human interactions can result in changes to communities and loss of resilience. Sutherland and Woodroof (2009) argue for "horizon scanning" as a systematic search for potential threats and opportunities that are currently poorly recognized. For kelp forests, this may well be in the realm of dealing with the complexities of diffuse stressors from land, and increasingly within the context of a changing climate.

How will we respond and adapt with respect to the problems that beset kelp forests worldwide? In whatever ways we do, it will take societal buy-in and trade-offs of usages beyond what we currently have.

## THE FUTURE
### HABITAT PRESERVATION

All ecologists have dabbled in the dark arts of prediction, although probably with little more success than most other professions. History has proven the overwhelming influence of surprises which, by their nature, are not predictable. As Niels Bohr and other scientific luminaries have noted, "prediction is very difficult, especially if it's about the future." Several current trends, however, are worrisome and bode ill for many of the earth's species and ecosystems (e.g., Pimm et al. 2014). Fears of mass extinctions, loss of habitats, and a changing climate provide a cocktail of gloom in many quarters. An encouraging sign, therefore, is that giant kelp forests seem to be thriving, despite the many un-natural stressors to which they are subjected. In doing so, a dominant provider of temperate reef biogenic habitat is largely intact. It is still provisioning food and detrital webs and, even though some species in forest communities are overly exploited, we know of no recent extinctions of kelp forest organisms. Preservation of critical habitat is a major focal area in conservation, and much effort is being expended worldwide to ensure the future of kelp forest habitats, both for their own sake and for the goods and services they provide humanity.

From experiments to date, it appears that increasing ocean acidification will not directly affect *Macrocystis*, although calcifying organisms may undergo deleterious effects (see Chapter 13). However, the synergistic effects of climate change, especially as they affect temperature, nutrients, storms, rainfall, and terrestrial runoff, may well alter kelp distributions, particularly at the warm-water–low-nutrient margins of their current distributions. The areas where loss of populations seems most likely to occur are along Baja California, Mexico, parts of southern California, the low-latitude populations of northern Chile and southern Peru, and along southeastern Australia. Potential gains from the spread of giant kelp toward the poles are unlikely because populations may be constrained by day length. Storm frequency and intensity may well play a role in prevention of the expansion of the range of giant kelp in the colder regions. As indicated by the spread of *Centrostephanus* into Tasmania and the substantial changes to community structure seen in kelp beds in a heated embayment (Schiel et al. 2004), there will no doubt be surprises that, with our limited knowledge of the biology and ecology of the myriad organisms in kelp communities, will be difficult and most likely impossible to predict. It seems to us that the best that can be done is to reduce other, human-caused, effects and hope that this will reduce some of the synergistic impacts of climate change. As indicated above, however, the critical feature is the giant kelp habitat itself, without which giant kelp and most of the other species will be reduced to minor roles (Graham 2004).

Most of our knowledge about giant kelp and their communities has come from field observations and experiments, and focused laboratory experiments based on field studies. There is, however, a growing trend of monitoring studies to feed into better management and achieve a wider knowledge of long-term changes. Many studies make use of databases from monitoring programs using meta-analyses across usually disparate studies to seek patterns, often involving management initiatives such as MPAs and their effects. In our desire and enthusiasm for broad patterns and overarching trends, we frequently obfuscate the fact that these types of analyses are correlative, often with weak inference (Platt 1964) and usually factor out significant details, nuances, and complexities. The term meta-analysis was coined to refer to the "statistical analysis of a large collection of results from individual studies for the purpose of integrating the findings" (Glass 1976). It may be their purpose, therefore, to rise above the detail, and to integrate over longer times and a wider range of conditions than generally shorter-term experiments can achieve. However, these types of studies may not necessarily serve us well when management issues tend to be local or regional, or where detail, nuance, process, and regional conditions matter. In these circumstances, there is a continuing need for field work that includes experimentation, and field-driven laboratory studies that are solidly based on natural history and designed to test key hypotheses.

In our 1985 book, we highlighted the research needs in California as the following:

1. Basic
   a. More population studies (recruitment, growth, mortality, etc.) of "important" kelp forest organisms
   b. Studies of the physical processes that structure giant kelp ecosystems
   c. Studies of the functional role of biological processes such as competition and predation
2. Applied
   a. Better understanding of the biological and socioeconomic consequences of sea otter management alternatives
   b. Multispecies approaches to biological and economic modeling
   c. Studies on stock enhancement

Each of these has received attention. We have vastly increased our knowledge of giant kelp biology and kelp forest ecology, and this is now more broadly based on information from other regions, especially the significant amount of work done in Chile. In particular, there is now a good understanding of the physical and biological processes that structure kelp forests in several regions, and we hope this understanding will continue to broaden as more regions are investigated. Cooperative, cross-disciplinary investigations in more countries using approaches such as those of the Santa Barbara Coastal Long-Term Ecological Research (LTER) would be especially informative, given the great increase in knowledge about kelp forests that has come from the LTER. Population studies of key species will continue to be useful, as they increase our understanding of the roles of the less iconic species that together make up the diversity of kelp forests. Some population and community studies that would especially improve understanding of giant kelp forests are:

1. *Giant kelp biology and population dynamics at high latitudes.* Modeling studies indicate that light constrains populations at their current high latitude limits. Experimental investigations are needed to better test this hypothesis and examine alternatives, including the effects of competition with other kelps and the interaction of light and temperature.

2. *Microscopic stages.* Where do these stages of giant kelp occur in natural environments, how persistent are they, and how extensive is the role of delayed development of these stages in recruitment, especially after large impacts on adults?

3. *Ecomorphic variation in giant kelp.* With the unification of giant kelp into one species, how does the great morphological diversity seen across the ecomorphs occur, given the genetic similarity? How is this affected by environmental variation? What can finer-scale genetic analyses further reveal about these differences and connectivity among populations?

4. *Regional differences in the ability of sea urchins to cause deforestation.* The focus has been on the differences in predators and their effects on sea urchins. How important are differences in urchin recruitment, juvenile mortality, and shelter-oriented behavior among species across regions in forming and maintaining urchin-dominated sites? Are there regional differences in the abundance and temporal availability of drift kelp and detrital fall that produce differences in urchin behavior and grazing effects across regions?

5. *Remote giant kelp communities.* How do the structure and functioning of relatively uninvestigated giant kelp communities, especially in the southern hemisphere, compare with what is currently known?

6. *Connectivity.* We have an understanding of basic processes but not yet a wider body of research, except for some sites in southern California, on how connected giant kelp populations are, the relative roles of dispersal via spores and drifting adults, how density dependent these are on adult populations, and how widely separated populations can recover from large-scale disturbances. How might connectivity be altered by a changing climate, such as increased storm activity? These issues of connectivity extend to the key organisms that interact with giant kelp and affect its abundance and distribution.

7. *Production dynamics of giant kelp.* What are the photosynthetic constraints on giant kelp near the extremes of its distribution? How are the production

dynamics of giant kelp affected by the often compromised light and nutrient environment of nearshore waters, especially as they relate to climate change? What is the role of giant kelp as a source of production versus structure in defining the species composition of kelp forest communities?

8. *Nontrophic relationships in community modeling.* Can more complexity of interactions and giant kelp dynamics, other than just somewhat hierarchical predator–grazer controls, be incorporated into community models to provide more realistic predictions of community variation over space and time?

9. *Global changes in giant kelp distribution and abundance.* Satellite and other remote sensing technologies are now available, and regularly used to assess giant kelp dynamics in southern California. An international, standardized program that expands such monitoring to include the entire distribution of giant kelp would provide important information on regional differences and insights into the effects of global change.

Although many scientists do not distinguish between basic science and applied science, we distinguish them here as fundamental science and its application. The needs of applied research tend to be more place-specific than the more fundamental research outlined above. With reference to our 1985 list, sea otter management by restricting their expansion into some areas like those with invertebrate fisheries along coastal southern California was not successful and has been abandoned in favor of letting otter populations expand naturally, regardless of economic consequences. Otters remain under full legal protection and there are severe penalties for harassing or killing them. It is not clear that the few attempts at stock enhancement of abalone and fish populations in California have significantly improved natural stocks, although these efforts are continuing. It appears that the emphasis worldwide in applied coastal science is mostly in MPAs. As monitoring programs continue in the hundreds of protected areas worldwide and we get a longer-term view of effects and effectiveness, we will undoubtedly refine our understanding of the restorative value of MPAs. In particular, it would be interesting to see the effects of MPAs on organisms other than kelp and fished species, especially as the abundances and sizes of formerly fished species increase. We should also derive a greater understanding by direct comparisons, where possible, with still-exploited control areas. The treating of MPAs as large-scale manipulative experiments has already yielded significant insights into the structure and function of protected or less-exploited kelp systems. With the advent of connected networks of reserves (cf. Gaines, White, et al. 2010, Gaines, Lester, et al. 2010) we should see more cohesive coastal management policies that achieve the evolving goals of MPAs.

We end with a paean for science-based and science-informed management. "Science into management" has often been a one-way street, with the onus on scientists to demonstrate the relevance of their research and to work with managers from a wide range of local, state, and national departments. However, there has been little onus on managers to base their decisions on best science, and policy decisions are frequently deficient in sound scientific underpinnings. There is also a wide recognition of the numerous "stakeholders" who affect and are affected by any management decisions. Stakeholders include a wide range of commercial, sporting, recreational, cultural, and societal interests, all of whom tend to have different perspectives on what the problems are and how they are best fixed, usually involving some other group making concessions. To overcome some of the adversarial boundaries and barriers, there is an increasing impetus for stakeholders, managers, and scientists to work together in more progressive ways in identifying and formulating better management policy. This is challenging, to say the least. It is based on the assumptions that we are all in this together and that there are few uses that do not impinge in some way on other peoples' values or uses, and will undoubtedly involve trade-offs among usages. How this works in practice remains to be seen, but it is critical that science is viewed as a key to solving problems and better management. For example, in New Zealand, one of the world leaders in MPAs and coastal management, the government has instituted National Science Challenges to focus on complex issues by drawing scientists together across disciplines and institutions to achieve common goals of major benefit to the country. In the marine arena, this includes enhanced and sustainable use of marine resources within environmental and biological constraints, and understanding the role of the southern ocean in climate change. The imperative is for scientists to work with relevant stakeholders and environmental managers to achieve enduring outcomes. The view is long term and underpinning science is a key component. Just how this will work over the socio-politico-ecological realm remains to be seen.

In our view, a major achievement to date across the world's kelp forests and nearshore marine systems is the recognition that our efforts in solving coastal marine problems have taken us only so far and that new ways of acting are required. It seems as if we are at a crossroads in this regard, with government-driven initiatives on one side and more bottom-up stakeholder / science / management approaches on the other. There is a high degree of codependency, and how this plays out will be determined over the coming years. It is indeed fortunate that giant kelp forests have proven to be so remarkably robust and resilient to most impacts. So long as they continue to survive and thrive, there is every reason to believe that the diverse and beautiful communities they sustain will also be maintained in an increasingly stressed environment.

# AFTERWORD

Writing this book has been a pleasant and at times arduous journey through the scientific literature, which has revealed far more depth and quality of science than we were aware of before we began. We enjoyed rereading the discoveries of early kelp forest pioneers, from natural historians and taxonomists to the first diving scientists. We realized that our own careers collectively span the period from these firsthand observers (MSF was a student of M. Neushul) through generations of students and research scientists to the present. But most of all, we could time travel to the past to two young students, continents apart, who wanted to dive, explore, and make a difference in their research and vocation. It's been a fun journey with outstanding students and colleagues, and the satisfaction of diving in our magnificent giant kelp forests has never abated.

The authors (DRS and MSF) after a dive in southern California ca. 1982. (Cautionary note to young students: much of the gear and activities seen and implied in this photo are not currently condoned by OSHA).

# REFERENCES

Abbott, I.A. and Hollenberg, G.J. 1976. *Marine Algae of California.* Stanford University Press, Stanford, CA.

Abbott, D.P. and Haderlie, E.C. 1980. Prosobranchia: Marine snails. Pp. 230–307. In Morris, R.H., Abbott, D.P., and Haderlie, E.C. (eds), *Intertidal Invertebrates of California.* Stanford University Press, Stanford, CA.

Åberg, P. 1992. A demographic study of two populations of the seaweed *Ascophyllum nodosum. Ecology* 73:1473–1487.

Abesamis, R.A. and Russ, G.R. 2010. Patterns of recruitment of coral reef fishes in a monsoonal environment. *Coral Reefs* 29:911–921.

Adami, M.L. and Gordillo, S. 1999. Structure and dynamics of the biota associated with *Macrocystis pyrifera* (Phaeophyta) from the Beagle Channel, Tierra del Fuego. *Scientia Marina* 63:183–191.

Adl, S.M., Simpson, A.G.B., Farmer, M.A., Andersen, R.A., Anderson, O.R., Barta, J.R. et al. 2005. The new higher level classification of eukaryotes with emphasis on the taxonomy of protists. *Journal of Eukaryotic Microbiology* 52:399–451.

Agardh, C.A. 1820. *Species Algarum Rite Cognitae, Cum Synonymis, Differentiis Specifics et Descriptionibus Succinctis.* Berling, Lund, Sweden.

Agardy, T., Bridgewater, P., Crosby, M.P., Day, J., Dayton, R., Kenshington, R. et al. 2003. Dangerous targets? Unresolved issues and ideological clashes around marine protected areas. *Aquatic Conservation: Marine Freshwater Ecosystems* 13:353–367.

Airame, S., Dugan, J.E., Lafferty, K.D., Leslie, H., McArdle, D.A., and Warner, R.R. 2003. Applying ecological criteria to marine reserve design: a case study from the California Channel Islands. *Ecological Applications* 13:170–184.

Airoldi, L. 2003. The effects of sedimentation on rocky coast assemblages. *Oceanography and Marine Biology: An Annual Review* 41:161–236.

Alberto, F., Whitmer, A., Coehlo, N. C. Zippay, M., Varela-Alvarez, E., Raimondi, P. T. et al. 2009. Microsatellite markers for the giant kelp *Macrocystis pyrifera*. *Conservation Genetics* 10:1915–1917.

Alberto, F., Raimondi, P. T., Reed, D. C., Coehlo, N. C., Leblois, R. W., Whitmer, A. et al. 2010. Habitat continuity and geographic distance predict population genetic differentiation in giant kelp. *Ecology* 91:49–56.

Alberto, F., Raimondi, P. T., Reed, D. C., Watson, J. R., Siegel, D. A., Mitarai, S., et al. 2011. Isolation by oceanographic distance explains genetic structure for *Macrocystis pyrifera* in the Santa Barbara Channel. *Molecular Ecology* 20:2543–2554.

Albright, L. J., Chocair, J., Masuda, K., and Valdes, M. 1982. Degradation of the kelps *Macrocystis intergrifolia* and *Nereocystis luetkeana* in British Columbia coastal waters. Pp. 215–233. In Srivastava, L. M. (ed.), *Synthetic and Degradative Processes in Marine Macrophytes*. Walter de Gruyter, Berlin, Germany.

Aleem, A. A. 1956. Quantitative underwater study of benthic communities inhabiting kelp beds off California. *Science* 123:183–183.

Alexander, T. J., Barrett, N., Haddon, M., and Edgar, G. 2009. Relationships between mobile macroinvertebrates and reef structure in a temperate marine reserve. *Marine Ecology Progress Series* 389:31–44.

Allison, G. W., Lubchenco, J., and Carr, M. H. 1998. Marine reserves are necessary but not sufficient for marine conservation. *Ecological Applications* 8:79–92.

Allison, G. W., Gaines, S. D., Lubchenco, J., and Possingham, H. P. 2003. Ensuring persistence of marine reserves: catastrophes require adopting an insurance factor. *Ecological Applications* 13:S8–S24.

Alonso, M. K., Crespo, E. A., Pedraza, S. N., Garcia, N. A., and Coscarella, M. A. 2000. Food habits of the South American sea lion, *Otaria flavescens*, off Patagonia, Argentina. *Fishery Bulletin* 98:250–263.

Alonzo, S. H., Key, M., Ish, T., and McCall, A. D. 2004. *Status of the California Sheephead (Semicossyphus pulcher) Stock (2004)*. Center for Stock Assessment Research, California Department of Fish and Game, University of California Santa Cruz and National Marine Fisheries Service, Santa Cruz, CA.

Ambrose, R. F. 1986. Effects of octopus predation on motile invertebrates in a rocky subtidal community. *Marine Ecology Progress Series* 30:261–273.

Ambrose, R. F. 1994. Mitigating the effects of a coastal plant on a kelp forest community: rationale and requirements for an artificial reef. *Bulletin of Marine Science* 55:694–708.

Ambrose, R. F. and Nelson, B. V. 1982. Inhibition of giant kelp recruitment by an introduced brown alga. *Botanica Marina* 25:265–267.

Ambrose, R. F., Engle, J. M., Coyer, J. A., and Nelson, B. V. 1993. Changes in urchin and kelp densities at Anacapa Island, California. Pp. 199–209. In Hochberg, F. G. (ed.), *Third California Islands Symposium: Recent Advances in Research on the California Islands*. Santa Barbara Museum of Natural History, Santa Barbara, CA.

Ames, J. A. and Morejohn, G. V. 1980. Evidence of white shark, *Carcharodon carcharias*, attacks on sea otters, *Enhydra lutris*. *California Fish and Game* 66:196–209.

Amsler, C. D. and Neushul, M. 1989. Chemotactic effects of nutrients on spores of the kelps *Macrocystis pyrifera* and *Pterygophora californica*. *Marine Biology* 102:557–564.

Amsler, C. D. and Neushul, M. 1990. Nutrient stimulation of spore settlement in the kelps *Pterogophora californica* and *Macrocystis pyrifera*. *Marine Biology* 107:297–304.

Amsler, C. D. and Neushul, M. 1991. Photosynthetic physiology and chemical composition of spores of the kelps *Macrocystis pyrifera*, *Nereocystis leutkeana*, *Laminaria farlowii*, and *Pterygophora californica* (Phaeophyceae). *Journal of Phycology* 27:26–34.

Amster, E., Tiwary, A., and Schenker, M. B. 2007. Case report: potential arsenic toxicosis secondary to herbal kelp supplement. *Environmental Health Perspectives* 115:606–608.

Anderson, B. S. and Hunt, J. W. 1988. Bioassay methods for evaluating the toxicity of heavy metals, biocides and sewage effluent using microscopic stages of giant kelp *Macrocystis pyrifera* (Agardh): a preliminary report. *Marine Environmental Research* 26:113–134.

Anderson, B. S., Hunt, J. W., Turpen, S. L., Coulon, A. R., and Martin, M. 1990. Copper toxicity to microscopic stages of giant kelp *Macrocystis pyrifera*: interpopulation comparisons and temporal variability. *Marine Ecology Progress Series* 68:147–156.

Anderson, E. K. and North, W. J. 1966. In situ studies of spore production and dispersal in the giant kelp *Macrocystis pyrifera*. *Proceedings of the International Seaweed Symposium* 5:73–86.

Anderson, E. K. and North, W. J. 1967. Zoospore release rates in giant kelp *Macrocystis*. *Bulletin of the Southern California Academy of Sciences* 66:223–232.

Anderson, N. 1974. A mathematical model for the growth of giant kelp. *Simulation* 22:97–105.

Anderson, P. K. 1995. Competition, predation and the evolution and extinction of Stellar's sea cow, *Hydrodamalis gigas*. *Marine Mammal Science* 11:391–394.

Anderson, R. J., Rand, A., Rothman, M. D., Share, A., and Bolton, J. J. 2007. Mapping and quantifying the South African kelp resource. *African Journal of Marine Science* 29:369–378.

Anderson, T. J. and Yoklavich, M. M. 2007. Multiscale habitat associations of deepwater demersal fishes off central California. *Fishery Bulletin* 105:168–179.

Andrew, N. L. 1988. Ecological aspects of the common sea urchin, *Evechinus chloroticus*, in northern New Zealand: a review. *New Zealand Journal of Marine and Freshwater Research* 22:415–426.

Andrew, N. L. and Choat, J. H. 1982. The influence of predation and conspecific adults on the abundance of juvenile *Evechinus chloroticus* (Echinoidea, Echinometridae). *Oecologia* 54:80–87.

Andrew, N. L. and O'Neill, A. L. 2000. Large-scale patterns in habitat structure on subtidal rocky reefs in New South Wales. *Marine and Freshwater Research* 51:255–263.

Andrew, N. L. and Underwood, A. J. 1993. Density dependent foraging in the sea urchin *Centrostephanus rodgersii* on shallow subtidal reefs in New South Wales, Australia. *Marine Ecology Progress Series* 99:89–98.

Andrews, H. L. 1945. The kelp beds of the Monterey region. *Ecology* 26:24–37.

Arkema, K. K., Reed, D. C., and Schroeter, S. C. 2009. Direct and indirect effects of giant kelp determine benthic community structure and dynamics. *Ecology* 90:3126–3137.

Armstrong, F. A. and LaFond, E. C. 1966. Chemical nutrient concentrations and their relationship to internal waves and turbidity off southern California. *Limnology and Oceanography* 11:538–547.

Arnold, K. E. and Manley, S. L. 1985. Carbon allocation in *Macrocystis pyrifera* (Phaeophyta): intrinsic variability in photosynthesis and respiration. *Journal of Phycology* 21:154–167.

Astorga, M. P., Hernández, C. E., Valenzuela, C. P., Avaria-Llautureo, J., and Westermeier, R. 2012. Origin, diversification and historical biogeography of the giant kelp genus *Macrocystis*: evidences from Bayesian phylogenetic analysis. *Revista de Biologia y Oceanographia* 47:573–579.

Attwood, C. G., Lucas, M. I., Probyn, T. A., McQuaid, C. D., and Fielding, P. J. 1991. Production and standing stocks of the kelp *Macrocystis laevis* Hay at the Prince Edward Islands, Subantarctic. *Polar Biology* 11:129–133.

Babcock, R. C., Kelly, S., Shears, N. T., Walker, J. W., and Willis, T. J. 1999. Changes in community structure in temperate marine reserves. *Marine Ecology Progress Series* 89:125–134.

Babcock, R. C., Shears, N. T., Alcala, A. C., Barrett, N. S., Edgar, G. E., Lafferty, K. D. et al. 2010. Decadal trends in marine reserves reveal differential rates of change in direct and indirect effects. *Proceedings of the National Academy of Sciences* 107:18256–18261.

Balch, D. M. 1909. On the chemistry of certain algae of the Pacific Coast. *The Journal of Industrial and Engineering Chemistry* 1:777–787.

Ballantine, W. J. and Gordon, D. P. 1979. New Zealand's first marine reserve, Cape Rodney to Okakiri Point, Leigh. *Biological Conservation* 15:273–280.

Ballantine, W. J. and Langlois, T. J. 2008. Marine reserves: the need of systems. *Hydrobiologia* 606:35–44.

Barilotti, D. C. and Silverthorne, W. 1972. A resource management study of *Gelidium robustum*. *Proceedings of the International Seaweed Symposium* 7:255–261.

Barilotti, D. C. and Zertuche-González, J. A. 1990. Ecological effects of seaweed harvesting in the Gulf of California and Pacific Ocean off Baja California and California. *Hydrobiologia* 204:35–40.

Barilotti, D. C., McPeak, R. H., and Dayton, P. K. 1985. Experimental studies on the effects of commercial kelp harvesting in central and southern California *Macrocystis pyrifera* kelp beds. *California Fish and Game* 71:4–20.

Barrales, H. and Lobban, C. S. 1975. The comparative ecology of *Macrocystis pyrifera* with emphasis on the forests of Chubt, Argentina. *Journal of Ecology* 63:657–677.

Barrett, N. S., Buxton, C. D., and Edgar, G. J. 2009. Changes in invertebrate and macroalgal populations in Tasmanian marine reserves in the decade following protection. *Journal of Experimental Marine Biology and Ecology* 370:104–119.

Barry, J. P. and Tegner, M. J. 1990. Inferring demographic processes from size frequency distributions: simple models indicate specific patterns of growth and mortality. *Fishery Bulletin* 88:13–19.

Barsky, K. C. 2001. California spiny lobster. Pp. 98–100. In Leet, W. S., Dewees, C. M., Klingbeil, R., and Larson, E. J. (eds), *California's Living Marine Resources: A Status Report*. California Department of Fish and Game, Sacramento, CA.

Basch, L. V. and Tegner, M. J. 2007. Reproductive responses of purple sea urchin (*Strongylocentrotus purpuratus*) populations to environmental conditions across a coastal depth gradient. *Bulletin of Marine Science* 81:255–282.

Bascom, W. 1964. *Waves and Beaches*. Doubleday & Co., Garden City, New York.

Bauhin, J. 1651. *Historia Plantarum Universalis*. Tomus III. Yverdon, Switzerland.

Baum, J. K. and Worm, B. 2009. Cascading top-down effects of changing oceanic predator abundances. *Journal of Animal Ecology* 78:699–714.

Bay, S. and Greenstein, D. 1993. *Toxic Effects of Elevated Salinity and Desalination Waste Water*. Southern California Coastal Water Research Project Annual Report 1992/1993, Costa Mesa, CA.

Beas-Luna, R. and Ladah, L. B. 2014. Latitudinal, seasonal, and small-scale spatial differences of the giant kelp, *Macrocystis pyrifera*, and an herbivore at their southern range limit in the northern hemisphere. *Botanica Marina* 57:73–83.

Beckley, L. E. and Branch, G. M. 1992. A quantitative scuba diving survey of the sublittoral macrobenthos at sub-Antarctic Marion Island. *Polar Biology* 11:553–563.

Bedford, D. 2001. Giant kelp. Pp. 227–281. In Leet, W. S., Dewees, C. M., Klingbeil, R., and Larson, E. J. (eds), *California's Living Marine Resources: A Status Report*. California Department of Fish and Game, Sacramento, CA.

Behrens, M. D. and Lafferty, K. D. 2004. Effects of marine reserves and urchin disease on southern Californian rocky reef communities. *Marine Ecology Progress Series* 279:129–139.

Benech, S. V. 1980. *Observations of the Sea Otter, Enhydris lutris, Population between Point Buchen and Rattlesnake Creek, San Luis Obispo, California January through December 1980*. ECOMAR Report VII-5-80. ECOMAR Inc., Santa Barbara, CA.

Bernardino, A. F., Smith, C. R., Baco, A., Altamira, I., and Sumida, P. Y. G. 2010. Macrofaunal succession in sediments around kelp and wood falls in the deep NE Pacific and community overlap with other reducing habitats. *Deep-Sea Research Part I: Oceanographic Research Papers* 57:708–723.

Bernstein, B. B. and Jung, N. 1979. Selective pressures and coevolution in a kelp canopy community in southern California. *Ecological Monographs* 49:335–355.

Bess, R. and Rallapudi, R. 2007. Spatial conflicts in New Zealand fisheries: the rights of fishers and protection of the marine environment. *Marine Policy* 31:719–729.

Beverton, R. J. H. and Holt, S. J. 1957. *On the Dynamics of Exploited Fish Populations*. Chapman & Hall, New York.

Bird, T. K. and Benson, J. 1987. *Seaweed Cultivation for Renewable Resources*. Elsevier Science Ltd, Amsterdam, the Netherlands.

Birks, J. B. 1962. *Rutherford at Manchester*. Heywood and Co., London, UK.

Bixler, H. J. and Porse, H. 2011. A decade of change in the seaweed hydrocolloids industry. *Journal of Applied Phycology* 23:321–335.

Blackford, J. C. 2010. Predicting the impacts of ocean acidification: challenges from an ecosystem perspective. *Journal of Marine Systems* 81:12–18.

Blanchette, C. A., Wieters, E. A., Broitman, B. R., Kinlan, B. P., and Schiel, D. R. 2009. Trophic structure and diversity in rocky intertidal upwelling ecosystems: a comparison of community patterns across California, Chile, South Africa and New Zealand. *Progress in Oceanography* 83:107–116.

Blankly, W. O. 1982. Feeding ecology of three inshore fish species at Marion Island (Southern Ocean). *South African Journal of Zoology* 17:164–170.

Boalch, G. T. 1981. Do we really need to grow *Macrocystis* in Europe? *Proceedings of the International Seaweed Symposium* 10:657–667.

Bohnsack, J. A. 1996. Marine reserves, zoning, and the future of fishery management. *Fisheries* 21:14–16.

Bolin, R. L. 1938. Reappearance of the southern sea otter along the California coast. *Journal of Mammalogy* 19:301–303.

Bolton, J. J. and Levitt, G. J. 1987. The influence of upwelling on South African west coast seaweeds. *South African Journal of Marine Science* 5:319–325.

Boolootian, R. A. 1961. The distribution of the California sea otter. *California Fish and Game* 47:287–292.

Bradley, R. A. and Bradley, D. W. 1993. Wintering shorebirds increase after kelp (*Macrocystis*) recovery. *Condor* 95:372–376.

Brandt, R. P. 1923. *Potash from Kelp: Early Development and Growth of Giant Kelp, Macrocystis pyrifera*. United States Department of Agriculture Bulletin No. 1191. United States Government Printing Office, Washington, DC.

Bray, R. N. 1981. Influence of water currents and zooplankton densities on daily foraging movements of blacksmith, *Chromis punctipinnis*, a planktivorous reef fish. *Fishery Bulletin* 78:829–841.

Bray, R. N., Miller, A. C., and Geesey, G. G. 1981. The fish connection: a trophic link between planktonic and rocky reef communities. *Science* 214:204–205.

Bray, R. N., Miller, A. C., Johnson, S., Krause, P. R., Robertson, D. L., and Westcott, A. M. 1988. Ammonium excretion by macroinvertebrates and fishes on a subtidal rocky reef in southern California. *Marine Biology* 100:21–30.

Breda, V. A. and Foster, M. S. 1985. Composition, abundance and phenology of foliose red algae associated with two central California kelp forests. *Journal of Experimental Marine Biology and Ecology* 94:115–130.

Breen, P. A. 1987. Seaweeds, sea urchins and lobsters: comment. *Canadian Journal of Fisheries and Aquatic Sciences* 44:1806–1807.

Breen, P. A. and Mann, K. H. 1976. Changing lobster abundance and destruction of kelp beds by sea urchins. *Marine Biology* 34:137–142.

Breitburg, D. L. 1984. Residual effects of grazing: inhibition of competitor recruitment by encrusting coralline algae. *Ecology* 65:1136–1143.

Brierley, A. S. and Kingsford, M. J. 2009. Impacts of climate change on marine organisms and ecosystems. *Current Biology* 19:R602–R614.

Briggs, J. C. 1974. *Marine Zoogeography*. McGraw-Hill, New York.

Britton-Simmons, K. H. 2004. Direct and indirect effects of the introduced alga *Sargassum muticum* on benthic, subtidal communities of Washington State, USA. *Marine Ecology Progress Series* 277:61–78.

Britton-Simmons, K. H., Foley, G., and Okamoto, D. 2009. Spatial subsidy in the subtidal zone: utilization of drift algae by a deep subtidal sea urchin. *Aquatic Biology* 5:233–243.

Brostoff, W. N. 1988. Taxonomic studies of *Macrocystis pyrifera* (L) C Agardh (Phaeophyta) in southern California: holdfasts and basal stipes. *Aquatic Botany* 31:289–305.

Brown, M. T., Nyman, M. A., Keough, J. A., and Chin, N. K. M. 1997. Seasonal growth of the giant kelp *Macrocystis pyrifera* in New Zealand. *Marine Biology* 129:417–424.

Bruno, J. F., Stachowicz, J. J., and Bertness, M. D. 2003. Inclusion of facilitation into ecological theory. *Trends in Ecology and Evolution* 18:119–125.

Bruno, J. F., Sweatman, H., Precht, W. F., Selig, E. R., and Schutte, V. G. W. 2009. Assessing evidence of phase shifts from coral to macroalgal dominance on coral reefs. *Ecology* 90:1478–1484.

Brzezinski, M. A., Reed, D. C., and Amsler, C. D. 1993. Neutral lipids as major storage products in zoospores of the giant kelp *Macrocystis pyrifera* (Phaeophyceae). *Journal of Phycology* 29:16–23.

Brzezinski, M. A., Reed, D. C., Harrer, S., Rassweiler, A., Melack, J. M., Goodridge, B. M. et al. 2013. Multiple sources and forms of nitrogen sustain year-round kelp growth on the Inner Continental Shelf of the Santa Barbara Channel. *Oceanography* 26: 114–123.

Buggeln, R. G., Fensom, D. S., and Emerson, C. J. 1985. Translocation of C-11 photoassimilate in the blade of *Macrocystis pyrifera* (Phaeophyceae). *Journal of Phycology* 21:35–40.

Burgman, M. A. and Gerard, V. 1990. A stage-structured, stochastic population model for the giant kelp *Macrocystis pyrifera*. *Marine Biology* 105:15–23.

Burridge, T. R. and Bidwell, J. 2002. Review of the potential use of brown algal ecotoxicological assays in monitoring effluent discharge and pollution in Southern Australia. *Marine Pollution Bulletin* 45:140–147.

Buschmann, A. H., Vaquez, J., Osorio, P., Reyes, E., Filun, L., Hernandez-González, M. C. et al. 2004. The effect of water movement, temperature and salinity on abundance and reproductive patterns of *Macrocystis* spp. (Phaeophyta) at different latitudes in Chile. *Marine Biology* 145:849–862.

Buschmann, A. H., Moreno, C., Vásquez, J. A., and Hernández-González, M. C. 2006. Reproduction strategies of *Macrocystis pyrifera* (Phaeophyta) in Southern Chile: the importance of population dynamics. *Journal of Applied Phycology* 18:575–582.

Buschmann, A. H., Hernández-González, M. C., and Varela, D. 2008. Seaweed future cultivation in Chile: perspectives and challenges. *International Journal of Environment and Pollution* 33:432–456.

Buschmann, A. H., Varela, D., Hernández-González, M., and Huovinen, P. 2008. Opportunities and challenges for the development of an integrated seaweed-based aquaculture activity in Chile: determining the physiological capabilities of *Macrocystis* and *Gracilaria* as biofilters. *Journal of Applied Phycology* 20:571–577.

Buschmann, A. H., Pereda, S. V., Varela, D. A., Hernandez-Gonzalez, M. C., Rodriguez-Maulen, J., Lopez, A. et al. 2013. Ecophysiological responses of annual populations of *Macrocystis pyrifera* under winter light conditions. *Phycologia* 52:15.

Businessgreen. 2012. *Seaweeds offer kelping hand for biofuels.* www.businessgreen.com.

Bustamante, R. H., Branch, G. M., and Eekhout, S. 1995. Maintenance of an exceptional intertidal grazer biomass in South Africa: subsidy by subtidal kelps. *Ecology* 76:2314–2329.

Butler, J., Neuman, M., Pinkard, D., Kvitek, R., and Cochrane, G. 2006. The use of multibeam sonar mapping technique to refine population estimates of the endangered white abalone (*Haliotis sorenseni*). *Fishery Bulletin* 104:521–532.

Byrne, M. 2011. Impact of ocean warming and ocean acidification on marine invertebrate life history stages: vulnerabilities and potential for persistence in a changing ocean. *Oceanography and Marine Biology: An Annual Review* 49:1–42.

Byrnes, J. E., Stachowicz, J. J., Hultgren, K. M., Hughes, A. R., Olyarnik, S. V., and Thornber, C. S. 2006. Predator diversity strengthens trophic cascades in kelp forests by modifying herbivore behaviour. *Ecology Letters* 9:61–71.

Byrnes, J. E., Reed, D. C., Cardinale, B. J., Cavanaugh, K. C., Holbrook, S. J., and Schmitt, R. J. 2011. Climate-driven increases in storm frequency simplify kelp forest food webs. *Global Change Biology* 17:2513–2524.

Byrnes, J. E., Cardinale, B. J., and Reed, D. C. 2013. Interactions between sea urchin grazing and prey diversity on temperate rocky reef communities. *Ecology* 94:1636–1646.

Cailliet, G. M. and Lea, R. N. 1977. Abundance of rare zoarcid, *Maynea californica* Gilbert, 1915, in Monterey Canyon, Monterey Bay, California. *California Fish and Game* 63:253–261.

Caissie, B. E., Brigham-Grette, J., Lawrence, K. T., Herbert, T. D., and Cook, M. S. 2010. Last Glacial Maximum to Holocene sea surface conditions at Umnak Plateau, Bering Sea, as inferred from diatom, alkenone, and stable isotope records. *Paleoceanography* 25:PA1206. doi:10.1029/2008PA001671.

Cameron, F. K. 1915. *Potash from kelp.* Report No. 100, United States Government Printing Office, Washington, DC.

Camus, P. A. 1994. Recruitment of the intertidal kelp *Lessonia nigrescens* Bory in northern Chile: successional constraints and opportunities. *Journal of Experimental Marine Biology and Ecology* 184:171–181.

Camus, P. A. 2001. Marine biogeography of continental Chile. *Revista Chilena De Historia Natural* 74:587–617.

Carney, L. T. 2011. A multispecies laboratory assessment of rapid sporophyte recruitment from delayed kelp gametophytes. *Journal of Phycology* 47:244–251.

Carney, L. T. and Edwards, M. S. 2006. Delayed development in kelp forest systems. *Journal of Phycology* 42:28–28.

Carney, L. T. and Edwards, M. S. 2010. Role of nutrient fluctuations and delayed development in gametophyte reproduction by *Macrocystis pyrifera* (Phaeophyceae) in southern California. *Journal of Phycology* 46:987–996.

Carney, L. T., Waaland, J. R., Klinger, T., and Ewing, K. 2005. Restoration of the bull kelp *Nereocystis luetkeana* in nearshore rocky habitats. *Marine Ecology Progress Series* 302:49–61.

Carney, L. T., Bohonak, A. J., Edwards, M. S., and Alberto, F. 2013. Genetic and experimental evidence for a mixed-age, mixed-origin bank of kelp microscopic stages in southern California. *Ecology* 94:1955–1965.

Carr, M. H. 1989. Effects of macroalgal assemblages on the recruitment of temperate zone reef fishes. *Journal of Experimental Marine Biology and Ecology* 126:59–76.

Carr, M. H. 1991. Habitat selection and recruitment of an assemblage of temperate zone reef fishes. *Journal of Experimental Marine Biology and Ecology* 146:113–137.

Carr, M. H. 1994. Effects of macroalgal dynamics on recruitment of a temperate reef fish. *Ecology* 75:1320–1333.

Carr, M. H. 2000. Marine protected areas: challenges and opportunities for understanding and conserving coastal marine ecosystems. *Environmental Conservation* 27:106–109.

Carr, M. H. and Reed, D. C. 1993. Conceptual issues relevant to marine harvest refuges: examples from temperate reef fishes. *Canadian Journal of Fisheries and Aquatic Sciences* 50:2019–2028.

Carr, M. H. and Reed, D. C. In press. Shallow rocky reefs and kelp forests. In Zabaleta, E. (ed.), *Ecosystems of California*. University of California Press, Berkeley, CA.

Carr, M. H., Neigel, J. E., Estes, J. A., Andelman, S., Warner, R. R., and Largier, J. L. 2003. Comparing marine and terrestrial ecosystems: implications for the design of coastal marine reserves. *Ecological Applications* 13:S90–S107.

Carroll, J. C., Engle, J. M., Coyer, J. A., and Ambrose, R. F. 2000. Long-term changes and species interactions in a sea urchin dominated community at Anacapa Island, California. Pp. 370–378. In Browne, D. R., Mitchell, K. L., and Chaney, H. W. (eds), *Proceedings of the 5th California Channel Islands Symposium*. Santa Barbara Museum of Natural History, Santa Barbara, CA.

Carter, J. W., Jessee, W. N., Foster, M. S., and Carpenter, A. L. 1985. Management of artificial reefs designed to support natural communities. *Bulletin of Marine Science* 37:114–128.

Carter, J. W., Carpenter, A. L., Foster, M. S., and Jessee, W. N. 1985. Benthic succession on an artificial reef designed to support a kelp reef community. *Bulletin of Marine Science* 37: 86–113.

Casas, G., Scrosati, R., and Luz Piriz, M. 2004. The invasive kelp *Undaria pinnatifida* (Phaeophyceae, Laminariales) reduces native seaweed diversity in Nuevo Gulf (Patagonia, Argentina). *Biological Invasions* 6:411–416.

Caselle, J. E., Hamilton, S. L., Schroeder, D. M., Love, M. S., Standish, J. D., Rosales-Casian, J. et al. 2011. Geographic variation in density, demography, and life history traits of a harvested, sex-changing, temperate reef fish. *Canadian Journal of Fisheries and Aquatic Sciences* 68:288–303.

Castilla, J. C. 1985. Food webs and functional aspects of the kelp, *Macrocystis pyrifera*, community in the Beagle Channel, Chile. Pp. 407–414. In Siegfried, W. W., Condy, P. R., and Laws, R. M. (eds), *Antarctic Nutrient Cycles and Food Webs*. Springer-Verlag, Berlin, Germany.

Castilla, J. C. 1999. Coastal marine communities: trends and perspectives from human-exclusion experiments. *Trends in Ecology and Evolution* 14:280–283.

Castilla, J. C. 2000. Roles of experimental marine ecology in coastal management and conservation. *Journal of Experimental Marine Biology and Ecology* 250:3–21.

Castilla, J. C. and Duran, L. R. 1985. Human exclusion from the rocky intertidal zone of central Chile: the effects on *Concholepas concholepas* (Gastropoda). *Oikos* 45:391–399.

Castilla, J. C. and Fernandez, M. 1998. Small-scale benthic fisheries in Chile: on co-management and sustainable use of benthic invertebrates. *Ecological Applications* 8:S124–S132.

Castilla, J. C. and Gelcich, S. 2008. *Management of the Loco (Concholepas concholepas) as a Driver for Self-Governance of Small-Scale Benthic Fisheries in Chile*. FAO Fisheries Technical Paper 504. FAO, Rome, Italy.

Castilla, J. C. and Moreno, C. 1982. Sea urchins and *Macrocystis pyrifera*: experimental test of their ecological relations in southern Chile. Pp. 257–263. In Lawrence, J. M. (ed.), *Proceedings of the International Echinoderm Conference*, Tampa Bay, Florida.

Castilla, J. C., Uribe, M., Bahamonde, N., Clarke, M., Desqueyroux-Faundez, R., Kong, I. et al. 2005. Down under the southeastern Pacific: marine non-indigenous species in Chile. *Biological Invasions* 7:213-232.

Cavalier-Smith, T. 2010. Kingdoms Protozoa and Chromista and the eozoan root of the eukaryotic tree. *Biology Letters* 6:342–345.

Cavanaugh, K. C., Siegel, D. A., Kinlan, B. P., and Reed, D. C. 2010. Scaling giant kelp field measurements to regional scales using satellite observations. *Marine Ecology Progress Series* 403:13–27.

Cavanaugh, K. C., Siegel, D. A., Reed, D. C., and Dennison, P. E. 2011. Environmental controls of giant-kelp biomass in the Santa Barbara Channel, California. *Marine Ecology Progress Series* 429:1–17.

Cavanaugh, K. C., Kendall, B. E., Siegel, D. A., Reed, D. C., Alberto, F., and Assis, J. 2013. Synchrony in dynamics of giant kelp forests is driven by both local recruitment and regional environmental controls. *Ecology* 94:499–509.

CDFG (California Department of Fish and Game). 2005. *California Marine Life Protection Act Initiative Draft Master Plan Framework: A Recommendation to the California Fish and Game Commission.* CDFG, Sacramento, CA.

CDFG (California Department of Fish and Game). 2011. *Marine Management News.* Pp. 1–10. California Department of Fish and Game. http://www.dfg.ca.gov/marine/newsletter/archive.asp.

CDFW (California Department of Fish and Wildlife). 2013. *Marine Protected Areas Update.* California Department of Fish and Wildlife. https://nrm.dfg.ca.gov/FileHandler.ashx?DocumentID=46029&inline=true.

CDFW (California Department of Fish and Wildlife). 2014a. *California 14–15 Ocean Sport Fishing Regulations.* California Department of Fish and Wildlife. www.wildlife.ca.gov.

CDFW (California Department of Fish and Wildlife) (2014b) *California Commercial Landings Data.* California Department of Fish and Wildlife. http://www.dfg.ca.gov/marine/landings/landings11.asp.

Cerda, O., Karsten, U., Rothausler, E., Tala, F., and Thiel, M. 2009. Compensatory growth of the kelp *Macrocystis integrifolia* (Phaeophyceae, Laminariales) against grazing of *Peramphithoe femorata* (Amphipoda, Ampithoidae) in northern-central Chile. *Journal of Experimental Marine Biology and Ecology* 377:61–67.

Cerda, O., Hinojosa, I. A., and Thiel, M. 2010. Nest-building behavior by the amphipod *Peramphithoe femorata* (Kroyer) on the kelp *Macrocystis pyrifera* (Linnaeus) C. Agardh from Northern-Central Chile. *Biological Bulletin* 218:248–258.

Chadwick, N. E. 1991. Spatial distribution and the effects of competition on some temperate *Scleractinia* and *Corallimorpharia*. *Marine Ecology Progress Series* 70:39–48.

Chapman, A. R. O. 1981. Stability of sea-urchin dominated barren grounds following destructive grazing of kelp in St Margarets Bay, Eastern Canada. *Marine Biology* 62:307–311.

Chapman, A. R. O. 1984. Reproduction, recruitment and mortality in two species of *Laminaria* in southwest Nova Scotia. *Journal of Experimental Marine Biology and Ecology* 78:99–109.

Chapman, A. R. O. 1986. Age versus stage: an analysis of age-specific and size-specific mortality and reproduction in a population of *Laminaria longicruris*. *Journal of Experimental Marine Biology and Ecology* 97:113–122.

Chapman, A. R. O. and Craigie, J. S. 1977. Seasonal growth in *Laminaria longicruris*: relations with dissolved inorganic nutrients and internal reserves of nitrogen. *Marine Biology* 40:197–205.

Chapman, A. R. O. and Lindley, J. E. 1980. Seasonal growth of *Laminaria solidungula* in the Canadian high arctic in relation to irradiance and dissolved nutrient concentrations. *Marine Biology* 57:1–5.

Chapman, V. J. 1970. *Seaweeds and Their Uses*. Methuen and Co. Ltd., London, UK.

Chapman, V. J. and Chapman, D. J. 1980. *Seaweeds and Their Uses*. Chapman & Hall, New York.

Charters, A. C., Neushul, M., and Barilotti, D. C. 1969. The functional morphology of *Eisenia arborea*. *Proceedings of the International Seaweed Symposium* 6:89–105.

Cheng, T. H. 1969. Production of kelp: a major aspect of China's exploitation of the sea. *Economic Botany* 23:215–236.

Chien, P. K. 1972. Attachment processes and early development by kelp embryos: a light and electron microscope study. Pp. 108–122. In North, W. J. (ed.), *Kelp Habitat Improvement Project Annual Report 1 July, 1971–30 June, 1972*. W.M. Keck Laboratory of Environmental Health Engineering, California Institute of Technology, Pasadena, CA.

Chien, P. K. 1973. Attachment of *Macrocystis* to natural and artificial substrates. Pp. 89–106. In North, W. J. (ed.), *Kelp Habitat Improvement Project Annual Report 1 July, 1972–30 June, 1973*. W.M. Keck Laboratory of Environmental Health Engineering, California Institute of Technology, Pasadena, CA.

Chin, N. K. M., Brown, M. T., and Heads, M. J. 1991. The biogeography of Lessoniaceae, with special reference to *Macrocystis* Agardh, C. (Phaeophyta, Laminariales). *Hydrobiologia* 215:1–11.

Choat, J. H. and Andrew, N. L. 1986. Interactions amongst species in a guild of subtidal benthic herbivores. *Oecologia* 68:387–394.

Choat, J. H. and Ayling, A. M. 1987. The relationship between habitat structure and fish faunas on New Zealand reefs. *Journal of Experimental Marine Biology and Ecology* 110:257–284.

Choat, J. H. and Schiel, D. R. 1982. Patterns of distribution and abundance of large brown algae and invertebrate herbivores in subtidal regions of northern New Zealand. *Journal of Experimental Marine Biology and Ecology* 60:129–162.

Chung, I. K., Beardall, J., Mehta, S., Sahoo, D., and Stojkovic, S. 2011. Using marine macroalgae for carbon sequestration: a critical appraisal. *Journal of Applied Phycology* 23:877–886.

Cie, D. K. and Edwards, M. S. 2008. The effects of high irradiance on the settlement competency and viability of kelp zoospores. *Journal of Phycology* 44:495–500.

Clark, R. P., Edwards, M. S., and Foster, M. S. 2004. Effects of shade from multiple kelp canopies on an understory algal assemblage. *Marine Ecology Progress Series* 267:107–119.

Clarke, W. D. 1971. Mysids of the southern kelp region. Pp. 369–380. In North, W. J. (ed.), *The Biology of Giant Kelp Beds (Macrocystis) in California*. Nova Hedwigia 32. Verlag Von J. Cramer, Lehre, Germany.

Clarke, W. D. and Neushul, M. 1967. Subtidal ecology of the southern California coast. Pp. 29–42. In Olson, T. A. and Burgess, F. J. (eds), *Pollution and Marine Ecology*. Interscience, New York.

Clements, K. D. and Choat, J. H. 1993. Influence of season, ontogeny and tide on the diet of the temperate marine herbivorous fish *Odax pullus* (Odacidae). *Marine Biology* 117:213–220.

Clendenning, K.A. 1971a. Photosynthesis and general development in *Macrocystis*. Pp. 169–190. In North, W.J. (ed.), *The Biology of Giant Kelp Beds (Macrocystis) in California*. Nova Hedwigia 32. Verlag Von J. Cramer, Lehre, Germany.

Clendenning, K.A. 1971b. Organic productivity in kelp areas. Pp. 259–263. In North, W.J. (ed.), *The Biology of Giant Kelp Beds (Macrocystis) in California*. Nova Hedwigia 32. Verlag Von J. Cramer, Lehre, Germany.

Cloud, P. 1976. Beginnings of biospheric evolution and their biogeochemical consequences. *Paleobiology* 2:351–387.

Cock, J.M., Sterck, L., Rouze, P., Scornet, D., Allen, A.E., Amoutzias, G. et al. 2010. The *Ectocarpus* genome and the independent evolution of multicellularity in brown algae. *Nature* 465:617–621.

Cole, K. 1968. Gametophytic development and fertilization in *Macrocystis integrifolia*. *Canadian Journal of Botany* 46:777–782.

Coleman, F.C., Figueira, W.F., Ueland, J.S., and Crowder, L.B. 2004. The impact of United States recreational fisheries on marine fish populations. *Science* 305:1958–1960.

Connell, J.H. 1974. Ecology: field experiments in marine ecology. Pp. 21–54. In Mariscal, R.N. (ed.), *Experimental Marine Biology*. Academic Press, New York.

Connell, J.H. 1978. Diversity in tropical rain forests and coral reefs. *Science* 199:1302–1310.

Connell, J.H. and Sousa, W.P. 1983. On the evidence needed to judge ecological stability or persistence. *American Naturalist* 121:789–824.

Connell, S.D. 2005. Assembly and maintenance of subtidal habitat heterogeneity: synergistic effects of light penetration and sedimentation. *Marine Ecology Progress Series* 289: 53–61.

Connell, S.D. 2007. Water quality and the loss of coral reefs and kelp forests: alternative states and the influence of fishing. Pp. 556–568. In Connell, S.D. and Gillanders, B.M. (eds), *Marine Ecology*. Oxford University Press, Sydney, Australia.

Connell, S.D. and Irving, A.D. 2008. Integrating ecology with biogeography using landscape characteristics: a case study of subtidal habitat across continental Australia. *Journal of Biogeography* 35:1608–1621.

Contreras, D., Schlatter, R., and Ramirez, C. 1983. Flora ficologica de las Islas Diego Ramirez (Chile). *INACH Ser. Client (Scientific Series of the Chilean Antarctic Institute)* 30:13–26.

Conversi, A. and McGowan, J.A. 1994. Natural versus human caused variability of water clarity in the southern California bight. *Limnology and Oceanography* 39:632–648.

Cook, P.A. and Gordon, H.R. 2010. World abalone supply, markets, and pricing. *Journal of Shellfish Research* 29:569–571.

Cooke, S.J. and Cowx, I.G. 2004. The role of recreational fishing in global fish crises. *Bioscience* 54:857–859.

Cooke, S.J. and Cowx, I.G. 2006. Contrasting recreational and commercial fishing: searching for common issues to promote unified conservation of fisheries resources and aquatic environments. *Biological Conservation* 128:93–108.

Cooper, J., Wieland, M., and Hines, A. 1977. Subtidal abalone populations in an area inhabited by sea otters. *Veliger* 20:163–167.

Costa, D. 1978. The sea otter: its interaction with man. *Oceanus* 21:24–30.

Costa, D. P. and Kooyman, G. L. 1984. Contribution of specific dynamic action to heat balance and the thermoregulation in the sea otter *Enhydra lutris*. *Physiological Zoology* 57:199–203.

Cotton, A. D. 1912. Marine algae. Clare Island Survey, Part 15. *Proceedings of the Royal Irish Academy B* 31:1–178.

Cousteau, J. and Schiefelbein, S. 2007. *The Human, the Orchid, and the Octopus: Exploring and Conserving Our Natural World*. Bloomsbury Publishing, New York.

Cowen, R. K. 1983. The effect of Sheephead (*Semicossyphus pulcher*) predation on red sea urchin (*Strongylocentrotus franciscanus*) populations: an experimental analysis. *Oecologia* 58:249–255.

Cowen, R. K. 1985. Large scale pattern of recruitment by the labrid, *Semicossyphus pulcher*: causes and implications. *Journal of Marine Research* 43:719–742.

Cowen, R. K., Agegian, C. R., and Foster, M. S. 1982. The maintenance of community structure in a central California giant kelp forest. *Journal of Experimental Marine Biology and Ecology* 64:189–201.

Cox, T. E. and Foster, M. S. 2013. The effects of storm-drains with periodic flows on intertidal algal assemblages in 'Ewa Beach (O'ahu), Hawai'i. *Marine Pollution Bulletin* 70:162–170.

Coyer, J. A. 1984. The invertebrate assemblage associated with the giant kelp, *Macrocystis pyrifera*, at Santa Catalina Island, California: a general description with emphasis on amphipods, copepods, mysids and shrimps. *Fishery Bulletin* 82:55–66.

Coyer, J. A., Ambrose, R. F., Engle, J. M., and Carroll, J. C. 1993. Interactions between corals and algae on a temperate zone rocky reef: mediation by sea urchins. *Journal of Experimental Marine Biology and Ecology* 167:21–37.

Coyer, J. A., Robertson, D. L., and Alberte, R. S. 1994. Genetic variability within a population and between diploid/haploid tissue of *Macrocystis pyrifera* (Phaeophyceae). *Journal of Phycology* 30:545–552.

Coyer, J. A., Smith, G. J., and Andersen, R. A. 2001. Evolution of *Macrocystis* spp. (Phaeophyceae) as determined by ITS1 and ITS2 sequences. *Journal of Phycology* 37:574–585.

Crandall, W. C. 1912. *The Kelps of the Southern California Coast*. United States 62nd Congress, 2nd Senate Session, Document 190, Appendix N., US Congressional Record, Washington, DC.

Crandall, W. C. 1915. The kelp beds from lower California to Puget Sound. In Cameron, F. K. (ed.), *Potash from Kelp*. U.S. Department of Agriculture, Washington, DC.

Crandall, W. C. 1918. A review of the kelp industry. *California Fish and Game* 4:105–107.

Cribb, A. B. 1954. *Macrocystis pyrifera* (L.) Ag. in Tasmanian waters. *Australian Journal of Marine and Freshwater Research* 5:1–34.

Critchley, A. T. and Dijkema, R. 1984. On the presence of the introduced brown algal *Sargassum muticum*, attached to commercially imported *Ostrea edulis* in the SW Netherlands. *Botanica Marina* 27:211–216.

Critchley, A. T., Farnham, W. F., Yoshida, T., and Norton, T. A. 1990. A bibliography of the invasive alga *Sargassum muticum* (Yendo) Fensholt (Fucales; Sargassaceae). *Botanica Marina* 33:551–562.

Darwin, C. 1839. *The Voyage of the Beagle*. (Reprinted in 1962, The Harvard Classics). P. F. Collier & Son Company, New York.

Darwin, C. 1890. *Journal of Researches into the Natural History and Geology of the Countries Visited during the Voyage Round the World of H.M.S. 'Beagle' under the Command of Captain Fitzroy.* R.N. John Murray, London, UK.

Daume, S., Brand-Gardner, S., and Woelkerling, W. J. 1999. Settlement of abalone larvae (*Haliotis laevigata* Donovan) in response to non-geniculate coralline red algae (Corallinales, Rhodophyta). *Journal of Experimental Marine Biology and Ecology* 234:125–143.

Daume, S., Brand-Gardner, S., and Woelkerling, W. J. 1999b. Preferential settlement of abalone larvae: diatom films vs. non-geniculate coralline red algae. *Aquaculture* 174:243–254.

Davenport, A. C. and Anderson, T. W. 2007. Positive indirect effects of reef fishes on kelp performance: the importance of mesograzers. *Ecology* 88:1548–1561.

Davis, G. E. 1989. Designated harvest refugia: the next stage of marine fishery management in California. *CalCOFI Reports* 30:53–58.

Davis, G. E., Haaker, P. L., and Richards, D. V. 1996. Status and threats of white abalone at the California Channel Islands. *Transactions of the American Fisheries Society* 125(1):42–48.

Davis, G. E. 2005. Science and society: marine reserve design for the California Channel Islands. *Conservation Biology* 19(6):1745–1751.

Davis, G. E., Haaker, P. L., and Richards, D. V. 1998. The perilous condition of white abalone *Haliotis sorenseni*, Bartsch, 1940. *Journal of Shellfish Research* 17:871–876.

Dawson, E. Y. 1951. A further study of upwelling and associated vegetation along Pacific Baja California, Mexico. *Journal of Marine Research* 10:39–58.

Dawson, E. Y., Neushul, M., and Wildman, R. D. 1960. Seaweeds associated with kelp beds along southern California and northwestern Mexico. *Pacific Naturalist* 1:1–81.

Day, J., Hockings, M., Holmes, G., Laffoley, D., Stolton, S., and Wells, S. 2012. *Guidelines for Applying the IUCN Protected Areas Management Categories to Marine Protected Areas.* IUCN, Gland, Switzerland.

Dayton, P. K. 1973. Dispersion, dispersal, and persistence of annual intertidal alga, *Postelsia palmaeformis* Ruprecht. *Ecology* 54:433–438.

Dayton, P. K. 1975. Experimental studies of algal canopy interactions in a sea otter dominated kelp community at Amchitka Island, Alaska. *Fishery Bulletin* 73:230–237.

Dayton, P. K. 1985a. Ecology of kelp communities. *Annual Review of Ecology and Systematics* 16:215–245.

Dayton, P. K. 1985b. The structure and regulation of some South American kelp communities. *Ecological Monographs* 55:447–468.

Dayton, P. K. 2003. The importance of the natural sciences to conservation. *American Naturalist* 162:1–13.

Dayton, P. K. and Tegner, M. J. 1984. Catastrophic storms, El Niño, and patch stability in a southern California kelp community. *Science* 224:283–285.

Dayton, P. K., Currie, V., Gerrodette, T., Keller, B. D., Rosenthal, R., and Ventresca, D. 1984. Patch dynamics and stability of some California kelp communities. *Ecological Monographs* 54:253–289.

Dayton, P. K., Seymour, R. J., Parnell, P. E., and Tegner, M. J. 1989. Unusual marine erosion in San Diego county from a single storm. *Estuarine, Coastal and Shelf Science* 29:151–160.

Dayton, P. K., Tegner, M. J., Parnell, P. E., and Edwards, P. B. 1992. Temporal and spatial patterns of disturbance and recovery in a kelp forest community. *Ecological Monographs* 62:421–445.

Dayton, P. K., Tegner, M. J., Edwards, P. B., and Riser, K. L. 1998. Sliding baselines, ghosts, and reduced expectations in kelp forest communities. *Ecological Applications* 8:309–322.

Dayton, P. K., Tegner, M. J., Edwards, P. B., and Riser, K. L. 1999. Temporal and spatial scales of kelp demography: the role of oceanographic climate. *Ecological Monographs* 69:219–250.

Dayton, P. K., Sala, E., Tegner, M. J., and Thrush, S. 2000. Marine reserves: parks, baselines, and fishery enhancement. *Bulletin of Marine Science* 66(3):617–634.

de Vernal, A., Eynaud, F., Henry, M., Hillaire-Marcel, C., Londeix, L., Mangin, S. et al. 2005. Reconstruction of sea-surface conditions at middle to high latitudes of the Northern Hemisphere during the Last Glacial Maximum (LGM) based on dinoflagellate cyst assemblages. *Quaternary Science Reviews* 24:897–924.

Dean, P. R. and Hurd, C. L. 2007. Seasonal growth, erosion rates, and nitrogen and photosynthetic ecophysiology of *Undaria pinnatifida* (heterokontophyta) in southern New Zealand. *Journal of Phycology* 43:1138–1148.

Dean, T. A. 1985. The temporal and spatial distribution of underwater quantum irradiance in a southern California kelp forest. *Estaurine, Coastal and Shelf Science* 21:835–844.

Dean, T. A. and Jacobsen, F. R. 1984. Growth of juvenile *Macrocystis pyrifera* (Laminariales) in relation to environmental factors. *Marine Biology* 83:301–311.

Dean, T. A. and Jacobsen, F. R. 1986. Nutrient-limited growth of juvenile kelp, *Macrocystis pyrifera*, during the 1982-1984 "El Niño" in southern California. *Marine Biology* 90:597–601.

Dean, T. A. and Deysher, L. E. 1983. The effects of suspended solids and thermal discharges on kelp. Pp. 114–135. In Bascom, W. (ed.), *The Effects of Waste Disposal on Kelp Communities*. Southern California Coastal Water Research Project, Long Beach, CA.

Dean, T. A., Deysher, L. E., Theis, K., Lagos, S., Jacobsen, F., and Bost, L. 1983. *The Effects of the San Onofre Nuclear Generating Station (Song) on the Giant Kelp Macrocystis Pyrifera: Final Preoperational Monitoring Report*. Marine Review Committee Inc., Encinitas, CA.

Dean, T. A., Schroeter, S. C., and Dixon, J. D. 1984. Effects of grazing by two species of sea urchins (*Strongylocentrotus franciscanus* and *Lytechinus anamesus*) on recruitment and survival of species of kelp (*Macrocystis pyrifera* and *Pterogophora californica*). *Marine Biology* 78:301–313.

Dean, T. A., Thies, K., and Lagos, S. L. 1989. Survival of juvenile giant kelp: the effects of demographic factors, competitors, and grazers. *Ecology* 70:483–495.

Dearn, S. L. 1987. *The Fauna of Subtidal Articulated Coralline Mats: Composition, Dynamics and Effects of Spatial Heterogeneity*. MSc thesis. California State University, Stanislaus, CA.

DeBoer, J. A. 1981. Nutrients. Pp. 356–392. In Lobban, C. S. and Wynne, M. J. (eds), *The Biology of Seaweeds*. University of California Press, Berkeley, CA.

Deiman, M., Iken, K., and Konar, B. 2012. Susceptibility of *Nerocystis luetkeana* (LamNiñari-ales, Ochrophyta) and *Eualaria fistulosa* (Laminariales, Ochrophyta) spores to sedimenta-tion. *Algae* 27:115–123.

Delille, D. and Perret, E. 1991. The influence of giant kelp *Macrocystis pyrifera* on the growth of sub-Antarctic marine bacteria. *Journal of Experimental Marine Biology and Ecology* 153:227–239.

DeMartini, E. E. and Roberts, D. A. 1990. Effects of giant kelp (*Macrocystis*) on the den-sity and abundance of fishes in a cobble-bottom kelp forest. *Bulletin of Marine Science* 46:287–300.

Demes, K. W., Graham, M. H., and Suskiewicz, T. S. 2009. Phenotypic plasticity reconciles incongruous molecular and morphological taxonomies: the giant kelp, *Macrocystis* (Lami-nariales, Phaeophyceae) is a monospecific genus. *Journal of Phycology* 45:1266–1269.

Denny, M. W. and Gaylord, B. 2010. Marine ecomechanics. *Annual Review of Marine Science* 2:89–114.

Denny, M. W. and Roberson, L. 2002. Blade motion and nutrient flux to the kelp, *Eisenia arborea*. *Biological Bulletin* 203:1–13.

Denny, M. W. and Shibata, M. F. 1989. Consequences of surf-zone turbulence for settlement and external fertilization. *American Naturalist* 134:859–889.

Denny, M. W., Gaylord, B., and Cowen, E. A. 1997. Flow and flexibility. II. The roles of size and shape in determining wave forces on the bull kelp *Nereocystis luetkeana*. *Journal of Experimental Biology* 200:3165–3183.

Denny, M. W., Gaylord, B., Helmuth, B., and Daniel, T. 1998. The menace of momentum: dynamic forces on flexible organisms. *Limnology and Oceanography* 43:955–968.

Denny, M. W., Hunt, L. J. H., Miller, L. P., and Harley, C. D. G. 2009. On the prediction of extreme ecological events. *Ecological Monographs* 79:397–421.

Department of Conservation and Ministry of Fisheries. 2005. *Marine Protected Areas: Policy and Implementation Plan*. Department of Conservation and Ministry of Fisheries, Wel-lington, New Zealand.

Devinny, J. S. and Volse, L. A. 1978. Effects of sediments on the development of *Macrocystis pyrifera* gametophytes. *Marine Biology* 48:343–348.

Deysher, L. E. 1993. Evaluation of remote-sensing techniques for monitoring giant kelp populations. *Hydrobiologia* 261:307–312.

Deysher, L. E. and Dean, T. A. 1984. Critical irradiance levels and the interactive effects of quantum irradiance and dose on gametogenesis in the giant kelp, *Macrocystis pyrifera*. *Journal of Phycology* 20:520–524.

Deysher, L. E. and Dean, T. A. 1986a. In situ recruitment of sporophytes of the giant kelp, *Macrocystis pyrifera* (L.) C. Agardh: effects of physical factors. *Journal of Experimental Marine Biology and Ecology* 103:41–63.

Deysher, L. E. and Dean, T. A. 1986b. Interactive effects of light and temperature on sporophyte production in the giant kelp *Macrocystis pyrifera*. *Marine Biology* 93: 17–20.

Dillehay, T. D., Ramirez, C., Pino, M., Collins, M. B., Rossen, J., and Pino-Navarro, J. D. 2008. Monte Verde: seaweed, food, medicine, and the peopling of South America. *Sci-ence* 320:784–786.

Dixon, J., Schroeter, S. C., and Kastendiek, J. 1981. Effects on the encrusting bryozoan, *Membranipora membranacea*, on the loss of blades and fronds by the giant kelp, *Macrocystis pyrifera* (Laminariales). *Journal of Phycology* 17:341–345.

Domning, D. 1972. Steller's sea cow and the origin of North Pacific aboriginal whaling. *Syesis* 5:187–189.

Domning, D. P. 1989. Kelp evolution: a comment. *Paleobiology* 15:53–56.

Doney, S. C., Fabry, V. J., Feely, R. A., and Kleypas, J. A. 2009. Ocean acidification: the other $CO_2$ problem. *Annual Review of Marine Science* 1:169–192.

Dong, C. M. and McWilliams, J. C. 2007. A numerical study of island wakes in the Southern California Bight. *Continental Shelf Research* 27:1233–1248.

Dong, C. M., Idica, E. Y., and McWilliams, J. C. 2009. Circulation and multiple-scale variability in the Southern California Bight. *Progress in Oceanography* 82:168–190.

Donner, S. D., Skirving, W. J., Little, C. M., Oppenheimer, M., and Hoegh-Guldberg, O. 2005. Global assessment of coral bleaching and required rates of adaptation under climate change. *Global Change Biology* 11:2251–2265.

Draget, K., Smidsrød, O., and Skjåk-Bræk, G. 2005. Alginates from algae. Pp. 1–30. In Steinbuchel, A. and Ree, S. (eds), *Polysaccharides and Polyamides in the Food Industry*. Wiley-VCH Verlag, Weinheim, Germany.

Dring, M. J. 1981. Chromatic adaptation of photosynthesis in benthic marine algae: an examination of its ecological significance using a theoretical model. *Limnology and Oceanography* 26:271–284.

Dring, M. J. 1987. Stimulation of light: saturated photosynthesis in brown algae by blue light. *British Phycological Journal* 22:302–302.

Dromgoole, F. I. 1988. Light fluctuations and the photosynthesis of marine algae. II. Photosynthetic response to frequency, phase ratio and amplitude. *Functional Ecology* 2:211–219.

Druehl, L. D. 1970. The pattern of Laminariales distribution in the northeast Pacific. *Phycologia* 9:237–247.

Druehl, L. D. 1973. Marine transplantation. *Science* 179:12.

Druehl, L. D. 1978. Distribution of *Macrocystis integrifolia* in British Columbia as related to environmental parameters. *Canadian Journal of Botany* 56:69–79.

Druehl, L. D. 1979. An enhancement scheme for *Macrocystis integrifolia* (Phaeophyceae). *Proceedings of the International Seaweed Symposium* 9:79–84.

Druehl, L. D. 1981. Geographical distribution. Pp. 306–325. In Lobban, C. S. and Wynne, M. J. (eds), *The Biology of Seaweeds*. University of California Press, Berkeley, CA.

Druehl, L. D. and Breen, P. A. 1986. Some ecological effects of harvesting *Macrocystis integrifolia*. *Botanica Marina* 29:97–103.

Druehl, L. D. and Kemp, L. 1982. Morphological and growth responses of geographically isolated *Macrocystis integrifolia* populations when grown in a common environment. *Canadian Journal of Botany* 60:1409–1413.

Druehl, L. D., Collins, J. D., Lane, C. E., and Saunders, G. W. 2005. An evaluation of methods used to assess intergeneric hybridization in kelp using Pacific Laminariales (Phaeophyceae). *Journal of Phycology* 41:250–262.

Dudgeon, S. R., Aronson, R. B., Bruno, J. F., and Precht, W. F. 2010. Phase shifts and stable states on coral reefs. *Marine Ecology Progress Series* 413:201–216.

Dugan, J. 1965. *Man under the Sea*. Collier Books, New York.

Dugan, J. E. and Davis, G. E. 1993. Applications of marine refugia to coastal fisheries management. *Canadian Journal of Fisheries and Aquatic Sciences* 50:2029–2042.

Dugan, J. E., Hubbard, D. M., McCrary, M. D., and Pierson, M. O. 2003. The response of macrofauna communities and shorebirds to macrophyte wrack subsidies on exposed sandy beaches of southern California. *Estuarine Coastal and Shelf Science* 58:25–40.

Dugan, J. E., Hubbard, D. M., Page, H. M., and Schimel, J. P. 2011. Marine macrophyte wrack inputs and dissolved nutrients in beach sands. *Estuaries and Coasts* 34:839–850.

Duggins, D. O. 1981. Sea urchin and kelp: the effects of short-term changes in urchin diet. *Limnology and Oceanography* 26:391–394.

Duggins, D. O. 1983. Starfish predation and the creation of mosaic patterns in a kelp-dominated community. *Ecology* 64:1610–1619.

Duggins, D. O., Eckman, J. E., and Sewell, A. T. 1990. Ecology of understory kelp environments II. Effects of kelp on recruitment of benthic invertebrates. *Journal of Experimental Marine Biology and Ecology* 143:27–45.

Duggins, D. O., Eckman, J. E., Siddon, C. E., and Klinger, T. 2003. Population, morphometric and biomechanical studies of three understory kelps along a hydrodynamic gradient. *Marine Ecology Progress Series* 265:57–76.

Dunton, K. H. 1990. Growth and production in *Laminaria solidungula* in relation to continuous underwater light levels in the Alaskan high arctic. *Marine Biology* 106:297–304.

Dunton, K. H. and Schell, D. M. 1986. Seasonal carbon budget and growth of *Laminaria solidungula* in the Alaskan high arctic. *Marine Ecology Progress Series* 31:57–66.

Dupont, S., Dorey, N., Stumpp, M., Melzner, F., and Thorndyke, M. 2013. Long-term and trans-life-cycle effects of exposure to ocean acidification in the green sea urchin *Strongylocentrotus droebachiensis*. *Marine Biology* 160:1835–1843.

Dupont, S. and Pörtner, H. O. 2013. A snapshot of ocean acidification research. *Marine Biology* 160:1765–1771.

Ebeling, A. W., Larson, R. J., and Alevizon, W. S. 1980. Habitat groups and island-mainland distribution of kelp-bed fishes off Santa Barbara, California. Pp. 404–431. In Power, D. M. (ed.), *Multidisciplinary Symposium on the California Islands*. Santa Barbara Museum of Natural History, Santa Barbara, CA.

Ebeling, A. W., Laur, D. R., and Rowley, R. J. 1985. Severe storm disturbances and reversal of community structure in a southern California kelp forest. *Marine Biology* 84:287–294.

Ebert, T. A. 1968. Growth rates of sea urchin *Strongylocentrotus purpuratus* related to food availability and spine abrasion. *Ecology* 49:1075–1091.

Ebert, T. A. 1983. Recruitment in echinoderms. Pp. 169–203. In Lawrence, J. M. (ed.), *Echinoderm Studies*, Vol. 1. A. A. Balkema, Rotterdam, the Netherlands.

Ebert, T. A. and Southon, J. R. 2003. Red sea urchins (*Strongylocentrotus franciscanus*) can live over 100 years: confirmation with A-bomb (14) carbon. *Fishery Bulletin* 101:915–922.

Ebert, T. A., Schroeter, S. C., and Dixon, J. D. 1993. Inferring demographic processes from size frequency distributions: effect of pulsed recruitment on simple models. *Fishery Bulletin* 91:237–243.

Edgar, G. J. 1987. Dispersal of faunal and floral propagules associated with drifting *Macrocystis pyrifera* plants. *Marine Biology* 95:599–610.

Edgar, G. J. and Barrett, N. S. 1999. Effects of the declaration of marine reserves on Tasmanian reef fishes, invertebrates and plants. *Journal of Experimental Marine Biology and Ecology* 242:107–144.

Edgar, G. J., Barrett, N. S., Morton, A. J., and Samson, C. R. 2004. Effects of algal canopy clearance on plant, fish and macroinvertebrate communities on eastern Tasmanian reefs. *Journal of Experimental Marine Biology and Ecology* 312:67–87.

Edmeades, D. C. 2002. The effects of liquid fertilisers derived from natural products on crop, pasture, and animal production: a review. *Australian Journal of Agriculture Research* 53:965–976.

Edwards, M. and Richardson, A. J. 2004. Impact of climate change on marine pelagic phenology and trophic mismatch. *Nature* 430:881–884.

Edwards, M. S. 1998. Effects of long-term kelp canopy exclusion on the abundance of the annual alga *Desmarestia ligulata* (Light F). *Journal of Experimental Marine Biology and Ecology* 228:309–326.

Edwards, M. S. 1999. Using *in situ* substratum steralization and fluorescence microscopy in studies of microscopic stages of marine macroalgae. *Hydrobiologia* 398–399:253–259.

Edwards, M. S. 2000. The role of alternate life-history stages of a marine macroalga: a seed bank analogue? *Ecology* 81:2404–2415.

Edwards, M. S. 2004. Estimating scale-dependency in disturbance impacts: El Niños and giant kelp forests in the northeast Pacific. *Oecologia* 138:436–447.

Edwards, M. S. and Hernández-Carmona, G. 2005. Delayed recovery of giant kelp near its southern range limit in the North Pacific following El Niño. *Marine Biology* 147:273–279.

Edwards, M. S. and Estes, J. A. 2006. Catastrophe, recovery and range limitation in E Pacific kelp forests: a large-scale perspective. *Marine Ecology Progress Series* 320:79–87.

Edyvane, K. 2003. *Conservation, Monitoring and Recovery of Threatened Giant Kelp (Macrocystis Pyrifera) Beds in Tasmania: Final Report.* Department of Primary Industries, Water and Environment, Hobart, Australia.

Elner, R. W. and Vadas, R. L. 1990. Inference in ecology: the sea urchin phenomenon in the Northwestern Atlantic. *American Naturalist* 136:108–125.

Elton, C. S. 1927. *Animal Ecology.* University of Chicago Press, Chicago, IL.

Engelen, A. and Santos, R. 2009. Which demographic traits determine population growth in the invasive brown seaweed *Sargassum muticum? Journal of Ecology* 97:675–684.

Erlandson, J. M., Rick, T. C., Estes, J. A., Graham, M. H., Braje, T. J., and Vellanoweth, R. L. 2005. Sea otters, shellfish, and humans: 10,000 years of ecological interaction on San Miguel Island, California. Pp. 58–68. In Garcecon, D. K. and Schwemm, C. A. (eds), *Proceedings of the Sixth California Islands Symposium.* Institute for Wildlife Studies, Aracata, CA.

Erlandson, J. M., Graham, M. H., Bourque, B. J., Corbett, D., Estes, J. A., and Steneck, R. S. 2007. The kelp highway hypothesis: marine ecology, the coastal migration theory, and the peopling of the Americas. *Journal of Island and Coastal Archaeology* 2:161–174.

Erlandson, J. M., Rick, T. C., Braje, T. J., Steinberg, A., and Vellanoweth, R. L. 2008. Human impacts on ancient shellfish: a 10,000 year record from San Miguel Island, California. *Journal of Archaeological Science* 35:2144–2152.

Eschmeyer, W. N., Herald, E. S., and Hamman, H. 1983. *A Field Guide to Pacific Coast Fishes of North America.* Houghton Mifflin Company, Boston, MA.

Estes, J. A. and Duggins, D. O. 1995. Sea otters and kelp forests in Alaska: generality and variation in a community ecological paradigm. *Ecological Monographs* 65:75–100.

Estes, J. A. and Harrold, C. 1988. Sea otters, sea urchins and kelp beds: some questions of scale. Pp. 116–150. In VanBlaricom, G. R. and Estes, J. A. (eds), *The Community Ecology of Sea Otters.* Springer-Verlag, Berlin, Germany.

Estes, J. A. and Palmisano, J. 1974. Sea otters: their role in structuring nearshore communities. *Science* 185:1058–1060.

Estes, J. A. and Steinberg, P. D. 1988. Predation, herbivory and kelp evolution. *Paleobiology* 14:19–36.

Estes, J. A. and Van Blaricom, G. R. 1985. Sea otters and shell fisheries. Pp. 187–235. In Beverton, R. H., Lavigne, D., and Beddington, J. (eds), *Conflicts between Marine Mammals and Fisheries.* Allen and Unwin, London, UK.

Estes, J. A., Smith, N. S., and Palmisano, J. F. 1978. Sea otter predation and community organization in western Aleutian Islands, Alaska. *Ecology* 59:822–833.

Estes, J. A., Jameson, R. J., and Rhode, E. B. 1982. Activity and prey selection in the sea otter: influence of population status on community structure. *The American Naturalist* 120:242–258.

Estes, J. A., Tinker, M. T., Williams, T. M., and Doak, D. F. 1998. Killer whale predation on sea otters linking oceanic and nearshore ecosystems. *Science* 282:473–476.

Estes, J. A., Hatfield, B. B., Ralls, K., and Ames, J. 2003. Causes of mortality in California sea otters during periods of population growth and decline. *Marine Mammal Science* 19:198–216.

Estes, J. A., Terborgh, J., Brashares, J. S., Power, M. E., Berger, J., Bond, W. J. et al. 2011. Trophic downgrading of planet earth. *Science* 333:301–306.

Fagerli, C. W., Norderhaug, K. M., and Christie, H. C. 2013. Lack of sea urchin settlement may explain kelp forest recovery in overgrazed areas in Norway. *Marine Ecology Progress Series* 488:119–132.

Fain, S. R. and Murray, S. N. 1982. Effects of light and temperature on net photosynthesis and dark respiration of gametophytes and embryonic sporophytes of *Macrocystis pyrifera. Journal of Phycology* 18:92–98.

Fankboner, P. V. and Druehl, L. D. 1976. In situ accumulation of marine algal exudate by a polychaete worm (*Schizobranchia insignis*). *Experientia* 32:1391–1392.

Fargione, P. V., Hill, J., Tilman, D., Polasky, S., and Hawthorne, P. 2008. Land clearing and the biofuel carbon debt. *Science* 319:1235–1238.

Feehan, C., Scheibling, R. E., and Lauzon-Guay, J. S. 2012. An outbreak of sea urchin disease associated with a recent hurricane: support for the "killer storm hypothesis" on a local scale. *Journal of Experimental Marine Biology and Ecology* 413:159–168.

Fejtek, S. M., Edwards, M. S., and Kim, K. Y. 2011. Elk kelp, *Pelagophycus porra,* distribution limited due to susceptibility of microscopic stages to high light. *Journal of Experimental Marine Biology and Ecology* 396:194–201.

Fernández, E., Cordova, C., and Tarazona, J. 1999. Condiciones de la pradera submareal de *Lessonia trabeculata* en La Isla Independence durante "El Niño 1997-98". Pp. 47–59.

In Tarazona, J. and Castillo, E. (eds), *El Niño 1997-98 y su impacto sobre los Ecosistemas Marino y Terrestre, Revista peruana de biologia volume estraordinario*. Facultad de Ciencias Biologicas UNMSM, Lima, Peru. (in Spanish)

Fernández, M. and Castilla, J. C. 2005. Marine conservation in Chile: historical perspective, lessons, and challenges. *Conservation Biology* 19:1752–1762.

Field, J. G. and Griffiths, C. L. 1991. Littoral and sublittoral ecosystems of southern Africa. Pp. 323–346. In Mathieson, A. C. and Nienhuis, P. H. (eds), *Ecosystems of the World 24: Intertidal and Littoral Ecosystems*. Elsevier, Amsterdam, the Netherlands.

Filbee-Dexter, K. and Scheibling, R. E. 2014. Sea urchin barrens as alternative stable states of collapsed kelp ecosystems. *Marine Ecology Progress Series* 495:1–25.

Fink, L. A. and Manley, S. L. 2011. The use of kelp sieve tube sap metal composition to characterize urban runoff in southern California coastal waters. *Marine Pollution Bulletin* 62:2619–2632.

Fisher, E. M. 1939. Habits of the southern sea otter. *Journal of Mammalogy* 20:21–36.

Fletcher, R. L. and Callow, M. E. 1992. The settlement, attachment and establishment of marine algal spores. *British Phycological Journal* 27:303–329.

Floerl, O., Inglis, G. J., and Hayden, B. J. 2005. A risk-based predictive tool to prevent accidental introductions of nonindigenous marine species. *Environmental Management* 35:765–778.

Flores-Aguilar, R. A., Gutierrez, A., Ellwanger, A., and Searcy-Bernal, R. 2007. Development and current status of abalone aquaculture in Chile. *Journal of Shellfish Research* 26:705–711.

Flowers, J. M., Schroeter, S. C., and Burton, R. S. 2002. The recruitment sweepstakes has many winners: genetic evidence from the sea urchin *Strongylocentrotus purpuratus*. *Evolution* 56:1445–1453.

Foley, M. M., Halpern, B. S., Micheli, F., Armsby, M. H., Caldwell, M. R., Crain, C. M. et al. 2010. Guiding ecological principles for marine spatial planning. *Marine Policy* 34:955–966.

Foster, M. S. 1972. The algal turf community in the nest of the ocean goldfish (*Hypsypops rubicunda*). *Proceedings of the International Seaweed Symposium* 7:55–60.

Foster, M. S. 1975a. Regulation of algal community development in a *Macrocystis pyrifera* forest. *Marine Biology* 32:331–342.

Foster, M. S. 1975b. Algal succession in a *Macrocystis pyrifera* forest. *Marine Biology* 32:313–329.

Foster, M. S. 1982. The regulation of macroalgal associations in kelp forests. Pp. 185–205. In Srivastava, L. M. (ed.), *Synthetic and Degradative Processes in Marine Macrophytes*. Walter de Gruyter, Berlin, Germany.

Foster, M. S. and Holmes, R. W. 1977. The Santa Barbara oil spill: an ecological disaster? Pp. 166–190. In Cairns, J., Dickson, K. L., and Herricks, E. E. (eds), *Recovery and Restoration of Damaged Ecosystems*. University Press of Virginia, Charlottesville, VA.

Foster, M. S. and Schiel, D. R. 1985. *The Ecology of Giant Kelp Forests in California: A Community Profile*. U.S. Fish and Wildlife Service, Biological Report 85 (7.2), U.S. Government Printing Office, Washington, DC.

Foster, M. S. and Schiel, D. R. 1988. Kelp communities and sea otters: keystone species or just another brick in the wall? Pp. 92–115. In VanBlaricom, G. R. and Estes, J. A. (eds), *The Community Ecology of Sea Otters*. Springer-Verlag, Berlin, Germany.

Foster, M.S. and Schiel, D.R. 1993. Zonation, El Niño disturbance, and the dynamics of sub-tidal vegetation along a 30 m depth gradient in two giant kelp forests. Pp. 151–162. In Battershill, C.N., Schiel, D.R., Jones, G.P., Creese, R.G., and MacDiarmid, A.B. (eds), *Proceedings of the Second International Temperate Reef Symposium*. NIWA Marine, Wellington, New Zealand.

Foster, M.S. and Schiel, D.R. 2010. Loss of predators and the collapse of southern California kelp forests (?): alternatives, explanations and generalizations. *Journal of Experimental Marine Biology and Ecology* 393:59–70.

Foster, M.S. and VanBlaricom, G.R. 2001. Spatial variation in kelp forest communities along the Big Sur coast of central California, USA. *Cryptogamie, Algologie* 22:173–186.

Foster, M.S., Neushul, M., and Charters, A.C. 1971. The Santa Barbara oil spill. I. Initial quantities and distribution of pollutant crude oil. *Environmental Pollution* 2:97–113.

Foster, M.S., Neushul, M., and Zingmark, R. 1971b. The Santa Barbara oil spill. II. Initial effects on littoral and kelp bed organisms. *Environmental Pollution* 2:115–134.

Foster, M.S., Schiel, D.R., and Carter, J.W. 1983. Ecology of kelp communities. Pp. 53–69. In Bascom, W. (ed.), *The Effects of Waste Disposal on Kelp Communities*. Southern California Coastal Water Research Project, Long Beach, CA.

Foster, M.S., DeVogelaere, A.P., Harrold, C., Pearse, J.S., and Thum, A.B. 1988. *Causes of Spatial and Temporal Patterns in Rocky Intertidal Communities of Central and Northern California*. Memoirs of the California Academy of Sciences Number 9. California Academy of Sciences, San Francisco, CA.

Foster, M.S., Edwards, M.S., Reed, D.C., Schiel, D.R., and Zimmerman, R.C. 2006. Top-down vs. bottom-up effects in kelp forests. *Science* 313:1737–1738.

Foster, M.S., Reed, D.C., Carr, M.H., Dayton, P.K., Malone, D.P., and Pearse, J.C. 2013. *Kelp Forests in California. Research and Discoveries: The Revolution of Science through SCUBA, Smithsonian Contributions in Marine Sciences*. Smithsonian Institution Scholarly Press, Washington, DC.

Fox, C.H. and Swanson, A.K. 2007. Nested PCR detection of microscopic life-stages of laminarian macroalgae and comparison with adult forms along intertidal height gradients. *Marine Ecology Progress Series* 332:1–10.

Fox, M.D. 2013. Resource translocation drives δ13C fractionation during recovery from disturbance in giant kelp, *Macrocystis pyrifera*. *Journal of Phycology* 49:811–815.

Fram, J.P., Stewart, H.L., Brzezinski, M.A., Gaylord, B., Reed, D.C., Williams, S.L. et al. 2008. Physical pathways and utilization of nitrate supply to the giant kelp, *Macrocystis pyrifera*. *Limnology and Oceanography* 53:1589–1603.

Frank, K.T., Petrie, B., Choi, J.S., and Leggett, W.C. 2005. Trophic cascades in a formerly cod-dominated ecosystem. *Science* 308:1621–1623.

Frantz, B.R., Foster, M.S., and Riosmena-Rodríguez, R. 2005. *Clathromorphum nereostratum* (Corallinales, Rhodophyta): the oldest alga? *Journal of Phycology* 41:770–773.

Fraser, C.I., Nikula, R., Spencer, H.G., and Waters, J.M. 2009. Kelp genes reveal effects of subantarctic sea ice during the Last Glacial Maximum. *Proceedings of the National Academy of Sciences* 106:3249–3253.

Fraser, C. I., Winter, D. J., Spencer, H. G., and Waters, J. M. 2010. Multigene phylogeny of the southern bull-kelp *Durvillaea* (Phaeophyceae: Fucales). *Molecular Phylogenetics and Evolution* 57:1301–1311.

Fritsch, F. E. 1945. *The Structure and Reproduction of the Algae.* Cambridge University Press, Cambridge, UK.

Frusher, S. D. 1997. *Stock Assessment Report: Rock Lobster.* Government of Tasmania, Australia, Internal report No. 35. Tasmanian Department of Primary Industry and Fisheries, Hobart, Australia.

Fukuhara, Y., Mizuta, H., and Yasui, H. 2002. Swimming activities of zoospores in *Laminaria japonica* (Phaeophyceae). *Fisheries Science* 68:1173–1181.

Fyfe, J. C. and Saenko, O. A. 2005. Human-induced change in the Antarctic Circumpolar Current. *Journal of Climate* 18:3068–3073.

Gaines, S. D. and Lubchenco, J. 1982. A unified approach to marine plant-herbivore interactions 2. Biogeography. *Annual Review of Ecology and Systematics* 13:111–138.

Gaines, S. D. and Roughgarden, J. 1987. Fish in offshore kelp forests affect recruitment to intertidal barnacle populations. *Science* 235:479–481.

Gaines, S. D., Gaylord, B., and Largier, J. L. 2003. Avoiding current oversights in marine reserve design. *Ecological Applications* 13(S1):S32–S46.

Gaines, S. D., White, C., Carr, M. H., and Palumbi, S. R. 2010. Designing marine reserve networks for both conservation and fisheries management. *Proceedings of the National Academy of Sciences* 107:18286–18293.

Gaines, S. D., Lester, S. E., Grorud-Colvert, K., Costello, C., and Pollnac, R. 2010. Evolving science of marine reserves: new developments and emerging research frontiers. *Proceedings of the National Academy of Sciences* 107:18251–18255.

Garbary, D. J., Kim, K. Y., Klinger, T., and Duggins, D. 1999. Red algae as hosts for endophytic kelp gametophytes. *Marine Biology* 135:35–40.

Garland, C. D., Cooke, S. L., Grant, J. F., and McMeekin, T. A. 1985. Ingestion of the bacteria on and the cuticle of crustose (non-articulated) coralline algae by post-larval and juvenile abalone (*Haliotis ruber* Leach) from Tasmanian waters. *Journal of Experimental Marine Biology and Ecology* 91:137–149.

Gaylord, B. and Denny, M. W. 1997. Flow and flexibility. I. Effects of size, shape and stiffness in determining wave forces on the stipitate kelps *Eisenia arborea* and *Pterogophora californica*. *Journal of Experimental Biology* 200:3141–3164.

Gaylord, B. and Gaines, S. D. 2000. Temperature or transport? Range limits in marine species mediated solely by flow. *American Naturalist* 155:769–789.

Gaylord, B., Reed, D. C., Raimondi, P. T., Washburn, L., and McLean, S. R. 2002. A physically based model of macroalgal spore dispersal in the wave and current-dominated nearshore. *Ecology* 83:1239–1251.

Gaylord, B., Denny, M. W., and Koehl, M. A. R. 2003. Modulation of wave forces on kelp canopies by alongshore currents. *Limnology and Oceanography* 48:860–871.

Gaylord, B., Reed, D. C., Washburn, L., and Raimondi, P. T. 2004. Physical-biological coupling in spore dispersal of kelp forest macroalgae. *Journal of Marine Systems* 49: 19–39.

Gaylord, B., Reed, D. C., Raimondi, P. T., and Washburn, L. 2006a. Macroalgal spore dispersal in coastal environments: mechanistic insights revealed by theory and experiment. *Ecological Monographs* 76:481–502.

Gaylord, B., Reed, D. C., Raimondi, P. T., and Washburn, L. 2006b. Ecomechanics of macroalgal spore dispersal in coastal environments: insights from theory and experiment. *Integrative and Comparative Biology* 46:E47.

Gaylord, B., Rosman, J. H., Reed, D. C., Koseff, J. R., Fram, J., MacIntyre, S. et al. 2007. Spatial patterns of flow and their modification within and around a giant kelp forest. *Limnology and Oceanography* 52:1838–1852.

Gaylord, B., Denny, M. W., and Koehl, M. A. R. 2008. Flow forces on seaweeds: field evidence for roles of wave impingement and organism inertia. *Biological Bulletin* 215:295–308.

Gaylord, B., Nickols, K. J., and Jurgens, L. 2012. Roles of transport and mixing processes in kelp forest ecology. *Journal of Experimental Biology* 215:997–1007.

Geange, S. W. 2014. Growth and reproductive consequences of photosynthetic tissue loss in the surface canopies of *Macrocystis pyrifera* (L.) C. Agardh. *Journal of Experimental Marine Biology and Ecology* 453:70–75.

Gell, F. R. and Roberts, C. M. 2003. Benefits beyond boundaries: the fishery effects of marine reserves. *Trends in Ecology and Evolution* 18:448–455.

Gerard, V. A. 1976. *Some Aspects of Material Dynamics and Energy Flow in a Kelp Forest in Monterey Bay, California.* PhD thesis. University of California, Santa Cruz.

Gerard, V. A. 1982a. Growth and utilization of internal nitrogen reserves by the giant kelp *Macrocystis pyrifera* in a low-nutrient environment. *Marine Biology* 66:27–35.

Gerard, V. A. 1982b. In situ rates of nitrate uptake by giant kelp *Macrocystis pyrifera* (L) C Agardh: tissue differences, environmental effects, and predictions of nitrogen limited growth. *Journal of Experimental Marine Biology and Ecology* 62:211–224.

Gerard, V. A. 1982c. In situ water motion and nutrient uptake by the giant kelp *Macrocystis pyrifera*. *Marine Biology* 69:51–54.

Gerard, V. A. 1984. The light environment in a giant kelp forest: influence of *Macrocystis pyrifera* on spatial and temporal variability. *Marine Biology* 84:189–195.

Gerard, V. A. and Kirkman, H. 1984. Ecological observations on a branched, loose-lying form of *Macrocystis pyrifera* (L.) C. Agardh in New Zealand. *Botanica Marina* 27:105–109.

Gerard, V. A. and North, W. J. 1984. Measuring growth, production, and yield of the giant kelp, *Macrocystis pyrifera*. *Hydrobiologia* 116:321–324.

Gerrodette, T. 1979. Equatorial submergence in a solitary coral, *Balanophyllia elegens*, and the critical life stage excluding the species from shallow-water in the south *Marine Ecology Progress Series* 1:227–235.

Ghelardi, R. J. 1971. "Species" structure of the animal community that lives in *Macrocystis pyrifera* holdfasts. Pp. 381–420. In North, W. J. (ed.), *The Biology of Giant Kelp Beds (Macrocystis) in California*. Nova Hedwigia 32. Verlag Von J. Cramer, Lehre, Germany.

Gilles, K. W. and Pearse, J. S. 1986. Studies on disease in sea urchins, *Strongylocentrotus purpuratus*: experimental infection and bacterial virulence. *Diseases of Aquatic Organisms* 1:105–114.

Glass, G. V. 1976. Primary, secondary, and meta-analysis of research. *Educational Researcher* 5:3–8.

Godoy, N., Gelcich, S., Vásquez, J. A., and Castilla, J. C. 2010. Spearfishing to depletion: evidence from temperate reef fishes in Chile. *Ecological Applications* 20:1504–1511.

Goldberg, N. A. and Foster, M. S. 2002. Settlement and post-settlement processes limit the abundance of the geniculate coralline alga *Calliarthron* on subtidal walls. *Journal of Experimental Marine Biology and Ecology* 278:31–45.

Gordon, B. L. 1974. *Monterey Bay Area: Natural History and Cultural Imprints.* Boxwood Press, Pacific Grove, CA.

Gorman, D., Russell, B. D., and Connell, S. D. 2009. Land-to-sea connectivity: linking human-derived terrestrial subsidies to subtidal habitat change on open rocky coasts. *Ecological Applications* 19:1114–1126.

Grabitske, H. A. and Slavin, J. L. 2008. Low-digestible carbohydrates in practice. *Journal of the American Dietetic Association* 108:1677–1681.

Graham, L. E. and Wilcox, L. W. 2000. *Algae.* Prentice Hall, Upper Saddle River, NJ.

Graham, L. E., Graham, J. E., and Wilcox, L. W. 2009. *Algae.* 2nd ed. Benjamin Cummings (Pearson), San Francisco, CA.

Graham, M. H. 1997. Factors determining the upper limit of giant kelp, *Macrocystis pyrifera* Agardh, along the Monterey Peninsula, central California, USA. *Journal of Experimental Marine Biology and Ecology* 218:127–149.

Graham, M. H. 1999. Identification of kelp zoospores from *in situ* plankton samples. *Marine Biology* 135:709–720.

Graham, M. H. 2002. Prolonged reproductive consequences of short-term biomass loss in seaweeds. *Marine Biology* 140:901–911.

Graham, M. H. 2003. Coupling propagule output to supply at the edge and interior of a giant kelp forest. *Ecology* 84:1250–1264.

Graham, M. H. 2004. Effects of local deforestation on the diversity and structure of southern California giant kelp forest food webs. *Ecosystems* 7:341–357.

Graham, M. H. and Mitchell, B. G. 1999. Obtaining absorption spectra from individual macroalgal spores using microphotometry. *Hydrobiologia* 399:231–239.

Graham, M. H., Harrold, C., Lisin, S., Light, K., Watanabe, J. M., and Foster, M. S. 1997. Population dynamics of giant kelp *Macrocystis pyrifera* along a wave exposure gradient. *Marine Ecology Progress Series* 148:269–279.

Graham, M. H., Dayton, P. K., and Erlandson, J. M. 2003. Ice ages and ecological transitions on temperate coasts. *Trends in Ecology and Evolution* 18:33–40.

Graham, M. H., Vásquez, J. A., and Buschmann, A. H. 2007. Global ecology of the giant kelp *Macrocystis*: from ecotypes to ecosystems. *Oceanography and Marine Biology: An Annual Review* 45:39–88.

Graham, M. H., Kinlan, B. P., Druehl, L. D., Garske, L. E., and Banks, S. 2007. Deep-water kelp refugia as potential hotspots of tropical marine diversity and productivity. *Proceedings of the National Academy of Sciences* 104:16576–16580.

Graham, M. H., Halpern, B. S., and Carr, M. H. 2008. Diversity and dynamics of Californian subtidal kelp forests. Pp. 103–134. In McClanahan, T. R. and Branch, G. M. (eds), *Food Webs and the Dynamics of Marine Reefs.* Oxford University Press, Oxford, UK.

Graham, M. H., Kinlan, B. P., and Grosberg, R. K. 2010. Post-glacial redistribution and shifts in productivity of giant kelp forests. *Proceedings of the Royal Society B: Biological Sciences* 277:399–406.

Grall, J., Le Loc'h, F., Guyonnet, B., and Riera, P. 2006. Community structure and food web based on stable isotopes ($\delta^{15}$ N and $\delta^{13}$ C) analysis of a North Eastern Atlantic maerl bed. *Journal of Experimental Marine Biology and Ecology* 338:1–15.

Grenville, D. J., Peterson, R. L., Barrales, H. L., and Gerrath, J. F. 1982. Structure and development of the secretory cells and duct system in *Macrocystis pyrifera* (L.) Agardh, C.A. *Journal of Phycology* 18:232–240.

Grigg, R. W. 1975. Age structure of a longevous coral: relative index of habitat suitability and stability. *American Naturalist* 109:647–657.

Grig, R. W. and Kiwala, R. S. 1970. Some ecological effects of discharged wastes on marine life. *California Fish and Game* 56:145–155.

Grove, R. S., Zabloudil, K., Norall, T., and Deysher, L. 2002. Effects of El Niño events on natural kelp beds and artificial reefs in southern California. *ICES Journal of Marine Science* 59:S330–S337.

Guenther, C. M., Lenihan, H. S., Grant, L. E., Lopez-Carr, D., and Reed, D. C. 2012. Trophic cascades induced by lobster fishing are not ubiquitous in southern California kelp forests. *PLOS ONE* 7:e49396.

Guiry, M. D. and Guiry, G. M. 2012. *AlgaeBase*. Worldwide Electronic Publication, National University of Ireland, Galway. Accessed December 18, 2012. http://www.algaebase.org.

Gutierrez, A. Correa, T., Munoz, V., Santibanez, A., Marcos, R., Caceres, C. et al. 2006. Farming of the giant kelp *Macrocystis pyrifera* in southern Chile for development of novel food products. *Journal of Applied Phycology* 18:259–267.

Haaker, P.L., Karpov, K., Rogers-Bennett, L., Taniguchi, I, Friedman, C.S., and Tegner, M.J. 2001. Abalone. Pp. 89–97. In Leet, W. S., Dewees, C. M., Klingbeil, R., and Larson, E. J. (eds), *California's Living Marine Resources: A Status Report*. California Department of Fish and Game, Sacramento, CA.

Hairston, N. G., Smith, F. E., and Slobodkin, L. B. 1960. Community structure, population control, and competition. *American Naturalist* 94:421–425.

Hales, J. M. and Fletcher, R. L. 1989. Studies on the recently introduced brown alga *Sargassum muticum* (Yendo) Fensholt. 4. The effect of temperature, irradiance and salinity on germling growth. *Botanica Marina* 32:167–176.

Halewood, E. R., Carlson, C. A., Brzezinski, M. A., Reed, D. C., and Goodman, J. 2012. Annual cycle of organic matter partitioning and its availability to bacteria across the Santa Barbara Channel continental shelf. *Aquatic Microbial Ecology* 67:189–209.

Halpern, B. S. 2003. The impact of marine reserves: do reserves work and does reserve size matter? *Ecological Applications* 13:117–137.

Halpern, B. S. and Cottenie, K. 2007. Little evidence for climate effects on local-scale structure and dynamics of California kelp forest communities. *Global Change Biology* 13:236–251.

Halpern, B. S. and Warner, R. R. 2002. Marine reserves have rapid and lasting effects. *Ecology Letters* 5:361–366.

Halpern, B.S. and Warner, R.R. 2003. Matching marine reserve design to reserve objectives. *Proceedings of the Royal Society B: Biological Sciences* 270:1871–1878.

Halpern, B.S., Cottenie, K., and Broitman, B.R. 2006. Strong top-down control in southern California kelp forest ecosystems. *Science* 312:1230–1232.

Halpern, B.S., Lester, S.E., and Kellner, J.B. 2010. Spillover from marine reserves and the replenishment of fished stocks. *Environmental Conservation* 36:268–276.

Hamilton, S.L., Caselle, J.E., Standish, J.D., Schroeder, D.M., Love, M.S., Rosales-Casian, J.A. et al. 2007. Size-selective harvesting alters life histories of a temperate sex-changing fish. *Ecological Applications* 17:2268–2280.

Hamilton, S.L., Caselle, J.E., Malone, D.P., and Carr, M.H. 2010. Incorporating biogeography into evaluations of the Channel Islands marine reserve network. *Proceedings of the National Academy of Sciences* 107:18272–18277.

Hamilton, S.L., Newsome, S.D., and Caselle, J.E. 2014. Dietary niche expansion of a kelp forest predator recovering from intense commercial exploitation. *Ecology* 95: 164–172.

Hamilton, S. L. and Caselle, J. E. 2015. Exploitation and recovery of a sea urchin predator has implications for the resilience of southern California kelp forests. *Proceedings of the Royal Society B* 282. http://dx.doi.org/10.1098/rspb.2014.1817.

Hamon, K.G., Thebaud, O., Frusher, S., and Richard-Little, L. 2009. A retrospective analysis of the effects of adopting individual transferable quotas in the Tasmanian red rock lobster, *Jasus edwardsii*, fishery. *Aquatic Living Resources* 22:549–558.

Hansen, J. and Nazarenko, L. 2004. Soot climate forcing via snow and ice albedos. *Proceedings of the National Academy of Sciences* 101:423–428.

Hanski, I. 1999. *Metapopulation Ecology*. Oxford University Press, Oxford, UK.

Harley, C.D.G., Anderson, K.M., Demes, K.W., Jorve, J.P., Kordas, R.L., Coyle, T.A. et al. 2012. Effects of climate change on global seaweed communities. *Journal of Phycology* 48:1064–1078.

Harper, J.L. 1977. *The Population Biology of Plants*. Academic Press, London, UK.

Harper, J.L. and White, J. 1974. The demography of plants. *Annual Review of Ecology and Systematics* 5:419–463.

Harrer, S.L., Reed, D.C., Holbrook, S.J., and Miller, R.J. 2013. Patterns and controls of the dynamics of net primary production by understory macroalgal assemblages in giant kelp forests. *Journal of Phycology* 49:248–257.

Harris, L.G., Ebeling, A.W., Laur, D.R., and Rowley, R.J. 1984. Community recovery after storm damage: a case of facilitation in primary succession. *Science* 224:1336–1338.

Harrold, C. and Lisin, S. 1989. Radio-tracking rafts of giant kelp: local production and regional transport. *Journal of Experimental Marine Biology and Ecology* 130:237–251.

Harrold, C. and Pearse, J.S. 1987. The ecological role of echinoderms in kelp forests. Pp. 137–233. In Jangoux, M. and Lawrence, J.M. (eds), *Echinoderm Studies*. A.A. Balkema Press, Rotterdam, the Netherlands.

Harrold, C. and Reed, D.C. 1985. Food availability, sea-urchin grazing and kelp forest community structure. *Ecology* 66:1160–1169.

Harrold, C., Watanabe, J., and Lisin, S. 1988. Spatial variation in the structure of kelp forest communities along a wave exposure gradient. *Marine Ecology* 9:131–156.

Harrold, C., Light, K., and Lisin, S. 1998. Organic enrichment of submarine-canyon and continental-shelf benthic communities by macroalgal drift imported from nearshore kelp forests. *Limnology and Oceanography* 43:669–678.

Hart, M. W. and Scheibling, R. E. 1988. Heat waves, baby booms, and the destruction of kelp beds by sea urchins. *Marine Biology* 99:167–176.

Hatfield, B. and Tinker, T. 2012. Spring 2012 *Mainland California Sea Otter Census Results*. USGS Western Ecological Research Center, Santa Cruz Field Station (2012-08-20). USGS. http://www.werc.usgs.gov/ProjectSubWebPage.aspx?SubWebPageID=22&ProjectID=91.

Haxo, F. T. and Blinks, L. R. 1950. Photosynthetic action spectra of marine algae. *The Journal of General Physiology* 33:389–422.

Hay, C. H. 1986. A new species of *Macrocystis* C. Ag. (Phaeophyta) from Marion Island, southern Indian Ocean. *Phycologia* 25:241–252.

Hay, C. H. 1990. The distribution of *Macrocystis* (Phaeophyta: Laminariales) as a biological indicator of cool sea surface temperature, with special reference to New Zealand waters. *Journal of the Royal Society of New Zealand* 20:313–336.

Hedgpeth, J. W. 1957. Marine biogeography. Pp. 359–382. In Hedgpeth, J. W. and Ladd, H. S. (eds), *Treatise of Marine Ecology and Paleoecology*. Geological Society of America, Memphis, TN.

Heine, J. N. 1983. Seasonal productivity of two red algae in a central California kelp forest. *Journal of Phycology* 19:146–152.

Helmuth, B., Veit, R. R., and Holberton, R. 1994. Long-distance dispersal of a sub-Antarctic brooding bivalve (*Giamardia trapesina*) by kelp rafting. *Marine Biology* 120:421–426.

Helmuth, B., Mieszkowska, N., Moore, P., and Hawkins, S. J. 2006. Living on the edge of two changing worlds: forecasting the responses of rocky intertidal ecosystems to climate change. *Annual Review of Ecology Evolution and Systematics* 36:373–404.

Henríquez, L. A. Buschmann, A. H., Maldonado, M. A., Graham, M. H., Hernandez-González, M. C., Pereda, S. V. et al. 2011. Grazing on giant kelp microscopic phases and the recruitment success of annual populations of *Macrocystis pyrifera* (Laminariales, Phaeophyta) in Southern Chile. *Journal of Phycology* 47:252–258.

Henry, E. C. and Cole, K. M. 1982a. Ultrastructure of swarmers in the Laminariales (Phaeophyceae). I. *Zoospores. Journal of Phycology* 18:550–569.

Henry, E. C. and Cole, K. M. 1982b. Ultrastructure of swarmers in the Laminariales (Phaeophyceae) 2. Sperm. *Journal of Phycology* 18:570–579.

Hepburn, C. D. and Hurd, C. L. 2005. Conditional mutualism between the giant kelp *Macrocystis pyrifera* and colonial epifauna. *Marine Ecology Progress Series* 302:37–48.

Hepburn, C. D., Pritchard, D. W., Cornwall, C. E., McLeod, R. J., Beardall, J., Raven, J. A. et al. 2011. Diversity of carbon use strategies in a kelp forest community: implications for a high $CO_2$ ocean. *Global Change Biology* 17:2488–2497.

Hernández-Carmona, G., Garcia, O., Robledo, D., and Foster, M. 2000. Restoration techniques for *Macrocystis pyrifera* (Phaeophyceae) populations at the southern limit of their distribution in Mexico. *Botanica Marina* 43:273–284.

Hernández-Carmona, G., Robledo, D., and Serviere-Zaragoza, E. 2001. Effect of nutrient availability on *Macrocystis pyrifera* recruitment and survival near its southern limit off Baja California. *Botanica Marina* 44:221–229.

Hernández-Carmona, G., Hughes, B., and Graham, M. H. 2006. Reproductive longevity of drifting kelp *Macrocystis pyrifera* (Phaeophyceae) in Monterey Bay, USA. *Journal of Phycology* 42:1199–1207.

Hernández-Carmona, G., Riosmena-Rodriguez, R., Serviere-Zaragoza, E., and Ponce-Diaz, G. 2011. Effect of nutrient availability on understory algae during El Niño Southern Oscillation (ENSO) conditions in Central Pacific Baja California. *Journal of Applied Phycology* 23:635–642.

Hewitt, C. L. Campbell, M. L., McEnnulty, F., Moore, K. M., Murfet, N. B., Robertson, B. et al. 2005. Efficacy of physical removal of a marine pest: the introduced kelp *Undaria pinnatifida* in a Tasmanian Marine Reserve. *Biological Invasions* 7:251–263.

Hilborn, R., Stokes, K., Maguire, J., Smith, T., Botsford, L. W., Mangel, M. et al. 2004. When can marine reserves improve fisheries management? *Ocean and Coastal Management* 47:197–205.

Hines, A. H. 1982. Coexistence in a kelp forest: size, population dynamics and resource partitioning in a guild of spider crabs (Brachyura, Majidae). *Ecological Monographs* 52:179–198.

Hines, A. H. and Loughlin, T. R. 1980. Observations of sea otters digging for clams at Monterey Bay, California. *Fishery Bulletin* 78:159–163.

Hines, A. H. and Pearse, J. S. 1982. Abalones, shells, and sea otters: dynamics of prey populations in central California. *Ecology* 63:1547–1560.

Hixon, M. A. 1980. Competitive interactions between California reef fishes of the genus *Embiotoca*. *Ecology* 61:918–931.

Hixon, M. A. 2006. Competition. Pp. 449–465. In Allen, L. G., Pondella, D. J., and Horn, M. H. (eds), *The Ecology of Marine Fishes: California and Adjacent Waters*. University of California Press, Berkeley, CA.

Hobday, A. J. 2000a. Age of drifting *Macrocystis pyrifera* (L.) C. Agardh rafts in the Southern California Bight. *Journal of Experimental Marine Biology and Ecology* 253:97–114.

Hobday, A. J. 2000b. Abundance and dispersal of drifting kelp *Macrocystis pyrifera* rafts in the Southern California Bight. *Marine Ecology Progress Series* 195:101–116.

Hobday, A. J. 2000c. Persistence and transport of fauna on drifting kelp (*Macrocystis pyrifera* (L.) C. Agardh) rafts in the Southern California Bight. *Journal of Experimental Marine Biology and Ecology* 253:75–96.

Hobday, A. J., Tegner, M. J., and Haaker, P. L. 2001. Over-exploitation of a broadcast spawning marine invertebrate: decline of the white abalone. *Fish Biology and Fisheries* 10: 493–514.

Hockey, P. A. R. 1988. Kelp gulls *Larus dominicanus* as predators in kelp *Macrocystis pyrifera* beds. *Oecologia* 76:155–157.

Hoegh-Guldberg, O. and Bruno, J. F. 2010. The impact of climate change on the world's marine ecosystems. *Science* 328:1523–1528.

Hoegh-Guldberg, O., Mumby, P. J., Hooten, A. J., Steneck, R. S., Greenfield, P., Gomez, E. et al. 2007. Coral reefs under rapid climate change and ocean acidification. *Science* 318:1737–1742.

Hoffmann, A. and Santelices, B. 1997. *Flora marina de Chile central*. Universidad Catolica de Chile, Santiago, Chile.

Hoffmann, A. J. and Santelices, B. 1991. Banks of algal microscopic forms: hypotheses on their functioning and comparisons with seed banks. *Marine Ecology Progress Series* 79:185–194.

Hofmann, G. E., Barry, J. P., Edmunds, P. J., Gates, R. D., Hutchins, D. A., Klinger, T. et al. 2010. The effect of ocean acidification on calcifying organisms in marine ecosystems: an organism-to-ecosystem perspective. *Annual Review of Ecology, Evolution, and Systematics* 41:127–147.

Holbrook, S. J., Schmitt, R. J., and Ambrose, R. F. 1990. Biogenic habitat structure and characteristics of temperate reef fish assemblages. *Australian Journal of Ecology* 15:489–503.

Holbrook, S. J., Schmitt, R. J., and Stephens, J. S. 1997. Changes in an assemblage of temperate reef fishes associated with a climate shift. *Ecological Applications* 7:1299–1310.

Hooker, J. D. 1847. *The Botany of the Antarctic Voyage of H.M. Discovery Ships Erebus and Terror in the Years 1839–1843. I. Flora Antarctica, Part 2.* Reeve Brothers, London, UK.

Houk, J. L. and Geibel, J. J. 1974. Observations of underwater tool use by the sea otter, *Enhydra lutris* Linnaeus. *California Fish and Game* 60:207–208.

Howard, M. D., Sutula, M., Caron, D. A., Chao, Y., Farrara, J. D., Frenzel, H. et al. 2014. Anthropogenic nutrient sources rival natural sources on small scales in the coastal waters of the Southern California Bight. *Limnology and Oceanography* 59:285–297.

Hruby, T. 1978. *Impacts of Large-Scale Aquatic Biomass Systems.* Woods Hole Oceanographic Institution Technical Report, Woods Hole Oceanographic Institution, Woods Hole, MA.

Hsieh, C. H., Kim, H. J., Watson, W., Di Lorenzo, E., and Sugihara, G. 2009. Climate-driven changes in abundance and distribution of larvae of oceanic fishes in the southern California region. *Global Change Biology* 15:2137–2152.

Hughes, A. D., Brunner, L., Cook, E. J., Kelly, M. S., and Wilson, B. 2012. Echinoderms display morphological and behavioural phenotypic plasticity in response to their trophic environment. *PLOS ONE* 7:e41243.

Hughes, B. B., Eby, R., Van Dye, E., Tinker, M. T., Marks, C. I., Johnson, K. S. et al. 2013. Recovery of a top predator mediates negative eutrophic effects on seagrass. *Proceedings of the National Academy of Sciences* 110:15313–15318.

Hurd, C. L. 2000. Water motion, marine macroalgal physiology and production. *Journal of Phycology* 36:453–472.

Hurd, C. L. and Pilditch, C. A. 2011. Flow-induced morphological variations affect diffusion boundary-layer thickness of *Macrocystis pyrifera* (Heterokontophyta, Laminariales). *Journal of Phycology* 47:341–351.

Hurd, C. L. and Stevens, C. L. 1997. Flow visualization around single-and multiple-bladed seaweeds with various morphologies. *Journal of Phycology* 33:360–367.

Hurd, C. L., Harrison, P. J., and Druehl, L. D. 1996. Effect of seawater velocity on inorganic nitrogen uptake by morphologically distinct forms of *Macrocystis integrifolia* from wave-sheltered and exposed sites. *Marine Biology* 126:205–214.

Hurd, C. L., Stevens, C. L., Laval, B. E., Lawrence, G. A., and Harrison, P. J. 1997. Visualization of seawater flow around morphologically distinct forms of the giant kelp *Macrocystis integrifolia* from wave-sheltered and exposed sites. *Limnology and Oceanography* 42:156–163.

Hurd, C. L., Hepburn, C. D., Currie, K. I., Raven, J. A., and Hunter, K. A. 2009. Testing the effects of ocean acidification on algal metabolism: considerations for experimental designs. *Journal of Phycology* 45:1236–1251.

Hymanson, Z. P., Reed, D. C., Foster, M. S., and Carter, J. W. 1990. The validity of using morphological-characteristics as predictors of age in the kelp *Pterogophora californica* (Laminariales, Phaeophyta). *Marine Ecology Progress Series* 59:295–304.

Inderjit, Chapman, D., Ranelletti, M., and Kaushik, S. 2006. Invasive marine algae: an ecological perspective. *Botanical Review* 72:153–178.

Inglis, G. 1989. The colonization and degradation of stranded *Macrocystis pyrifera* (L) C Ag. by the macrofauna of a New Zealand sandy beach. *Journal of Experimental Marine Biology and Ecology* 125:203–217.

IPCC (Intergovernmental Panel on Climate Change). 2013. Summary for policymakers. In Stocker, T. F., Qin, D., Plattner, G. K., Tignor, M., Allen, S. K., Boschung, J. et al. (eds), *Climate Change 2013: The Physical Science Basis. Contribution of Working Group I to the Fifth Assessment Report of the Intergovernmental Panel on Climate Change.* Cambridge University Press, Cambridge, UK/New York.

Isaacs, J. D. 1976. *Wave Climate Modification and Monitoring.* Pp. 87–90. University of California Sea Grant College Program Annual Report 1975-1976, University of California, La Jolla, CA.

Jackson, G. A. 1977. Nutrients and production of giant kelp, *Macrocystis pyrifera*, off southern California. *Limnology and Oceanography* 22:979–995.

Jackson, G. A. 1983. The physical and chemical environment of a kelp community. Pp. 11–37. In Bascom, W. (ed.), *The Effects of Waste Disposal on Kelp Communities.* Southern California Coastal Water Research Project, Long Beach, CA.

Jackson, G. A. 1987. Modeling the growth and harvest yield of the giant kelp *Macrocystis pyrifera. Marine Biology* 95:611–624.

Jackson, G. A. and Winant, C. D. 1983. Effect of a kelp forest on coastal currents. *Continental Shelf Research* 20:75–80.

Jackson, J. B. C., Kirby, M. X., Berger, W. H., Bjorndal, K. A., Botsford, L. W., Bourque, B. J. et al. 2001. Historical overfishing and the recent collapse of coastal ecosystems. *Science* 293:629–638.

James, D. E., Stull, J. K., and North, W. J. 1990. Toxicity of sewage-contaminated sediment cores to *Macrocystis pyrifera* (Laminariales, Phaeophyta) gametophytes determined by digital image analysis. *Hydrobiologia* 204:483–489.

Jangoux, M. 1984. Diseases of echinoderms. *Helgolander Wissenschaftliche Meeresuntersuchungen* 37:207–216.

Jenkins, S. R., Hawkins, S. J., and Norton, T. A. 2004. Long term effects of *Ascophyllum nodosum* canopy removal on mid-shore community structure. *Journal of the Marine Biological Association of the United Kingdom* 84:327–329.

Jerlov, N. G. 1968. *Optical Oceanography.* Elsevier Publishing Co., New York.

Johnson, C. R., Valentine, J., Pederson, H., Heinzeller, T., and Nebelsick, J. 2003. A most unusual barrens: complex interactions between lobsters, sea urchins and algae facilitates spread of an exotic kelp in eastern Tasmania. *Echinoderms: München* 1:213–220.

Johnson, C. R., Ling, S., Ross, D., Shepherd, S., and Miller, K. 2005. *Establishment of the Long-Spined Sea Urchin (Centrostephanus Rodgersii) in Tasmania: First Assessment of Potential Threats to Fisheries*. FRDC Project No. 2001/044. Tasmanian Aquaculture and Fisheries Institute. Fisheries Research and Development Corporation, Australian Government.

Johnson, C. R., Banks, S. C., Barrett, N. S., Cazassus, F., Dunstan, P. K., Edgar, G. J. et al. 2011. Climate change cascades: shifts in oceanography, species' ranges and subtidal marine community dynamics in eastern Tasmania. *Journal of Experimental Marine Biology and Ecology* 400:17–32.

Johnson, D. L. 1980. Episodic vegetation stripping, soil erosion, and landscape modification in prehistoric and recent historic time, San Miguel Island, California. Pp. 103–121. In Power, D. M. (ed.), *The California Islands: Proceedings of a Multidisciplinary Symposium*. Santa Barbara Museum of Natural History, Santa Barbara, CA.

Jones, G. M. and Scheibling, R. E. 1985. *Paramoeba* sp (Amebida, Paramoebidae) as the possible causative agent of sea-urchin mass mortality in Nova Scotia. *Journal of Parasitology* 71:559–565.

Jones, G. P. 1988. Ecology of rocky reef fish of north-eastern New Zealand: a review. *New Zealand Journal of Marine and Freshwater Research* 22:445–462.

Jones, H. P. and Schmitz, O. J. 2009. Rapid recovery of damaged ecosystems. *PLOS ONE* 4:e5653.

Jones, L. G. 1971. Studies of selected small herbivorous invertebrates inhabiting *Macrocystis* canopies and holdfasts in southern California kelp beds. Pp. 343–367. In North, W. J. (ed.), *The Biology of Giant Kelp Beds (Macrocystis) in California*. Nova Hedwigia 32. Verlag Von J. Cramer, Lehre, Germany.

Jones, N. S. and Kain, J. M. 1967. Subtidal algal colonization following removal of *Echinus*. *Helgolander Wissenschaftliche Meeresuntersuchungen* 15:460–466.

Kain, J. M. 1979. A view of the genus *Laminaria*. *Oceanography and Marine Biology: An Annual Review* 17:101–161.

Kamenos, N. A., Burdett, H. L., Aloisio, E., Findlay, H. S., Martin, S., Longbone, C. et al. 2013. Coralline algal structure is more sensitive to rate, rather than the magnitude, of ocean acidification. *Global Change Biology* 19:3621–3628.

Karpov, K., Haaker, P., Taniguchi, I., and Rogers-Bennett, L. 2000. Serial depletion and the collapse of the California abalone (*Haliotis* spp.) fishery. Pp. 11–24. In Campbell, A. (ed.). *Workshop on Rebuilding Abalone Stocks in British Columbia*. Canadian Special Publication of Fisheries and Aquatic Sciences. NRC Research Press, Ottawa, Canada.

Kastendiek, J. E. 1982. Competitor mediated coexistence: interactions among three species of benthic macroalgae. *Journal of Experimental Marine Biology and Ecology* 62:201–210.

Kato, S. 1972. Sea urchins: new fishery develops in California. *Marine Fisheries Review* 34:23–30.

Kay, M. C., Lenihan, H. S., Guenther, C. M., Wilson, J. R., Miller, C. J., and Shrout, S. W. 2012. Collaborative assessment of California spiny lobster population and fishery responses to a marine reserve network. *Ecological Applications* 22:322–335.

Keats, D. 1986. Comment on 'Seaweeds, sea urchins, and lobsters: a reappraisal' by RJ Miller. *Canadian Journal of Fisheries and Aquatic Sciences* 43:1675–1676.

Kelsey, H. 1986. *Juan Rodríguez Cabrillo*. Huntington Library, San Marino, CA.

Kennelly, S. J. and Underwood, A. J. 1984. Underwater microscopic sampling of a sublittoral kelp community. *Journal of Experimental Marine Biology and Ecology* 76:67–78.

Kenner, M. C. 1992. Population dynamics of the sea urchin *Strongylocentrotus purpuratus* in a central California kelp forest: recruitment, mortality, growth and diet. *Marine Biology* 112:107–118.

Kenyon, K. W. 1969. The sea otter in the eastern Pacific Ocean. *North American Fauna* 68:1–252.

Kim, S. L. 1992. The role of drift kelp in the population ecology of a *Diopatra ornata* Moore (Polychaeta, Onuphidae) ecotone. *Journal of Experimental Marine Biology and Ecology* 156:253–272.

Kim, Y. S. and Nam, K. W. 1997. Maturation of gametophytes of *Undaria pinnatifida* (Harvey) Suringar in Korea. *Journal of the Korean Fisheries Society* 30:505–510.

Kimura, R. S. and Foster, M. S. 1984. The effects of harvesting *Macrocystis pyrifera* on the algal assemblage in a giant kelp forest. *Hydrobiologia* 116–117:425–428.

Kingsford, M. J. 1995. Drift algae: a contribution to near-shore habitat complexity in the pelagic environment and an attractant for fish. *Marine Ecology Progress Series* 116:297–301.

Kingsford, M. J. and Choat, J. H. 1985. The fauna associated with drift algae captured with a plankton-mesh purse seine net. *Limnology and Oceanography* 30:618–630.

Kingsford, M. J., Schiel, D. R., and Battershill, C. N. 1989. Distribution and abundance of fish in a rocky reef environment at the sub-Antarctic Auckland Islands, New Zealand. *Polar Biology* 9:179–186.

Kinlan, B. P. and Gaines, S. D. 2003. Propagule dispersal in marine and terrestrial environments: a community perspective. *Ecology* 84:2007–2020.

Kinlan, B. P., Graham, M. H., Sala, E., and Dayton, P. K. 2003. Arrested development of giant kelp (*Macrocystis pyrifera*, Phaeophyceae) embryonic sporophytes: a mechanism for delayed recruitment in perennial kelps? *Journal of Phycology* 39:47–57.

Kinlan, B. P., Gaines, S. D., and Lester, S. E. 2005. Propagule dispersal and the scales of marine community process. *Diversity and Distributions* 11:139–148.

Kirtman, B., Power, S. B., Adedoyin, J. A., Boer, G. J., Bojariu, R., Camilloni, I., et al. 2013. Near-term climate change: projections and predictability. In Stocker, T. F., Qin, D., Plattner, G. K., Tignor, M., Allen, S. K., Boschung, J., et al. (eds), *The Physical Science Basis: Contribution of Working Group I to the Fifth Assessment Report of the Intergovernmental Panel on Climate Change*. Cambridge University Press, Cambridge, UK and New York.

Kitching, J. A. and Ebling, F. J. 1961. The ecology of Lough XI. The control of algae by *Paracentrotus lividus* (Echinoidea). *Journal of Animal Ecology* 30:373–383.

Kitching, J. A., Mccan, T. T., and Hilson, H. C. 1934. Studies in sublittoral ecology. I. A submarine gully in Wembury Bay, South Devon. *Journal of the Marine Biological Association of the United Kingdom* 19:677–705.

Klein, C. J. Chan, A., Cundiff, L., Gardner, A. J., Hrovat, Y., Scholz, A., et al. 2008. Striking a balance between biodiversity conservation and socioeconomic viability in the design of marine protected areas. *Conservation Biology* 22:691–700.

Koehl, M. A. R. 1984. How do benthic organisms withstand moving water. *American Zoologist* 24:57–70.

Koehl, M. A. R. 1986. Mechanical design of spicule-reinforced connective tissues. *American Zoologist* 26:A38.

Koehl, M. A. R. 1999. Ecological biomechanics of benthic organisms: life history mechanical design and temporal patterns of mechanical stress. *Journal of Experimental Biology* 202:3469–3476.

Koehl, M. A. R. and Wainwright, S. A. 1977. Mechanical adaptations of a giant kelp. *Limnology and Oceanography* 22:1067–1071.

Koehl, M. A. R., Silk, W. K., Liang, H., and Mahadevan, L. 2008. How kelp produce blade shapes suited to different flow regimes: a new wrinkle. *Integrative and Comparative Biology* 48:834–851.

Konar, B. 2000. Seasonal inhibitory effects of marine plants on sea urchins: structuring communities the algal way. *Oecologia* 125:208–217.

Konar, B. and Foster, M. S. 1992. Distribution and recruitment of subtidal geniculate coralline algae. *Journal of Phycology* 28:273–280.

Konar, B. and Roberts, C. 1996. Large-scale landslide effects on two exposed rocky subtidal areas in California. *Botanica Marina* 39:517–524.

Konar, B., Edwards, M. S., and Estes, J. A. 2014. Biological interactions maintain the boundaries between kelp forests and urchin barrens in the Aleutian Archipelago. *Hydrobiologia* 724:91–107.

Konotchick, T., Parnell, P. E., Dayton, P. K., and Leichter, J. J. 2012. Vertical distribution of *Macrocystis pyrifera* nutrient exposure in southern California. *Estuarine Coastal and Shelf Science* 106:85–92.

Konotchick, T., Dupont, C. L., Valas, R. E., Badger, J. H., and Allen, A. E. 2013. Transcriptomic analysis of metabolic function in the giant kelp, *Macrocystis pyrifera*, across depth and season. *New Phytologist* 198:398–407.

Kopczak, C. D. 1994. Variability of nitrate uptake capacity in *Macrocystis pyrifera* (Laminariales, Phaeophyta) with nitrate and light availability. *Journal of Phycology* 30:573–580.

Kopczak, C. D., Zimmerman, R. C., and Kremer, J. N. 1991. Variation in nitrogen physiology and growth among geographically isolated populations of the giant kelp, *Macrocystis pyrifera* (Phaeophyta). *Journal of Phycology* 27:149–158.

Korb, R. E. and Gerard, V. A. 2000. Nitrogen assimilation characteristics of polar seaweeds from differing nutrient environments. *Marine Ecology Progress Series* 198:83–92.

Kovalenko, I., Zdyrko, B., Magasinski, A., Hertzberg, B., Milicev, Z., Burtovyy, R. et al. 2011. A major constituent of brown algae for use in high-capacity Li-Ion batteries. *Science* 334:75–79.

Kramer, D. E. and Nordin, D. M. A. 1979. *Studies on the Handling and Processing of Sea Urchin Roe. I. Fresh Product.* Fisheries and Marine Service Technical Report No. 870, Department of Fisheries and Oceans, BC, Canada.

Kroeker, K. J., Kordas, R. L., Crim, R. N., and Singh, G. G. 2010. Meta-analysis reveals negative yet variable effects of ocean acidification on marine organisms. *Ecology Letters* 13:1419–1434.

Kroeker, K. J., Kordas, R. L., Crim, R., Hendriks, I. E., Ramajo, L., Singh, G. S. et al. 2013. Impacts of ocean acidification on marine organisms: quantifying sensitivities and interaction with warming. *Global Change Biology* 19:1884–1896.

Krumhansl, K. A. and Scheibling, R. E. 2012. Production and fate of kelp detritus. *Marine Ecology Progress Series* 467:281–302.

Kühnemann, O. 1970. Algunas considerationes sobre los bosques de *Macrocystis pyrifera*. *Physis* 29:273–296. (in Spanish)

Küpper, F. C., Carpenter, L. J., McFiggans, G. B., Palmer, C. J., Waite, T. J., Boneberg, E.-M. et al. 2008. Iodide accumulation provides kelp with an inorganic antioxidant impacting atmospheric chemistry. *Proceedings of the National Academy of Sciences* 105:6954–6958.

Kuwabara, J. S. 1982. Micronutrients and kelp cultures: evidence for cobalt and manganese deficiency in southernCalifornia deep seawater. *Science* 216:1219–1221.

Kuwabara, J. S. and North, W. J. 1980. Culturing microscopic stages of *Macrocystis pyrifera* (Phaeophyta) in Aquil, a chemically defined medium. *Journal of Phycology* 16:546–549.

Ladah, L. B. and Zertuche-González, J. A. 2004. Giant kelp (*Macrocystis pyrifera*) survival in deep water (25–40 m) during El Niño of 1997–1998 in Baja California, Mexico. *Botanica Marina* 47:367–372.

Ladah, L. B. and Zertuche-González, J. A. 2007. Survival of microscopic stages of a perennial kelp (*Macrocystis pyrifera*) from the center and the southern extreme of its range in the Northern Hemisphere after exposure to simulated El Niño stress. *Marine Biology* 152:677–686.

Ladah, L. B., Zertuche-González, J. A., and Hernández-Carmona, G. 1999. Giant kelp (*Macrocystis pyrifera*, Phaeophyceae) recruitment near its southern limit in Baja California after mass disappearance during ENSO 1997-1998. *Journal of Phycology* 35:1106–1112.

Lafferty, K. D. 2004. Fishing for lobsters indirectly increases epidemics in sea urchins. *Ecological Applications* 14:1566–1573.

Lafferty, K. D. and Kuris, A. M. 1993. Mass mortality of abalone *Haliotis cracherodii* on California Channel Islands: tests of epidemiologic hypotheses. *Marine Ecology Progress Series* 96:239–248.

Lane, C. E., Mayes, C., Druehl, L. D., and Saunders, G. W. 2006. A multi-gene molecular investigation of the kelp (Laminariales, Phaeophyceae) supports substantial taxonomic re-organization. *Journal of Phycology* 42:493–512.

Langford, T. 1990. *Ecological Effects of Thermal Discharges*. Elsevier, Essex, UK.

Larivière, S. 1998. *Lontra felina*. *Mammalian Species* 575:1–5.

Larson, R. J. 1980. Competition, habitat selection and the bathymetric segregation of two rockfish (*Sebastes*) species. *Ecological Monographs* 50:221–239.

Larson, R. J. and DeMartini, E. E. 1984. Abundance and vertical distribution of fishes in a cobble-bottom kelp forest off San Onofre, California. *Fishery Bulletin* 82:37–53.

Lastra, M., Page, H. M., Dugan, J. E., Hubbard, D. M., and Rodil, I. F. 2008. Processing of allochthonous macrophyte subsidies by sandy beach consumers: estimates of feeding rates and impacts on food resources. *Marine Biology* 154:163–174.

Lauck, T., Clark, C. W., Mangel, M., and Munro, G. R. 1998. Implementing the precautionary principle in fisheries management through marine reserves. *Ecological Applications* 8:72–78.

Laur, D., Ebeling, A., and Coon, D. 1988. Effects of sea otter foraging on subtidal reef communities off central California. Pp. 151–168 . In VanBlaricom, G. and Estes, J. A. (eds), *The Community Ecology of Sea Otters*. Springer-Verlag, Berlin, Germany.

Lauzon-Guay, J. S. and Scheibling, R. E. 2007a. Behaviour of sea urchin *Strongylocentrotus droebachiensis* grazing fronts: food-mediated aggregation and density-dependent facilitation. *Marine Ecology Progress Series* 329:191–204.

Lauzon-Guay, J. S. and Scheibling, R. E. 2007b. Seasonal variation in movement, aggregation and destructive grazing of the green sea urchin (*Strongylocentrotus droebachiensis*) in relation to wave action and sea temperature. *Marine Biology* 151:2109–2118.

Lavoie, D. 1985. Population dynamics and ecology of beach wrack macroinvertebrates of the central California coast. *Bulletin of the Southern California Academy of Sciences* 84:1–22.

Lawrence, J. M. 1975. On the relationship between marine plants and sea urchins. *Oceanography and Marine Biology: An Annual Review* 13:213–286.

Lawton, J. H. 1994. What do species do in ecosystems. *Oikos* 71:367–374.

Lawton, J. H. 1999. Are there general laws in ecology? *Oikos* 84:177–192.

Leal, P. P., Hurd, C. L., and Roleda, M. Y. 2014. Meiospores produced in sori of nonsporophyllous laminae of *Macrocystis pyrifera* (Laminariales, Phaeophyceae) may enhance reproductive output. *Journal of Phycology* 50:400–405.

Leet, W. S., Dewees, C. M., Klingbeil, R., and Larson, E. J. 2001. *California's Living Marine Resources: A Status Report*. California Department of Fish and Game, Sacramento, CA.

Leighton, D. L. 1971. Grazing activities of benthic invertebrates in southern California kelp beds. Pp. 421–453. In North, W. J. (ed.), *The Biology of Giant Kelp Beds (Macrocystis) in California*. Nova Hedwigia 32. Verlag Von J. Cramer, Lehre, Germany.

Leleu, K., Remy-Zephyr, B., Grace, R., and Costello, M. J. 2012. Mapping habitats in a marine reserve showed how a 30 year trophic cascade altered ecosystem structure. *Biological Conservation* 155:193–201.

Leonard, G. H. 1994. Effect of the bat star *Asterina miniata* (Brandt) on recruitment of the giant kelp *Macrocystis pyrifera* C. Agardh. *Journal of Experimental Marine Biology and Ecology* 179:81–98.

Lester, S. A., Halpern, B. S., Grorud-Colvert, K., Lubchenco, J., Ruttenberg, B. I., Gaines, S. D. et al. 2009. Biological effects within no-take marine reserves: a global synthesis. *Marine Ecology Progress Series* 384:33–46.

Levenbach, S. 2009. Grazing intensity influences the strength of an associational refuge on temperate reefs. *Oecologia* 159:181–190.

Levin, S. A. and Paine, R. T. 1974. Disturbance, patch formation, and community structure. *Proceedings of the National Academy of Sciences* 71:2744–2747.

Levins, R. 1969. Some demographic and genetic consequences of environmental heterogeneity for biological control. *Bulletin for the Entomological Society of America* 15:237–240.

Levyns, M. 1933. Sexual reproduction in *Macrocystis pyrifera* Ag. *Annals of Botany* 47:349–353.

Lewbel, G. S., Wolfson, A., Gerrodette, T., Lippincott, W. H., Wilson, J. L., and Littler, M. M. 1981. Shallow-water benthic communities on California's outer continental shelf. *Marine Ecology Progress Series* 4:159–168.

Lewis, R. D. and McKee, K. K. 1989. *A Guide to the Artificial Reefs of Southern California*. California Department of Fish and Game, Sacramento, CA.

Lewis, R. J. and Neushul, M. 1994. Northern and southern hemisphere hybrids of *Macrocystis* (Phaeophyceae). *Journal of Phycology* 30:346–353.

Li, D. 2011. Five trillion basidiospores in a fruiting body of *Calvatia gigantea*. *Mycosphere* 2:457–462.

Limbaugh, C. 1955. *Fish Life in the Kelp Beds and the Effects of Kelp Harvesting*. Institute of Marine Resources, University of California, La Jolla, CA.

Lindberg, D. R. 1991. Marine biotic interchange between the northern and southern hemispheres. *Paleobiology* 17:308–324.

Ling, S. D. 2008. Range expansion of a habitat-modifying species leads to loss of taxonomic diversity: a new and impoverished reef state. *Oecologia* 156:883–894.

Ling, S. D., Johnson, C. R., Frusher, S. D., and Ridgway, K. R. 2009. Overfishing reduces resilience of kelp beds to climate-driven catastrophic phase shift. *Proceedings of the National Academy of Sciences* 106:22341–22345.

Ling, S. D., Ibbott, S., and Sanderson, J. C. 2010. Recovery of canopy-forming macroalgae following removal of the enigmatic grazing sea urchin *Heliocidaris erythrogramma*. *Journal of Experimental Marine Biology and Ecology* 395:135–146.

Lissner, A. L. and Dorsey, J. H. 1986. Deep-water biological assemblages of a hard-bottom bank-ridge complex of the southern California continental borderland. *Bulletin of the Southern California Academy of Sciences* 85:87–101.

Living Color. 2012. *Artificial grass, kelp, algae*. www.livingcolor.com.

Lobban, C. S. 1978. Growth and death of the *Macrocystis* sporophyte (Phaeophyceae, Laminariales). *Phycologia* 17:196–212.

Lotze, H. K., Erlandson, J. M., Hardt, M. J., Norris, R. D., Roy, K., Smith, T. D. et al. 2011. Uncovering the ocean's past. Pp. 137–162. In Jackson, J. B. C., Alexander, K. E., and Sala, E. (eds), *Shifting Baselines: The Past and the Future of Ocean Fisheries*. Island Press, Washington, DC.

Lowry, L. F. and Pearse, J. S. 1973. Abalones and sea urchins in an area inhabited by sea otters. *Marine Biology* 23:213–219.

Lubchenco, J., Palumbi, S. R., Gaines, S. D., and Andelman, S. 2003. Plugging a hole in the ocean: the emerging science of marine reserves. *Ecological Applications* 13(1):3–7.

Lucas, A. J., Dupont, C. L., Tai, V., Largier, J. L., Palenik, B., and Franks, P. J. S. 2011. The green ribbon: multiscale physical control of phytoplankton productivity and community structure over a narrow continental shelf. *Limnology and Oceanography* 56:611–626.

Lüning, K. 1980. Critical levels of light and temperature regulating the gametogenesis of 3 *Laminaria* species (Phaeophyceae). *Journal of Phycology* 16:1–15.

Lüning, K. 1981. Photobiology of seaweeds: ecophysiological aspects. *Proceedings of the International Seaweed Symposium* 10:35–55.

Lüning, K. 1993. Environmental and internal control of seasaonal growth in seaweeds. *Hydrobiologia* 260–261:1–14.

Lüning, K. and Dring, M. J. 1975. Reproduction, growth and photosynthesis of gametophytes of *Laminaria saccharina* grown in blue and red-light. *Marine Biology* 29:195–200.

Lüning, K. and Müller, D. G. 1978. Chemical interaction in sexual reproduction of several Laminariales: release and attraction of spermatozoids. *Zeitschrift fur Pflanzenphysiologie* 89:333–341.

Lüning, K. and Neushul, M. 1978. Light and temperature demands for growth ad reproduction of Laminarian gametophytes in southern and central California. *Marine Biology* 45:297–309.

Lüning, K. and Dieck, T. I. 1990. Distribution and evolution of the Laminariales; North Pacific-Atlantic relationships. Pp. 187–204. In *Evolutionary Biogeography of the Marine Algae of the North Atlantic*. Springer Verlag, Berlin, Germany.

Macaya, E. C. and Zuccarello, G. C. 2010a. DNA barcoding and genetic divergence in the giant kelp *Macrocystis* (Laminariales). *Journal of Phycology* 46:736–742.

Macaya, E. C. and Zuccarello, G. C. 2010b. Genetic structure of the giant kelp *Macrocystis pyrifera* along the southeastern Pacific. *Marine Ecology Progress Series* 420:103–112.

Macaya, E. C., Boltana, S., Hinojosa, I. A., Macchiavello, J. E., Valdivia, N. A., Vasquez, N. R. et al. 2005. Presence of sporophylls in floating kelp rafts of *Macrocystis* spp. (Phaeophyceae) along the Chilean Pacific coast. *Journal of Phycology* 41:913–922.

Mackie, J. A., Darling, J. A., and Geller, J. B. 2012. *Ecology of Cryptic Invasions: Latitudinal Segregation among Watersipora (Bryozoa) Species*. Scientific Reports 2. 871; doi:1038 /srep00871.

Madariaga, D. J., Ortiz, M., and Thiel, M. 2013. Demography and feeding behavior of the kelp crab *Taliepus marginatus* in subtidal habitats dominated by the kelps *Macrocystis pyrifera* or *Lessonia trabeculata*. *Invertebrate Biology* 132:133–144.

Maestre, A. and Pitt, R. 2005. *The National Stormwater Quality Database, Version 1.1: A Compilation and Analysis of Npdes Stormwater Monitoring Information*. United States Environmental Protection Agency, Washington, DC.

Maloney, E., Fairey, R., Lyman, A., Walton, Z., and Sigala, M. 2007. *Introduced Aquatic Species in California's Open Coastal Waters 2007: Final Report*. California Department of Fish and Game, Sacramento, CA.

Malthus, T. 1798. *An Essay on the Principle of Population*. London, England. 126 pp.

Manley, S. L. 1983. Composition of sieve tube sap from *Macrocystis pyrifera* (Phaeophyta) with emphasis on the inorganic constituents *Journal of Phycology* 19:118–121.

Manley, S. L. and Lowe, C. G. 2012. Canopy-forming kelps as California's coastal dosimeter: I-131 from damaged Japanese reactor measured in *Macrocystis pyrifera*. *Environmental Science and Technology* 46:3731–3736.

Manley, S. L. and North, W. J. 1984. Phosphorus and the growth of juvenile *Macrocystis pyrifera* (Phaeophyta) sporophytes. *Journal of Phycology* 20:389–393.

Mann, K. H. and Breen, P. A. 1972. Relation between lobster abundance, sea urchins and kelp beds. *Journal of the Fisheries Research Board of Canada* 29:603–605.

Mansilla, A. and Ávila, M. 2011. Using *Macrocystis pyrifera* (L.) C. Agardh from southern Chile as a source of applied biological compounds. *Revista Brasileira de Farmacognosia* 21:262–267.

Marsden, I. D. 1991. Kelp-sandhopper interactions on a sand beach in New Zealand. I. drift composition and distribution. *Journal of Experimental Marine Biology and Ecology*, 152:61–74.

Martin, K. 2008. Beach grooming. *Coastal Services* 11:5–9.

Martínez, E. A., Cardenas, L., and Pinto, R. 2003. Recovery and genetic diversity of the intertidal kelp *Lessonia nigrescens* (Phaeophyceae) 20 years after El Niño 1982/83. *Journal of Phycology* 39:504–508.

Mason, J. E. 2004. Historical patterns from 74 years of commercial landings from California waters. *CalCOFI Reports*, 45:180–190.

Mattison, J. E., Trent, J. D., Shanks, A. L., Akin, T. B., and Pearse, J. S. 1977. Movement and feeding activity of red sea urchins (*Strongylocentrotus franciscanus*) adjacent to a kelp forest. *Marine Biology* 39:25–30.

Maxwell, S. M., Hazen, E. L., Bograd, S. J., Halpern, B. S., Breed, G. A., Nickel, B. et al. 2013. Cumulative human impacts on marine predators. *Nature Communications* 4:2688.

MBC (MBC Applied Environmental Sciences). 2013. *Status of the Kelp Beds 2012: Ventura, Los Angeles, Orange and San Diego Counties.* MBC Applied Environmental Sciences, Costa Mesa, CA.

McClanahan, T. R. and Mangi, S. 2000. Spillover of exploitable fishes from a marine park and its effect on the adjacent fishery. *Ecological Applications* 10:1792–1805.

McClanahan, T. R., Baird, A. H., Marshall, P. A., and Toscano, M. A. 2004. Comparing bleaching and mortality responses of hard corals between southern Kenya and the Great Barrier Reef, Australia. *Marine Pollution Bulletin* 48:327–335.

McConnico, L. A. and Foster, M. S. 2005. Population biology of the intertidal kelp, *Alaria marginata* Postels and Ruprecht: a non-fugitive annual. *Journal of Experimental Marine Biology and Ecology* 324:61–75.

McFarland, W. N. and Prescott, J. 1959. Standing crop, chlorophyll content and *in situ* metabolism of a giant kelp community in southern California. *Publication of the Institute of Marine Science University Texas* 6:109–132.

McHugh, D. J. 2003. *A Guide to the Seaweed Industry.* FAO Fisheries Technical Paper No. 441. FAO, Rome, Italy.

McLachlan, A. and Brown, A. C. 2006. *The Ecology of Sandy Shores.* Academic Press, Burlington, MA.

McLean, J. H. 1962. Sublittoral ecology of kelp beds of open coast area near Carmel, California. *Biological Bulletin* 122:95–114.

McNair, J. N., Newbold, J. D., and Hart, D. D. 1997. Turbulent transport of suspended particles and dispersing benthic organisms: how long to hit bottom? *Journal of Theoretical Biology* 188:29–52.

McPeak, R. H. and Glantz, D. A. 1984. Harvesting California's kelp forests. *Oceanus* 27:19–26.

McPeak, R. H., Glantz, D. A., and Shaw, C. R. 1988. *The Amber Forest.* Watersport Publishing, San Diego, CA.

McPhee-Shaw, E. E., Siegel, D. A., Washburn, L., Brzezinski, M. A., Jones, J. L., Leydecker, A. et al. 2007. Mechanisms for nutrient delivery to the inner shelf: observations from the Santa Barbara Channel. *Limnology and Oceanography* 52:1748–1766.

Medlin, L. K., Kooistra, W., Potter, D., Saunders, G. W., and Andersen, R. A. 1997. Phylogenetic relationships of the 'golden algae' (haptophytes, heterokont chromophytes) and their plastids. *Plant Systematics and Evolution*, 187–219.

Menge, B. A. 1992. Community regulation: under what conditions are bottom-up factors important on rocky shores. *Ecology* 73:755–765.

Menge, B. A. 2008. Marine food webs: conceptual development of a central research paradigm. Pp. 3–28. In McClanahan, T. R. and Branch, G. M. (eds), *Food Webs and the Dynamics of Marine Reefs.* Oxford University Press, Oxford, UK.

Micheli, F. and Halpern, B. S. 2005. Low functional redundancy in coastal marine assemblages. *Ecology Letters* 8:391–400.

Micheli, F., Dunbar, B., Watanabe, J., Shelton, A. O., Bushinsky, S. M., Chin, A. L. et al. 2008. Persistence of depleted abalones in marine reserves of central California. *Biological Conservation* 141:1078–1090.

Michelou, V. K., Caporaso, J. G., Knight, R., and Palumbi, S. R. 2013. The ecology of microbial communities associated with *Macrocystis pyrifera*. *PLOS ONE* 8:e67480.

Miller, D. J. and Giebel, J. J. 1973. *Summary of Blue Rockfish and Lingcod Life Histories; A Reef Ecology Study; and Giant Kelp, Macrocystis Pyrifera, Experiments in Monterey Bay, California.* Fisheries Bulletin 158, California Department of Fish and Game, Sacramento, CA, 1–137.

Miller, K. A. and Engle, J. M. 2009. The natural history of *Undaria pinnatifida* and *Sargassum filicinum* at the California Channel Islands: non-native seaweeds with different invasion styles. Pp. 131–140. In Diamiani, C. C. and Garcelon, D. K. (eds), *Proceedings of the 7th California Islands Symposium.* Institute for Wildlife Studies, Arcata, CA.

Miller, K. A., Aguilar-Rosas, L. E., and Pedroche, F. F. 2011. A review of non-native seaweeds from California, USA and Baja California, México. *Hidrobiológica* 21:365–379.

Miller, R. J. 1985. Seaweeds, sea urchins and lobsters: a reappraisal. *Canadian Journal of Fisheries and Aquatic Sciences* 42:2061–2072.

Miller, R. J. 1986. Reply to comment by D. Keats. *Canadian Journal of Fisheries and Aquatic Sciences* 43:1676.

Miller, R. J. 1987. Reply to comment by P. A. Breen. *Canadian Journal of Fisheries and Aquatic Sciences* 44:1807–1809.

Miller, R. J. and Colodey, A. G. 1983. Widespread mass mortalities of the green sea urchin in Nova Scotia, Canada. *Marine Biology* 73:263–267.

Miller, R. J. and Etter, R. J. 2008. Shading facilitates sessile invertebrate dominance in the rocky subtidal Gulf of Maine. *Ecology* 89:452–462.

Miller, R. J. and Page, H. M. 2012. Kelp as a trophic resource for marine suspension feeders: a review of isotope-based evidence. *Marine Biology* 159:1391–1402.

Miller, R. J., Reed, D. C., and Brzezinski, M. A. 2011. Partitioning of primary production among giant kelp (*Macrocystis pyrifera*), understory macroalgae, and phytoplankton on a temperate reef. *Limnology and Oceanography* 56:119–132.

Mitchell, C. T. and Hunter, J. R. 1970. Fishes associated with drifting kelp, *Macrocystis pyrifera*, off the coast of southern California and northern Baja California. *California Fish and Game* 56:288–297.

Moitoza, D. J. and Phillips, D. W. 1979. Prey defense, predator preference and non-random diet: the interactions between *Pycnopodia helianthoides* and two species of sea urchins. *Marine Biology* 53:299–304.

Mooney, H., Larigauderie, A., Cesario, M., Elmquist, T., Hoegh-Guldberg, O., Lavorel, S. et al. 2009. Biodiversity, climate change, and ecosystem services. *Current Opinion in Environmental Sustainability* 1:46–54.

Moore, L. B. 1943. Observations on the growth of *Macrocystis* in New Zealand with a description of a free-living form. *Transactions and Proceedings of the Royal Society of New Zealand* 72:333–341.

Mora, C. and Sale, P. F. 2011. Ongoing global biodiversity loss and the need to move beyond protected areas: a review of the technical and practical shortcomings of protected areas on land and sea. *Marine Ecology Progress Series* 434:251–266.

Morelissen, B., Dudley, B. D., Geange, S. W., and Phillips, N. E. 2013. Gametophyte reproduction and development of *Undaria pinnatifida* under varied nutrient and irradiance conditions. *Journal of Experimental Marine Biology and Ecology* 448:197–206.

Moreno, C. A. and Jara, H. F. 1984. Ecological studies on fish fauna associated with *Macrocystis pyrifera* belts in the south of Fueguian Islands, Chile. *Marine Ecology Progress Series* 15:99–107.

Moreno, C. A. and Sutherland, J. P. 1982. Physical and biological processes in a *Macrocystis pyrifera* community near Valdivia, Chile. *Oecologia* 55:1–6.

Moritz, C. and Agudo, R. 2013. The future of species under climate change: resilience or decline? *Science* 341:504–508.

Morris, R. H., Abbott, D. P., and Haderlie, E. C. 1980. *Intertidal Invertebrates of California.* Stanford University Press, Stanford, CA.

Morse, D. E., Hooker, N., Duncan, H., and Jensen, L. 1979. Gamma-aminobutyric acid, a neurotransmitter, induces planktonic abalone larvae to settle and begin metamorphosis. *Science* 204:407–410.

Müller, D. G. 1981. Sexuality and sexual attraction. Pp. 661–676. In Lobban, C. S. and Wynne, M. J. (eds), *The Biology of Seaweeds. Botanical Monographs* 17. University of California Press, Berkeley, CA.

Müller, D. G. 1988. Demonstration of sexual pheromones in Laminariales and Desmarestiales. Pp. 243–250. In Lobban, C. S., Chapman, D. J., and Cremer, B. (eds), *Experimental Phycology: A Laboratory Manual.* Cambridge University Press, Cambridge, UK.

Müller, D. G., Maier, I., and Gassmann, G. 1985. Survey on sexual pheromone specificity in Laminariales (Phaeophyceae). *Phycologia* 24:475–477.

Muñoz, V., Hernández-González, M. C., Buschmann, A. H., Graham, M. H., and Vásquez, J. A. 2004. Variability in per capita oogonia and sporophyte production from giant kelp gametophytes (*Macrocystis pyrifera*, Phaeophyceae). *Revista Chilena De Historia Natural* 77:639–647.

Murdoch, W. W., Mechalas, B., and Fay, R. C. 1989. *Final Report of the Marine Review Committee to the California Coastal Commission on the Effects of the San Onofre Nuclear Generating Station on the Marine Environment.* California Coastal Commission, San Francisco, CA.

Murray, S. N., Littler, M. M., and Abbott, I. A. 1980. Biogeography of the California marine algae with emphasis on the southern California islands. Pp. 325–339. In Power, D. M. (ed.), *The California Islands: Proceedings of a Multidisciplinary Symposium.* Santa Barbara Museum of Natural History, Santa Barbara, CA.

Murray, S. N., Ambrose, R. F., Bohnsack, J. A., Botsford, L. W., Carr, M. H., Davis, G. E. et al. 1999. No-take reserve networks: sustaining fishery populations and marine ecosystems. *Fisheries* 24:11–25.

Muth, A. F. 2012. Effects of zoospore aggregation and substrate rugosity on kelp recruitment success. *Journal of Phycology* 48:1374–1379.

NASA JPL (National Aeronautics and Space Administration Jet Propulsion Laboratory). 2013. *MODIS Aqua Level 3 SST thermal IR monthly 4km nighttime.* http://podaac.jpl.nasa.gov /dataset/MODIS_AQUA_L3_SST_THERMAL_MONTHLY_4KM_NIGHTTIME.

Neuman, M., Tissot, B., and VanBlaricom, G. 2010. Overall status and threats assessment of black abalone (*Haliotis cracherodii* Leach, 1814) populations in California. *Journal of Shellfish Research* 29:577–586.

Neushul, M. 1959. *Studies on the Growth and Reproduction of the Giant Kelp, Macrocystis*. PhD thesis. University of California, Los Angeles.

Neushul, M. 1963. Studies on giant kelp, *Macrocystis*. 2. Reproduction. *American Journal of Botany* 50:354–359.

Neushul, M. 1965. Scuba diving studies of the vertical distribution of benthic marine plants. *Botanica Gothoburgensia* 3:161–176

Neushul, M. 1971a. The species of *Macrocystis* with particular reference to those of North and South America. Pp. 211–222. In North, W. J. (ed.), *The Biology of Giant Kelp Beds (Macrocystis) in California*. Nova Hedwigia 32. Verlag Von. J. Cramer, Lehre, Germany.

Neushul, M. 1971b. Submarine illumination in *Macrocystis*. Pp. 241–254. In North, W. J. (ed.), *The Biology of Giant Kelp Beds (Macrocystis) in California*. Nova Hedwigia 32. Verlag Von J. Cramer, Lehre, Germany.

Neushul, M. 1972. Functional interpretation of benthic marine algal morphology. Pp. 47–74. In Abbott, I. A. and Kurogi, M. (eds), *Contributions to the Systematics of Benthic Marine Algae of the North Pacific*. Japanese Society of Phycology, Kobe, Japan.

Neushul, M. 1987. Energy from marine biomass: a historical record. Pp. 1–37. In Bird, T. K. and Benson, P. H. (eds), *Seaweed Cultivation for Renewable Resources*. Elsevier Science Publishers, Amsterdam, the Netherlands.

Neushul, M. and Harger, B. W. G. 1985. Studies of biomass yield from a nearshore macroalgal test farm. *Journal of Solar Energy Engineering* 107:93–97.

Neushul, M. and Haxo, F. T. 1963. Studies on giant kelp, *Macrocystis*. 1. Growth of young plants. *American Journal of Botany* 50:349–353.

Neushul, M., Clarke, W. D., and Brown, D. W. 1967. Subtidal plant and animal communities of the southern California islands. Pp. 37–55. In Philbrick, R. (ed.), *Proceedings of the Symposium on the Biology of the California Islands*. Santa Barbara Botanic Garden, Santa Barbara, CA.

Neushul, M., Foster, M. S., Coon, D. A., Woessner, J. W., and Harger, B. W. W. 1976. In situ study of recruitment, growth and survival of subtidal marine algae: techniques and preliminary results. *Journal of Phycology* 12:397–408.

Neushul, P. 1989. Seaweed for war: California's World War I kelp industry. *Technology and Culture* 30:561–583.

Newcombe, E. M., Cardenas, C. A., and Geange, S. W. 2012. Green sea urchins structure invertebrate and macroalgal communities in the Magellan Strait, southern Chile. *Aquatic Biology* 15:135–144.

Nicholson, N. L. 1979. Evolution within *Macrocystis*: northern and southern hemisphere taxa. Pp. 433–442. In *Proceedings of the International Symposium for Marine Biogeography and Evolution in the Southern Hemisphere*. NZ Department of Science and Industrial Research, Auckland, New Zealand.

Nisbet, R. M. and Bence, J. R. 1989. Alternative dynamic regimes in canopy-forming kelp: a variant on density-vague population regulation. *American Naturalist* 134:377–408.

NOAA 2012. *Restoring kelp on Palos Verdes Shelf.* www.montroserestoration.gov.

North, W. J. 1964. *An Investigation of the Effects of Discharged Wastes on Kelp.* The Resources Agency of California, State Water Quality Control Board, Sacramento, CA.

North, W. J. 1966. Kelp habitat improvement project: annual report 1965-1966. W.M. Keck Laboratory of Environmental Health Engineering, California Institute of Technology, Pasadena, CA.

North, W. J. 1969. *Kelp Habitat Improvement Project, Annual Report 1968-1969.* W.M. Keck Laboratory of Environmental Health Engineering, California Institute of Technology, Pasadena, CA.

North, W. J. 1971a. *The Biology of Giant Kelp Beds (Macrocystis) in California.* Nova Hedwigia 32. Verlag Von J. Cramer, Lehre, Germany.

North, W. J. 1971b. Introduction and background. Pp. 1–97. In North, W. J. (ed.), *The Biology of Giant Kelp Beds (Macrocystis) in California.* Nova Hedwigia 32. Verlag Von J. Cramer, Lehre, Germany.

North, W. J. 1976. Aquaculture techniques for creating and restoring beds of giant kelp, *Macrocystis* spp. *Journal of the Fisheries Research Board of Canada* 33:1015–1023.

North, W. J. 1980. Trace metals in giant kelp, *Macrocystis. American Journal of Botany* 67:1097–1101.

North, W. J. 1983. Separating the effects of temperature and nutrients. Pp. 243–255. In Bascom, W. (ed.), *The Effects of Waste Disposal on Kelp Communities.* Southern California Coastal Water Research Project, Long Beach, CA.

North, W. J. 1987a. Biology of the Macrocystis resource in North America. Pp. 265–311. In Doty, M. S., Caddy, J. F., and Santelices, B. (eds), *Case Studies of Seven Seaweed Resources.* FAO Fisheries Technical Paper No. 281. FAO, Rome, Italy.

North, W. J. 1987b. Oceanic farming of *Macrocystis,* the problems and non-problems. Pp. 39–67. In Bird, T. K. and Benson, P. H. (eds), *Seaweed Cultivation for Renewable Resources.* Elsevier, Amsterdam, the Netherlands.

North, W. J. T. Liu, J. Chen, and Suo, R. 1988. Cultivation of *Laminaria* and *Macrocystis* (Laminariales, Phaeophyta) in the Peoples Republic of China. *Phycologia* 27: 298–299.

North, W. J. 1994. Review of *Macrocystis* biology. Pp. 447–527. In Akatsuka, I. (ed.), *Biology of Economic Algae.* SPB Academic Publishing, The Hague, the Netherlands.

North, W. J. 1998. Interview by S. Erwin. Oral History Project, California Institute of Technology. accessed October 15, 2013, http://oralhistories.library.caltech.edu/34/1/OH_North_W.pdf.

North, W. J. and Hubbs, C. L. 1968. *Utilization of kelp-bed resources in southern California. The Resources Agency,* California Department of Fish and Game, Sacramento, CA.

North, W. J. and Pearse, J. S. 1970. Sea urchin population explosion in southern California coastal waters. *Science* 167:209–210.

North, W. J. and Schaefer, M. B. 1964. *An Investigation of the Effects of Discharged Wastes on Kelp.* The Resources Agency of California, State Water Quality Control Board Publication, Sacramento, CA.

North, W. J., Neushul, M., and Clendenning, K. A. 1964. Successive biological changes observed in a marine cove exposed to a large spillage of mineral oil. Pp. 335–354. In *Pollution Marines par les Produits Petroliers, Symposium de Monaco,* Monaco.

North, W. J., Gerard, V., and McPeak, R. 1981. Experimental fertilizing of coastal *Macrocystis* beds. *Proceedings of the International Seaweed Symposium* 10:613–618.

North, W. J., Gerard, V., and Kuwabara, J. S. 1982. Farming *Macrocystis* at coastal and oceanic sites. Pp. 247–264. In Srivastava, L. M. (ed.), *Synthetic and Degradative Processes in marine macrophytes*. Walter de Gruyter, Berlin, Germany.

Norton, T. A., Mathieson, A. C., and Neushul, M. 1981. Morphology and environment. Pp. 421–451. In Lobban, C. S. and Wynne, M. J. (eds), *The Biology of Seaweeds*. University of California Press, Berkeley, CA.

Norton, T. A., Mathieson, A. C., and Neushul, M. 1982. A review of some aspects of form and function in seaweeds. *Botanica Marina* 25:501–510.

Nowlis, J. S. 2000. Short-and long-term effects of three fishery-management tools on depleted fisheries. *Bulletin of Marine Science* 66:651–662.

NRC (National Research Council). 2003. *Oil in the Sea. III*. The National Academies Press, Washington, DC.

Nyman, M. A., Murray, T. B., Neushul, M., and Keogh, J. A. 1990. *Macrocystis pyrifera* in New Zealand: testing two mathematical models for whole plant growth. *Journal of Applied Phycology* 2:249–257.

O'Connor, K. C. and Anderson, T. W. 2010. Consequences of habitat disturbance and recovery to recruitment and the abundance of kelp forest fishes. *Journal of Experimental Marine Biology and Ecology* 386:1–10.

Occhipinti-Ambrogi, A. 2007. Global change and marine communities: alien species and climate change. *Marine Pollution Bulletin* 55:342–352.

Ojeda, F. P. and Santelices, B. 1984. Invertebrate communities in holdfasts of the kelp *Macrocystis pyrifera* from southern Chile. *Marine Ecology Progress Series* 16:65–73.

Okamoto, D. K., Stekoll, M. S., and Eckert, G. L. 2013. Coexistence despite recruitment inhibition of kelps by subtidal algal crusts. *Marine Ecology Progress Series* 493:103–112.

Okey, T. A. 2003. Macrobenthic colonist guilds and renegades in Monterey Canyon (USA) drift algae: partitioning multidimensions. *Ecological Monographs* 73:415–440.

Oppliger, L. V., Correa, J. A., Engelen, A. H., Tellier, F., and Vieira, V. 2012. Temperature effects on gametophyte life-history traits and geographic distribution of two cryptic kelp species. *PLOS ONE* 7:e39289.

Ostarello, G. l. 1973. Natural history of hydrocoral *Allopora californica* Verill (1866). *Biological Bulletin* 145:548–564.

Ostfeld, R. S. 1982. Foraging strategies and prey switching in the California sea otter. *Oecologia* 53:170–178.

Pace, D. 1981. Kelp community development in Barkley Sound, British Columbia following sea urchin removal. Pp. 457–463. In Fogg, G. E. and Eifion Jones, W. (eds), *Proceedings of the 8th International Seaweed Symposium*. The Marine Sciences Laboratories, Menai Bridge, UK.

Page, H. M., Reed, D. C., Brzezinski, M. A., Melack, J. M., and Dugan, J. E. 2008. Assessing the importance of land and marine sources of organic matter to kelp forest food webs. *Marine Ecology Progress Series* 360:47–62.

Page, H. M., Brooks, A. J., Kulbicki, M., Galzin, R., Miller, R. J., Reed, D. C. et al. 2013. Stable isotopes reveal trophic relationships and diet of consumers in temperate kelp forest and coral reef ecosystems. *Oceanography* 26:180–189.

Paine, R. T. 1980. Food webs: linkage, interaction strength and community infrastructure. *Journal of Animal Ecology* 49:667–685.

Pandolfi, J. M., Connolly, S. R., Marshall, D. J., and Cohen, A. L. 2011. Projecting coral reef futures under global warming and ocean acidification. *Science* 333:418–422.

Papenfuss, G. 1942. Studies of South African Phaeophyceae. I. *Ecklonia maxima, Laminaria pallida, Macrocystis pyrifera. American Journal of Botany* 29:15–24.

Parfrey, L. W., Barbero, E., Lasser, E., Dunthorn, M., Bhattacharya, D., Patterson, D. J. et al. 2006. Evaluating support for the current classification of eukaryotic diversity. *PLOS Genetics* 2:2062–2073.

Parker, B. C. 1965. Translocation in the giant kelp *Macrocystis* I. rates, direction, quantity of C14 labelled products and fluorescein. *Journal of Phycology* 1:41–46.

Parker, B. C. 1971. The internal structure of *Macrocystis*. Pp. 99–121. In North, W. J. (ed.), *The Biology of Giant Kelp Beds (Macrocystis) in California.* Nova Hedwigia 32. Verlag Von J. Cramer, Lehre, Germany.

Parker, B. C. and Dawson, E. Y. 1965. Fleshy seaweeds from California Miocene deposits. *American Journal of Botany* 52:643.

Parnell, P. E., Dayton, P. K., Lennert-Cody, C. E., Rasmussen, L. L., and Leichter, J. J. 2006. Marine reserve design: optimal size, habitats, species affinities, diversity, and ocean microclimate. *Ecological Applications* 16:945–962.

Parnell, P. E., Miller, E. F., Lennert-Cody, C. E., Dayton, P. K., Carter, M. L., and Stebbins, T. D. 2010. The response of giant kelp (*Macrocystis pyrifera*) in southern California to low-frequency climate forcing. *Limnology and Oceanography* 55:2686–2702.

Pauly, D., Christensen, V., Dalsgaard, J., Froese, R., and Torres, F. J. 1998. Fishing down marine food webs. *Science* 279:860–863.

Pearse, J. S. and Hines, A. H. 1977. Ecological studies in a kelp forest inhabited by sea otters. *Journal of Phycology* 13:S52.

Pearse, J. S. and Hines, A. H. 1979. Expansion of a central California kelp forest following mass mortality of sea urchins. *Marine Biology* 51:83–91.

Pearse, J. S. and Hines, A. H. 1987. Long-term population dynamics of sea urchins in a central California kelp forest: rare recruitment and rapid decline. *Marine Ecology Progress Series* 39:275–283.

Pearse, J. S. and Lowry, L. F. 1974. *An Annotated Species List of the Benthic Algae and Invertebrates in the Kelp Forest Community At Point Cabrillo, Pacific Grove, California.* Technical Report Number 1, Coastal Marine Laboratory (now Institute of Marine Sciences), University of California, Santa Cruz, CA.

Pearse, J. S., Clark, M. E., Leighton, D. L., Mitchell, C. T., and North, W. J. 1970. Marine waste disposal and sea urchin ecology. In *Kelp Habitat Improvement Project, Annual Report, 1 July 1969-30 June 1970.* California Institute of Technology. Pasadena. CA.

Pearse, J. S., Costa, D. P., Yellin, M. B., and Agegian, C. R. 1977. Localized mass mortality of red sea urchin, *Strongylocentrotus franciscanus*, near Santa Cruz, California. *Fishery Bulletin* 75:647–648.

Pearse, J. S., Mooi, R., Lockhart, S. J., and Brandt, A. 2009. Brooding and species diversity in the Southern Ocean: selection for brooders or speciation within brooding clades? Pp. 181–196. In Krupnik, I., Lang, M. A., and Miller, S. E. (eds), *Smithsonian at the Poles:*

*Contributions to International Polar Year Science.* Smithsonian Institution Scholarly Press, Washington, DC.

Pearse, J. S., Carr, M. H., Baxter, C. H., Watanabe, J. M., Foster, M. S., Steller, D. L. et al. 2013. Kelpbeds as classrooms: perspectives and lessons learned. Pp. 133–142. In Lang, M. A., Marinelli, R. L., Roberts, S. J., and Taylor, P. R. (eds), *Research and Discoveries: The Revolution of Science through Scuba.* Smithsonian Contributions to the Marine Sciences, Washington, DC.

Pearson, G. A., Serrao, E. A., and Brawley, S. H. 1998. Control of gamete release in fucoid algae: sensing hydrodynamic conditions via carbon acquisition. *Ecology* 79:1725–1739.

Pendleton, L. H. and Rooke, J. 2006. *Understanding the Potential Economic Impact of Marine Recreational Fishing: California.* http://www.dfg.ca.gov/mlpa/pdfs/binder3di.pdf.

Pequegnat, W. E. 1964. The epifauna of California siltstone reef. *Ecology* 45:272–283.

Pérez-Matus, A. and Shima, J. S. 2010. Density- and trait-mediated effects of fish predators on amphipod grazers: potential indirect benefits for the giant kelp *Macrocystis pyrifera*. *Marine Ecology Progress Series* 417:151–U168.

Pérez-Matus, A., Ferry-Graham, L. A., Cea, A., and Vásquez, J. A. 2007. Community structure of temperate reef fishes in kelp-dominated subtidal habitats of northern Chile. *Marine and Freshwater Research* 58:1069–1085.

Perissinotto, R. and McQuaid, C. D. 1992. Deep occurrence of the giant kelp *Macrocystis laevis* in the Southern Ocean. *Marine Ecology Progress Series* 81:89–95.

Perry, A. L., Low, P. J., Ellis, J. R., and Reynolds, J. D. 2005. Climate change and distribution shifts in marine fishes. *Science* 308:1912–1915.

Peters, A. F. and Breeman, A. M. 1993. Temperature tolerance and latitudinal range of brown algae from temperate Pacific South America. *Marine Biology* 115:143–150.

Peters, R. H. 1988. Some general problems for ecology illustrated by food web theory. *Ecology* 69:1673–1676.

Peterson, N. M., Peterson, M. J., and Peterson, T. R. 2005. Conservation and the myth of consensus. *Conservation Biology* 19:762–767.

Pidwirny, M. 2006. Earth-Sun Relationships and Insolation. In Pidwirny, M. (ed.) *Fundamentals of Physical Geography.* 2nd ed., University of British Columbia. http://www.physicalgeography.net/fundamentals/chapter7.html.

Pimm, S., Jenkins, C., Abell, R., Brooks, T., Gittleman, J., Joppa, L., et al. 2014. The biodiversity of species and their rates of extinction, distribution, and protection. *Science* 344:1246752.

Pinnegar, J. K., Polunin, N. V. C., Francour, P., Badalamenti, F., Chemello, R., Harmelin-Vivien, M. L. et al. 2000. Trophic cascades in benthic marine ecosystems: lessons for fisheries and protected-area management. *Environmental Conservation* 27:179–200.

Pirker, J. G. 2002. *Demography, Biomass Production and Effects of Harvesting Giant Kelp Macrocystis pyrifera (Linnaeus) in Southern New Zealand.* PhD thesis. University of Canterbury, Christchurch, New Zealand.

Pirker, J. G., Schiel, D. R., and Lees, H. 2000. *Seaweed Products for Barrel Culture Paua Farming. TBGF Project: Contract SR1401.* University of Canterbury, Christchurch, New Zealand.

Platt, J. R. 1964. Strong inference. *Science* 146:347–353.

Polis, G. A. 1991. Complex trophic interactions in deserts: an empirical critique of food web theory. *American Naturalist* 138:123–155.

Polis, G. A. 1999. Why are parts of the world green? Multiple factors control productivity and the distribution of biomass. *Oikos* 86:3–15.

Polis, G. A., Anderson, W. B., and Holt, R. D. 1997. Toward an integration of landscape and food web ecology: the dynamics of spatially subsidized food webs. *Annual Review of Ecology and Systematics* 28:289–316.

Polis, G. A., Sears, A. L. W., Huxel, G. R., Strong, D. R., and Maron, J. 2000. When is a trophic cascade a trophic cascade? *Trends in Ecology and Evolution* 15:473–475.

Ponce-Díaz , G., Lluch-Cota, S. E., Bautista-Romero, J. J., and Lluch-Belda, D. 2003. Multiscale characterization of the sea temperature in an area of abalone banks (*Haliotis* spp.) at Bahia Asuncion, Baja California Sur, Mexico. *Ciencias Marinas* 29:291–303.

Porzio, L., Buia, M. C., and Hall-Spencer, J. M. 2011. Effects of ocean acidification on macroalgal communities. *Journal of Experimental Marine Biology and Ecology* 400:278–287.

Power, M. E., Tilman, D., Estes, J. A., Menge, B. A., Bond, W. J., Mills, L. S. et al. 1996. Challenges in the quest for keystones. *Bioscience* 46:609–620.

Prescott, J. H. 1959. Rafting of jack rabbit on kelp. *Journal of Mammalogy* 40:443–444.

Price, M. L. 2008. *A Biography of Conrad Limbaugh.* University of California, San Diego, La Jolla, CA.

Quast, J. C. 1971a. Some physical aspects of the inshore environment particularly as it affects kelp bed fishes. Pp. 229–240. In North, W. J. (ed.), *The Biology of Giant Kelp Beds (Macrocystis) in California.* Nova Hedwigia 32. Verlag Von J. Cramer, Lehre, Germany.

Quast, J. C. 1971b. Fish fauna of the rocky inshore zone. Pp. 481–507. In North, W. J. (ed.), *The Biology of Giant Kelp Beds (Macrocystis) in California.* Nova Hedwigia 32. Verlag Von J. Cramer, Lehre, Germany.

Quast, J. C. 1971c. Observations on the food of the kelp bed fishes Pp. 541–579. In North, W. J. (ed.), *The Biology of Giant Kelp Beds (Macrocystis) in California.* Nova Hedwigia 32. Verlag Von J. Cramer, Lehre, Germany.

Raimondi, P., Cailliet, G., and Foster, M. S. 2005. *Diablo Canyon Power Plant: Independent Scientist's Recommendations to the Regional Board Regarding "Mitigation" for Cooling Water Impacts.* California central coast regional water quality control Board, San Luis Obispo, CA.

Raimondi, P. T., Reed, D. C., Gaylord, B., and Washburn, L. 2004. Effects of self-fertilization in the giant kelp, *Macrocystis pyrifera.* Ecology 85:3267–3276.

Raimondi, P. T., Wilson, M. C., Ambrose, R. F., Engle, J. M., and Minchinton, T. E. 2002. Continued declines of black abalone along the coast of California: are mass mortalities related to El Niño events? *Marine Ecology Progress Series* 242:143–152.

Rainey, C. 1998. *Wet Suit Pursuit: Hugh Bradner's Development of the First Wet Suit. Archives of the Scripps Institution of Oceanography.* University of California, San Diego, La Jolla, CA.

Rassweiler, A., Arkema, K. K., Reed, D. C., Zimmerman, R. C., and Brzezinski, M. A. 2008. Net primary production, growth, and standing crop of *Macrocystis pyrifera* in Southern California. *Ecology* 89:2068–2068.

Reed, D. C. 1987. Factors affecting the production of sporophylls in the giant kelp *Macrocystis pyrifera. Journal of Experimental Marine Biology and Ecology* 113:61–69.

Reed, D.C. 1990a. The effects of variable settlement and early competition on patterns of kelp recruitment. *Ecology* 71:776–787.

Reed, D.C. 1990b. An experimental evaluation of density dependence in a subtidal algal population. *Ecology* 71:2286–2296.

Reed, D.C. and Brzezinski, M.A. 2009. Kelp forests. Pp. 31–37. In Laffoley, D. and Grimsditch, G. (eds), *The Management of Natural Coastal Carbon Sinks*. IUCN, Gland, Switzerland.

Reed, D.C. and Foster, M.S. 1984. The effects of canopy shading on algal recruitment and growth in a giant kelp forest. *Ecology* 65:937–948.

Reed, D.C., Laur, D.R., and Ebeling, A.W. 1988. Variation in algal dispersal and recruitment: the importance of episodic events. *Ecological Monographs* 58:321–335.

Reed, D.C., Neushul, M., and Ebeling, A.W. 1991. Role of settlement density on gametophyte growth and reproduction in the kelps *Pterogophora californica* and *Macrocystis pyrifera* (Phaeophyceae). *Journal of Phycology* 27:361–366.

Reed, D.C., Amsler, C.D., and Ebeling, A.W. 1992. Dispersal in kelps: factors affecting spore swimming and competency. *Ecology* 73:1577–1585.

Reed, D.C., Lewis, R.J., and Anghera, M. 1994. Effects of an open coast oil production outfall on patterns of giant kelp (*Macrocystis pyrifera*) recruitment. *Marine Biology* 120:25–31.

Reed, D.C., Ebeling, A.W., Anderson, T.W., and Anghera, M. 1996. Differential reproductive responses to fluctuating resources in two seaweeds with different reproductive strategies. *Ecology* 77:300–316.

Reed, D.C., Anderson, T.W., Ebeling, A.W., and Anghera, M. 1997. The role of reproductive synchrony in the colonization potential of kelp. *Ecology* 78:2443–2457.

Reed, D.C., Brzezinski, M.A., Coury, D.A., Graham, W.M., and Petty, R.L. 1999. Neutral lipids in macroalgal spores and their role in swimming. *Marine Biology* 133:737–744.

Reed, D.C., Raimondi, P.T., Carr, M.H., and Goldwasser, L. 2000. The role of dispersal and disturbance in determining spatial heterogeneity in sedentary organisms. *Ecology* 81:2011–2026.

Reed, D.C., Schroeter, S.C., and Raimondi, P.T. 2004. Spore supply and habitat availability as sources of recruitment limitation in the giant kelp *Macrocystis pyrifera* (Phaeophyceae). *Journal of Phycology* 40:275–284.

Reed, D.C., Schroeter, S.C., and Huang, D. 2006. *An Experimental Investigation of the Use of Artificial Reefs to Mitigate the Loss of Giant Kelp Forest Habitat. A Case Study of the San Onofre Nuclear Generating Station's Artificial Reef Project*. California Sea Grant College Program, University of California, San Diego, CA.

Reed, D.C., Schroeter, S.C., Huang, D., Anderson, T.W., and Ambrose, R.F. 2006. Quantitative assessment of different artificial reef designs in mitigating losses to kelp forest fishes. *Bulletin of Marine Science* 78:133–150.

Reed, D.C., Kinlan, B.P., Raimondi, P.T., Washburn, L., Gaylord, B., and Drake, P.T. 2006. A metapopulation perspective on patch dynamics and connectivity of giant kelp. Pp. 352–386. In Kritzer, J.P. and Sale, P.F. (eds), *Marine Metapopulations*. Academic Press, San Diego, CA.

Reed, D.C., Rassweiler, A., and Arkema, K.K. 2008. Biomass rather than growth rate determines variation in net primary production by giant kelp. *Ecology* 89:2493–2505.

Reed, D. C, Rassweiler, A., and Arkema, K. 2009. Density derived estimates of standing crop and net primary production in the giant kelp *Macrocystis pyrifera*. *Marine Biology* 156:2077–2083.

Reed, D. C., Rassweiler, A., Carr, M. H., Cavanaugh, K. C., Malone, D. P., and Siegel, D. A. 2011. Wave disturbance overwhelms top-down and bottom-up control of primary production in California kelp forests. *Ecology* 92:2108–2116.

Revelle, R. and Wheelock, C. D. 1954. *An Oceanographic Investigation of Conditions in the Vicinity of Whites Point and Hyperion Sewage Outfall, Los Angeles, California*. Scripps Institution of Oceanography, La Jolla, CA.

Reynolds, R. A. and Wilen, J. E. 2000. The sea urchin fishery: harvesting, processing and the market. *Marine Resources Economics* 15:115–126.

Ridgway, K. 2007. Long-term trend and decadal variability of the southward penetration of the East Australian Current. *Geophysical Research Letters* 34:L13613.

Riedman, M. L. and Estes, J. A. 1988. A review of the history, distribution and foraging ecology of sea otters. Pp. 4–21. In Van Blaricom, G. R. and Estes, J. A. (eds), *The Community Ecology Of Sea Otters*. Springer-Verlag, Berlin, Germany.

Riedman, M. L., Hines, A. H., and Pearse, J. S. 1981. Spatial segregation of four species of turban snails (Gastropoda, Tegula) in central California. Veliger 24:97–102.

Rios, C., Arntz, W. E., Gerdes, D., Mutschke, E., and Montiel, A. 2007. Spatial and temporal variability of the benthic assemblages associated to the holdfasts of the kelp *Macrocystis pyrifera* in the Straits of Magellan, Chile. *Polar Biology* 31:89–100.

Riosmena-Rodriguez, R., Boo, G. H., Lopez-Vivas, J. M., Hernández-Velasco, A., Saenz-Arroyo, A., and Boo, S. M. 2012. The invasive seaweed *Sargassum filicinum* (Fucales, Phaeophyceae) is on the move along the Mexican Pacific coastline. *Botanica Marina* 55: 547–551.

Roberson, L. M. and Coyer, J. A. 2004. Variation in blade morphology of the kelp *Eisenia arborea*: incipient speciation due to local water motion? *Marine Ecology Progress Series* 282:115–128.

Roberts, C. M. 2000. Selecting marine reserve locations: optimality versus opportunism. *Bulletin of Marine Science* 66:581–592.

Roberts, C. M. and Polunin, N. V. C. 1991. Are marine reserves effective in management of reef fisheries? *Reviews in Fish Biology and Fisheries* 1:65–91.

Roberts, D. A., Johnston, E. L., and Knott, N. A. 2010. Impacts of desalination plant discharges on the marine environment: a critical review of published studies. *Water Research* 44:5117–5128.

Rodriguez, G. E., Rassweiler, A., Reed, D. C., and Holbrook, S. J. 2013. The importance of progressive senescence in the biomass dynamics of giant kelp (*Macrocystis pyrifera*). *Ecology* 94:1848–1858.

Roesijadi, G. S., Jones, S. B., Snowden-Swan, L. J., and Zhu, Y. 2010. *Macroalgae as a Biomass Feedstock: A Preliminary Analysis*. Pacific Northwest National Laboratory, Report PNNL-19944. US National Technical Information Service, Springfield, VA.

Rogers-Bennett, L. 2013. *Strongylocentrotus franciscanus* and *Strongylocentrotus purpuratus*. Pp. 413–435. In Lawrence, J. M. (ed.), *Sea Urchins: Biology and Ecology*. Elsevier, Amsterdam, the Netherlands.

Rogers-Bennett, L., Haaker, P. L., Karpov, K. A., and Kushner, D. J. 2002. Using spatially explicit data to evaluate marine protected areas for abalone in southern California. *Conservation Biology* 16:1308–1317.

Roleda, M. Y., Morris, J. N., McGraw, C. M., and Hurd, C. L. 2012. Ocean acidification and seaweed reproduction: increased $CO_2$ ameliorates the negative effect of lowered pH on meiospore germination in the giant kelp *Macrocystis pyrifera* (Laminariales, Phaeophyceae). *Global Change Biology* 18:854–864.

Roleda, M. Y., Boyd, P. W., and Hurd, C. L. 2012. Before ocean acidification: calcifier chemistry lessons. *Journal of Phycology* 48:840–843.

Rosell, K. G. and Srivastava, L. M. 1985. Seasonal variations in total nitrogen, carbon and amino acids in *Macrocystis integrifolia* and *Nereocystis luetkeana* (Phaeophyta). *Journal of Phycology* 21:304–309.

Rosenberg, G., Littler, D. S., Littler, M. M., and Oliveira, E. C. 1995. Primary production and photosynthetic quotients of seaweeds from Sao Paolo State, Brazil. *Botanica Marina* 38:369–377.

Rosenthal, R. J., Clarke, W. D., and Dayton, P. K. 1974. Ecology and natural history of a stand of giant kelp, *Macrocystis pyrifera*, off Del-Mar, California. *Fishery Bulletin* 72:670–684.

Rosman, J. H., Koseff, J. R., Monismith, S. G., and Grover, J. 2007. A field investigation into the effects of a kelp forest (*Macrocystis pyrifera*) on coastal hydrodynamics and transport. *Journal of Geophysical Research: Oceans* 112:C2.

Rothäusler, E., Gomez, I., Hinojosa, I. A., Karsten, U., Miranda, L., Tala, F. et al. 2011. Kelp rafts in the Humboldt Current: interplay of abiotic and biotic factors limit their floating persistence and dispersal potential. *Limnology and Oceanography* 56:1751–1763.

Rowley, R. J. 1989. Settlement and recruitment of sea urchins (*Strongylocentrotus* spp) in a sea urchin barren ground and a kelp bed: are populations regulated by settlement or post-settlement processes. *Marine Biology* 100:485–494.

Rowley, R. J. 1994. Marine reserves in fisheries management. *Aquatic Conservation: Marine and Fresh Water Ecosystems* 4:233–254.

The Royal Society. 2005. *Ocean Acidification due to Increasing Atmospheric Carbon Dioxide.* Policy document 12 / 05. The Royal Society, London, UK.

Russell, L. K., Hepburn, C. D., Hurd, C. L., and Stuart, M. D. 2008. The expanding range of *Undaria pinnatifida* in southern New Zealand: distribution, dispersal mechanisms and the invasion of wave-exposed environments. *Biological Invasions* 10:103–115.

Sagarin, R. D., Barry, J. P., Gilman, S. E., and Baxter, C. H. 1999. Climate-related change in an intertidal community over short and long time scales. *Ecological Monographs* 69:465–490.

Saito, Y. 1975. *Undaria.* Pp. 304–320. In Tokida, J. and Hirose, H. (eds), *Advances in Phycology in Japan.* Dr W Junk, The Hague, the Netherlands.

Sala, E., Boudouresque, C. F., and Harmelin-Vivien, M. 1998. Fishing, trophic cascades, and the structure of algal assemblages: evaluation of an old but untested paradigm. *Oikos* 82:425–439.

Sala, E. and Dayton, P. K. 2011. Predicting strong community impacts using experimental estimates of per capita interaction strength: benthic herbivores and giant kelp recruitment. *Marine Ecology* 32:300–312.

Sala, E. and Graham, M. H. 2002. Community-wide distribution of predator-prey interaction strength in kelp forests. *Proceedings of the National Academy of Sciences* 99:3678–3683.

Sala, E., Boudouresque, C. F., and Harmelin-Vivien, M. 1998. Fishing, trophic cascades, and the structure of algal assemblages: evaluation of an old but untested paradigm. *Oikos* 82:425–439.

Sala, E., Ribes, M., Hereu, B., Zabala, M., Alva, V., Coma, R. et al. 1998. Temporal variability in abundance of the sea urchins *Paracentrotus lividus* and *Arbacia lixula* in the northwestern Mediterranean: comparison between a marine reserve and an unprotected area. *Marine Ecology Progress Series* 168:135–145.

Sale, P. F. 1978. Coexistence of coral reef fishes – a lottery for living space. *Environmental Biology of Fishes* 3:85–102.

Sale, P. F., Cowen, R. K., Danilowics, B. S., Jones, G. P., Kritzer, J. P., Lindeman, K. C. et al.. 2005. Critical science gaps impede use of no-take fishery reserves. *Trends in Ecology and Evolution* 20:74–80.

Sanbonsuga, Y. and Neushul, M. 1978. Hybridization of *Macrocystis* (Phaeophyta) with other float-bearing kelps. *Journal of Phycology* 14:214–224.

Sanbonsuga, Y. and Neushul, M. 1980. Hybridization and genetics of algae. Pp. 69–83. In Gantt, E. (ed.), *Handbook of Phycological Methods: Developmental and Cytological Methods.* Cambridge University Press, Cambridge, UK.

Santelices, B. 1990. Patterns of reproduction, dispersal and recruitment in seaweeds. *Oceanography and Marine Biology: An Annual Review* 28:177–276.

Santelices, B. 1991. Littoral and sublittoral communities of continental Chile. Pp. 347–369. In Mathieson, A. C. and Nienhuis, P. H. (eds), *Ecosystems of the World 24: Intertidal and Littoral Ecosystems.* Elsevier, Amsterdam, the Netherlands.

Santelices, B. and Ojeda, F. P. 1984a. Population dynamics of coastal forests of *Macrocystis pyrifera* in Puerto Toro, Isla Navarino, Southern Chile. *Marine Ecology Progress Series* 14: 175–183.

Santelices, B. and Ojeda, F. P. 1984b. Effects of canopy removal on the understory algal community structure of coastal forests of *Macrocystis pyrifera* from southern South America. *Marine Ecology Progress Series* 14:165–173.

Sargent, M. C. and Lantrip, L. W. 1952. Photosynthesis, growth and translocation in giant kelp. *American Journal of Botany* 39:99–107.

Saunders, G. W. and Druehl, L. D. 1992. Nucleotidesequence of the small subunit ribosomal RNA genes from selected Laminariales (Phaeophyta): implications for kelp evolution. *Journal of Phycology* 28:544–549.

Saupe, S. 2011. Afognak Island *Macrocystis pyrifera.* Pp. 41. *Kodiak Island Marine Science Symposium.* Alaska Sea Grant Program, Fairbanks, AK.

Scagel, R. F. 1947. *An investigation on marine plants near Hardy Bay, B.C.* B.C. Provincial Department of Fisheries, Report No. 1, Victoria, BC, Canada, 70 pp.

SCE (Southern California Edison Company). 2001. *Report of 2000 Data, Marine Environmental Analysis and Interpretation.* San Onofre Nuclear Generating Station, Southern California Edison Company, Rosemead, CA.

Schaal, G., Riera, P., and Leroux, C. 2012. Food web structure within kelp holdfasts (*Laminaria*): a stable isotope study. *Marine Ecology* 33:370–376.

Schaeffer, T., Foster, M., Landrau, M., and Walder, R. 1999. Diver disturbance in kelp forests. *California Fish and Game* 85:170–176.

Schaffelke, B. and Hewitt, C. L. 2007. Impacts of introduced seaweeds. *Botanica Marina* 50:397–417.

Scheffer, M. and Carpenter, S. R. 2003. Catastrophic regime shifts in ecosystems: linking theory to observation. *Trends in Ecology and Evolution* 18:648–656.

Scheffer, M., Carpenter, S., Foley, J. A., Folke, C., and Walker, B. 2001. Catastrophic shifts in ecosystems. *Nature* 413:591–596.

Scheffer, M., Carpenter, S., and de Young, B. 2005. Cascading effects of overfishing marine systems. *Trends in Ecology and Evolution* 20:579–581.

Scheibling, R. E. 1986. Increased macroalgal abundance following mass mortalities of sea urchins *Strongylocentrotus droebachiensis* along the Atlantic coast of Nova Scotia. *Oecologia* 68:186–198.

Scheibling, R. E. 1996. The role of predation in regulating sea urchin populations in eastern Canada. *Oceanologica Acta* 19:421–430.

Scheibling, R. E. and Stephenson, R. L. 1984. Mass mortality of *Strongylocentrotus droebachiensis* (Echinodermata, Echinoidea) off Nova Scotia, Canada. *Marine Biology* 78:153–164.

Scheibling, R. E. and Lauzon-Guay, J. S. 2010. Killer storms: North Atlantic hurricanes and disease outbreaks in sea urchins. *Limnology and Oceanography* 55:2331–2338.

Scheibling, R. E., Feehan, C., and Lauzon-Guay, J. S. 2010. Disease outbreaks associated with recent hurricanes cause mass mortality of sea urchins in Nova Scotia. *Marine Ecology Progress Series* 408:109–116.

Schiel, D. R. 1982. Selective feeding by the echnoid, *Evechinus chloroticus* and the removal of plants from subtidal algal stands in northern New Zealand. *Oecologia* 54:379–388.

Schiel, D. R. 1985. A short-term demographic study of *Cystoseira osmundacea* (Fucales, Cystoseiraceae) in central California. *Journal of Phycology* 21:99–106.

Schiel, D. R. 1988. Algal interactions on shallow subtidal reefs in Northern New Zealand: a review. *New Zealand Journal of Marine and Freshwater Research* 22:481–489.

Schiel, D. R. 1990. Macroalgal assemblages in New Zealand – structure, interactions and demography. *Hydrobiologia* 192:59–76.

Schiel, D. R. 1997. Review of abalone culture and research in New Zealand. *Molluscan Research* 18:289–298.

Schiel, D. R. 2006. Rivets or bolts? When single species count in the function of temperate rocky reef communities. *Journal of Experimental Marine Biology and Ecology* 338:233–252.

Schiel, D. R. 2009. Multiple stressors and disturbances: when change is not in the nature of things. Pp. 281–294. in Wahl, M. (ed.), *Marine Hard Bottom Communities: Patterns, Dynamics, Diversity and Change*. Springer, Berlin, Germany.

Schiel, D. R. 2011. Biogeographic patterns and long-term changes on New Zealand coastal reefs: non-trophic cascades from diffuse and local impacts. *Journal of Experimental Marine Biology and Ecology* 400:33–51.

Schiel, D. R. 2013. The other 93%: trophic cascades, stressors, and managing coastlines in non-marine protected areas. *New Zealand Journal of Marine and Freshwater Research* 47:374–391.

Schiel, D. R. and Choat, J. H. 1980. Effects of density on monospecific stands of marine algae. *Nature* 285:324–326.

Schiel, D. R. and Foster, M. S. 1986. The structure of subtidal algal stands in temperate waters. *Oceanography and Marine Biology: An Annual Review* 24:265–307.

Schiel, D. R. and Foster, M. S. 1992. Restoring kelp forests. Pp. 279–342. In Thayer, G. (ed.), *Restoring the Nation's Marine Environments*. Maryland Sea Grant College, College Park, MD.

Schiel, D. R. and Foster, M. S. 2006. The population biology of large brown seaweeds: ecological consequences of multiphase life histories in dynamic coastal environments. *Annual Review of Ecology, Evolution, and Systematics* 37:343–372.

Schiel, D. R. and Hickford, M. J. H. 2001. *Biological Structure of Nearshore Rocky Subtidal Habitats in Southern New Zealand*. Science for Conservation, Department of Conservation, Wellington, New Zealand.

Schiel, D. R. and Thompson, G. A. 2012. Demography and population biology of the invasive kelp *Undaria pinnatifida* on shallow reefs in southern New Zealand. *Journal of Experimental Marine Biology and Ecology* 434:25–33.

Schiel, D. R., Andrew, N. L., and Foster, M. S. 1995. The structure of subtidal algal and invertebrate assemblages at the Chatham Islands, New Zealand. *Marine Biology* 123:355–367.

Schiel, D. R., Steinbeck, J. R., and Foster, M. S. 2004. Ten years of induced ocean warming causes comprehensive changes in marine benthic communities. *Ecology* 85:1833–1839.

Schmitt, R. J. 1982. Consequences of dissimilar defenses against predation in a subtidal marine community. *Ecology* 63:1588–1601.

Schmitt, R. J. 1987. Indirect interactions between prey: apparent competition, predator aggregation and habitat segregation. *Ecology* 68:1887–1897.

Schmitt, R. J. 1996. Exploitation competition in mobile grazers: trade-offs in use of a limited resource. *Ecology* 77:408–425.

Schmitt, R. J. and Holbrook, S. J. 1990. Contrasting effects of giant kelp on dynamics of surfperch populations. *Oecologia* 84:419–429.

Schmitt, R. J., Osenberg, C. W., and Bercovitch, M. G. 1983. Mechanisms and consequences of shell fouling in the kelp snail, *Norrisia norrisi* (Sowerby) (Trochidae): indirect effects of octopus drilling. *Journal of Experimental Marine Biology and Ecology* 69:267–281.

Schmitz, K. and Srivastava, L. M. 1974. Fine structure and development of sieve tubes in *Laminaria groenlandica* Rosenv. *Cytobiologie* 10:66–87.

Schofield, O., Ducklow, H. W., Martinson, D. G., Meredith, M. P., Moline, M. A., and Fraser, W. R. 2010. How do polar marine ecosystems respond to rapid climate change? *Science* 328:1520–1523.

Schroeder, D. M. and Love, M. S. 2002. Recreational fishing and marine fish populations in California. *CalCOFI Report* 43:182–190.

Schroeter, R. E. and Moyle, P. B. 2006. Alien species. Pp. 611–620. In Allen, L. G., Pondella, D. J., and Horn, M. H. (eds), *The Ecology of Marine Fishes: California and Adjacent Waters*. University of California Press, Berkeley, CA.

Schroeter, S. C., Dixon, J. D., Kastendiek, J., and Smith, R. O. 1993. Detecting the ecological effects of environmental impacts: a case study of kelp forest invertebrates. *Ecological Applications* 3:331–350.

Schurz, W. L. 1917. The Manila galleon and California. *The Southwestern Historical Quarterly* 21:107–126.

Scuba Advisor. 2012. *Scuba destinations*. http://www.scubaadvisor.com.

Seaweed Energy Solutions. 2012. http://seaweedenergysolutions.com.

Selig, E. R. and Bruno, J. F. 2010. A global analysis of the effectiveness of marine protected areas in preventing coral loss. *PLOS ONE* 5:e9278.

Sellheim, K., Stachowicz, J. J., and Coates, R. C. 2010. Effects of a nonnative habitat-forming species on mobile and sessile epifaunal communities. *Marine Ecology Progress Series* 398:69–80.

Serrao, E. A., Kautsky, L., and Brawley, S. H. 1996. Distributional success of the marine seaweed *Fucus vesiculosus* L in the brackish Baltic Sea correlates with osmotic capabilities of Baltic gametes. *Oecologia* 107:1–12.

Setchell, W. A. 1893. On the classification and geographical distribution of the Laminariaceae. *Transactions of the Connecticut Academy of Arts and Science* 9:333–375.

Setchell, W. A. 1908. *Nereocystis* and *Pelagophycus*. *Botanical Gazette* 45:125–134.

Setchell, W. A. 1932. Macrocystis and its holdfasts. *University of California Publications in Botany* 16:445–492.

Seymour, R. J. 1983. *Extreme Waves in California During Winter 1983*. California Department of Boating and Waterways, Sacramento, CA.

Shaffer, J. A. 2000. Seasonal variation in understory kelp bed habitats of the Strait of Juan de Fuca. *Journal of Coastal Research* 16:768–775.

Shanks, A. L., Grantham, B. A., and Carr, M. H. 2003. Propagule dispersal distance and the size and spacing of marine reserves. *Ecological Applications* 13(S1):S159–S169.

Shears, N. T. and Babcock, R. C. 2002. Marine reserve demontrate top-down control of community structure on temperate reefs. *Oceologia* 132:131–142.

Shears, N. T. and Babcock, R. C. 2003. Continuing trophic cascade effects after 25 years of no-take marine reserve protection. *Marine Ecology Progress Series* 246:1–16.

Shears, N. T., Babcock, R. C., and Salomon, A. K. 2008. Context-dependent effects of fishing: variation in trophic cascades across environmental gradients. *Ecological Applications* 18:1860–1873.

Shears, N. T., Kushner, D. J., Katz, S. L., and Gaines, S. D. 2012. Reconciling conflict between the direct and indirect effects of marine reserve protection. *Environmental Conservation* 39:225–236.

Shephard, S. A. and Sanderson, C. 2013. The giant kelp *Macrocystis pyrifera*. Pp. 79–86. In Shepard, S. A. and Edgar, G. J. (eds), *Ecology of Australian Temperate Reefs: The Unique South*. CSIRO Publishing, Collingwood, Australia.

Sherlock, V., Pickmere, S., Currie, K., Hadfield, M., Nodder, S., and Boyd, P. W. 2007. Predictive accuracy of temperature-nitrate relationships for the oceanic mixed layer of the New Zealand region. *Journal of Geophysical Research-Oceans* 112:C06010. doi:10.1029/2006JC003562.

Shimek, S. J. 1977. The underwater foraging habits of the sea otter, *Enhydra lutris*. *California Fish and Game* 63:120–122.

Silva, P. C. 1962. Comparison of algal floristic patterns in the Pacific with those in the Atlantic and Indian Oceans, with special reference to *Codium*. Pp. 201–216. In *Proceedings of the 9th Pacific Science Congress*, Bangkok, Thailand.

Simberloff, D. 2004. Community ecology: is it time to move on? *American Naturalist* 163:787–799.

Simenstad, C. A., Estes, J. A., and Kenyon, K. W. 1978. Aleuts, sea otters and alternate stable state communities. *Science* 200:403–411.

Singer, M. M., George, S., Jacobson, S., Lee, I., Weetman, L. L., Tjeerdema, R. S. et al. 1995. Acute toxicity of oil dispersant Corexit- 9954 to marine organisms. *Ecotoxicology and Environmental Safety* 32:81–86.

Skottsberg, C. 1907. Zur kenntnis der Subantarktischen und Antarktischen Meeresalgen. I. Phaeophyceen. *Wissenschafliche Ergenbnisse Schwedisch Süpolar-Expedition* 4:1–172. (in German)

Skottsberg, C. 1911. The Wilds of Patagonia. Edward Arnold, London.

Smale, D. A., Burrows, M. T., Moore, P., O'Connor, N., and Hawkins, S. J. 2013. Threats and knowledge gaps for ecosystem services provided by kelp forests: a northeast Atlantic perspective. *Ecology and Evolution* 3:4016–4038.

Smith, J. M. B. and Bayliss-Smith, T. P. 1998. Kelp-plucking: coastal erosion facilitated by bull-kelp *Durvillaea antarctica* at subantarctic Macquarie Island. *Antarctic Science* 10:431–438.

Sousa, W. P. 2001. Natural disturbance and the dynamics of marine benthic communities. Pp. 85–130. In Bertness, M. D., Gaines, S., and Hay, M. E. (eds), *Marine Community Ecology*. Sinauer Assoc., Sunderland, MA.

Spalding, H., Foster, M. S., and Heine, J. N. 2003. Composition, distribution, and abundance of deep-water (> 30 m) macroalgae in central California. *Journal of Phycology* 39: 273–284.

Spalding, M. D., Fox, H. E., Halpern, B. S., McManus, M. A., Molnar, J., Allen, G. R. et al. 2007. Marine ecoregions of the world: a bioregionalization of coastal and shelf areas. *Bioscience* 57:573–583.

Springer, Y. P., Hays, C. G., Carr, M. H., and Mackey, M. R. 2010. Toward ecosystem-based management of marine macroalgae – the bull kelp, *Nereocystis luetkeana*. *Oceanography and Marine Biology: An Annual Review* 48:1–42.

Stachowicz, J. J. and Byrnes, J. E. 2006. Species diversity, invasion success, and ecosystem functioning: disentangling the influence of resource competition, facilitation, and extrinsic factors. *Marine Ecology Progress Series* 311:251–262.

Stanford, E. C. 1883. On algin: a new substance obtained from some of the commoner species of marine algae. *The Chemical News* 47:254–257.

Stanley, S. M. 2009. *Earth System History*. 3rd ed. W.H. Freeman, New York.

Steele, M. A., Schroeter, S. C., Carpenter, R. C., and Kushner, D. J. 2006. Top-down vs. bottom-up effects in kelp forests. *Science* 313:1738.

Steinbeck, J., Schiel, D. R., and Foster, M. S. 2005. Detecting long-term changes in complex communities: a case study for the rocky intertidal zone. *Ecological Applications* 15:1813–1832.

Steinberg, P. D., Estes, J. A., and Winter, F. C. 1995. Evolutionary consequences of food-chain length in kelp forest communities. *Proceedings of the National Academy of Sciences* 92:8145–8148.

Steneck, R. S. 1998. Human influences on coastal ecosystems: does overfishing create trophic cascades? *Trends in Ecology and Evolution* 13:429–430.

Steneck, R. S., Graham, M. H., Bourque, B. J., Corbett, D., Erlandson, J. M., Estes, J. A. et al. 2002. Kelp forest ecosystems: biodiversity, stability, resilience and future. *Environmental Conservation* 29:436–459.

Stephens, J. 2001. California sheephead. Pages 155-156. In Leet, W. S., Dewees, C. M., Kingbell, R., and Larson, E. J. (eds), *California's Living Marine Resources: A Status Report*. California Department of Fish and Game, Sacramento, CA.

Stephenson, A. 1977. Age determination and morphological variation of Ontario otters. *Canadian Journal of Zoology* 55:1577–1583.

Stevens, C. A., Hurd, C. L., and Smith, M. J. 2001. Water motion relative to subtidal kelp fronds. *Limnology and Oceanography* 46:668–678.

Stewart, H. L., Fram, J. P., Reed, D. C., Williams, S. L., Brzezinski, M. A., MacIntyre, S. et al. 2009. Differences in growth, morphology and tissue carbon and nitrogen of *Macrocystis pyrifera* within and at the outer edge of a giant kelp forest in California, USA. *Marine Ecology Progress Series* 375:101–112.

Stokstad, E. 2012. Engineered superbugs boost hopes of turning seaweed into fuel. *Science* 335:273.

Stokstad, E. 2014. Death of the stars. *Science* 344:464–467.

Strange, E., Allen, D., Mills, D., and Raimondi, P. 2004. *Research on Estimating the Environmental Benefits of Restoration to Mitigate or Avoid Environmental Impacts Caused by California Power Plant Cooling Water Intake Structures*. California Energy Commission Report 500-04-092. California Energy Commission, Sacramento, CA.

Strong, D. R. 1992. Are trophic cascades all wet? Differentiation and donor control in speciose ecosystems. *Ecology* 73:747–754.

Strong, D. R. 1997. Quick indirect interactions in intertidal food webs. *Trends in Ecology and Evolution* 12:173–174.

Sutherland, W. J. and Woodroof, H. J. 2009. The need for environmental horizon scanning. *Trends in Ecology and Evolution* 24:523–527.

Sverdrup, H. V., Johnson, M. W., and Fleming, R. H. 1942. *The Oceans*. Prentice-Hall Inc, New York.

Swartz, R. C., Cole, F. A., Schults, D. W., and Deben, W. A. 1986. Ecological changes in the southern California bight near a large sewage outfall: benthic conditions in 1980 and 1983. *Marine Ecology Progress Series* 31:1–13.

Sydeman, W. J., Garcia-Reyes, M., Schoeman, D. S., Rykaczewski, R. R., Thompson, S. A., Black, B. A. et al. 2014. Climate change and wind intensification in coastal upwelling ecosystems. *Science* 345:77–80.

Sykes, M. 1908. Anatomy and histology of *Macrocystis pyrifera* and *Laminaria saccharina*. *Annals of Botany* 22:291–325.

Syms, C. and Carr, M. H. 2001. *Marine protected areas: evaluating MPA effectiveness in an uncertain world*. Marinet: North American Commission for Environmental Cooperation. www.orchestrabycrossdraw.com/30/Posting.cfm?2B07183C303E151A0109 7F505D546A.

Tait, L. W. and Schiel, D. R. 2011. Legacy effects of canopy disturbance on ecosystem functioning in macroalgal assemblages. *PLOS ONE* 6: e26986.

Tarasoff, F. and Kooyman, G. 1973. Observations on the anatomy of the respiratory system of the river otter, sea otter, and harp seal. I. The topography, weight, and measurements of the lungs. *Canadian Journal of Zoology* 51:163–170.

Taylor, D. I. and Schiel, D. R. 2003. Wave-related morality in zygotes of habitat-forming algae from different exposures in southern New Zealand: the importance of "stickability". *Journal of Experimental Marine Biology and Ecology* 290:229–245.

Taylor, D. I. and Schiel, D. R. 2010. Algal populations controlled by fish herbivory across a wave exposure gradient on southern temperate shores. *Ecology* 91:201–211.

Taylor, D. I, Delaux, S., Stevens, C., Nokes, R., and Schiel, D. 2010. Settlement rates of macroalgal propagules: cross-species comparisons in a turbulent environment. *Limnology and Oceanography* 55:66–76.

Tegner, M. J. 1980. Multispecies considerations of resource management in southern California kelp beds. Canadian Technical Report of Fisheries and Aquatic Sciences 954:125–143.

Tegner, M. J. 1993. Southern California abalones: can stocks be rebuilt using marine harvest refugia. *Canadian Journal of Fisheries and Aquatic Sciences* 50:2010–2018.

Tegner, M. J. and Dayton, P. K. 1981. Population structure, recruitment and mortality of two sea urchins (*Strongylocentrotus franciscanus* and *S. purpuratus*) in a kelp forest. *Marine Ecology Progress Series* 5:255–268.

Tegner, M. J. and Dayton, P. K. 1987. El Niño effects on southern California kelp forest communities. *Advances in Ecological Research* 17:243–279.

Tegner, M. J. and Dayton, P. K. 2000. Ecosystem effects of fishing in kelp forest communities. ICES *Journal of Marine Science* 57:579–589.

Tegner, M. J. and Levin, L. 1982. Do sea urchins and abalones compete in California kelp forest communities. Pp. 265–271. In *Proceedings of the International Echinoderms Conference (Tampa Bay)*. AA Balkema, Rotterdam, the Netherlands.

Tegner, M. J. and Levin, L. A. 1983. Spiny lobsters and sea urchins: analysis of a predator prey interaction. *Journal of Experimental Marine Biology and Ecology* 73:125–150.

Tegner, M. J., Dayton, P. K., Edwards, P. B., and Riser, K. L. 1995. Sea urchin cavitation of giant kelp (*Macrocystis pyrifera* Agardh, C.) holdfasts and its effects on kelp mortality across a large California forest. *Journal of Experimental Marine Biology and Ecology* 191:83–99.

Tegner, M. J., Dayton, P. K., Edwards, P. B., Riser, K. L., Chadwick, D. B., Dean, T. A. et al. 1995. Effects of a large sewage spill on a kelp forest community: catastrophe or disturbance? *Marine Environmental Research* 40:181–224.

Tegner, M. J., Basch, L. V., and Dayton, P. K. 1996. Near extinction of an exploited marine invertebrate. *Trends in Ecology and Evolution* 11:278–208.

Tegner, M. J., Dayton, P. K., Edwards, P. B., and Riser, K. L. 1997. Large-scale, low-frequency oceanographic effect on kelp forest succession: a tale of two cohorts. *Marine Ecology Progress Series* 146:117–134.

Tegner, M. J., Haaker, P. L., Riser, K. L., and Vilchis, L. I. 2001. Climate variability, kelp forests, and the southern California red abalone fishery. *Journal of Shellfish Research* 20:755–764.

Tellier, F., Meynard, A. P., Correa, J. A., Faugeron, S., and Valero, M. 2009. Phylogeographic analyses of the 30° S south-east Pacific biogeographic transition zone establish the occurrence of a sharp genetic discontinuity in the kelp *Lessonia nigrescens*: vicariance or parapatry? *Molecular Phylogenetics and Evolution* 53:679–693.

Thiel, M. and Gutow, L. 2005a. The ecology of rafting in the marine environment. I. The floating substrata. *Oceanography and Marine Biology: An Annual Review* 42:181–264.

Thiel, M. and Gutow, L. 2005b. The ecology of rafting in the marine environment. II. The rafting organisms and community. *Oceanography and Marine Biology: An Annual Review* 43:279–418.

Thiel, M., Macaya, E. C., Acuna, E., Arntz, W. E., Bastias, H., Brokordt, K. et al. 2007. The Humboldt current system of northern and central Chile. *Oceanography and Marine Biology: An Annual Review* 45:195–344.

Thompson, G. A. and Schiel, D. R. 2012. Resistance and facilitation by native algal communities in the invasion success of *Undaria pinnatifida*. *Marine Ecology Progress Series* 468:95–105.

Thompson, W. C. 1959. Attachment of the giant kelp *Macrocystis pyrifera* in fine sediments and its biological and geological significance. P.586. In *Proceedings of the International Oceanographic Congress*. American Association for the Advancement of Science, Washington, DC.

Thomsen, M. S., Wernberg, T., Tuya, F., and Silliman, B. R. 2009. Evidence for impacts of non-indigenous macroalgae: a meta-analysis of experimental field studies. *Journal of Phycology* 45:812819.

Thornber, C. S., Kinlan, B. P., Graham, M. H., and Stachowicz, J. J. 2004. Population ecology of the invasive kelp *Undaria pinnatifida* in California: environmental and biological controls on demography. *Marine Ecology Progress Series* 268:69–80.

Thrush, S. F. and Dayton, P. K. 2010. What can ecology contribute to ecosystem-based management? *Annual Review of Marine Science* 2:419–441.

Tianjing Liu, T. J., Tianjing Liu, T. J., Suo, R. Y., Liu, X. Y., Hu, D. Q., Shi, Z. J., Liu, G. Y. et al. 1984. Studies on the artificial cultivation and propagation of giant kelp (*Macrocystis pyrifera*). *Hydrobiologia* 116:259–262.

Tinker, M. T., Bentall, G., and Estes, J. A. 2008. Food limitation leads to behavioral diversification and dietary specialization in sea otters. *Proceedings of the National Academy of Sciences* 105:560–565.

Todd, N. P. M. 2012. The 1961 Rutherford Jubilee Conference: perspectives from 2011. *Journal of Physics: Conference Series* 381:012127.

Dieck, T. I. 1993. Temperature tolerance and survival in darkness of kelp gametophytes (Laminariales, Phaeophyta): ecological and biogeographical implications. *Marine Ecology Progress Series* 100:253–264.

Tomczak, M. and Godfrey, J. S. 2003. *Regional Oceanography: An Introduction*. 2nd ed. Daya Publishing House, Delhi, India.

Torres-Moye, G., Edwards, M. S., and Montaño-Moctezuma, C. G. 2013. Benthic community structure in kelp forests from the Southern California Bight. *Ciencias Marinas* 39:239–252.

Tovey, D. J. and Moss, B. L. 1978. Attachment of haptera of *Laminaria digitata* (Huds) Lamour. *Phycologia* 17:17–22.

Towle, D. W. and Pearse, J. S. 1973. Production of the giant kelp, *Macrocystis*, estimated by in situ incorporation of 14C in polyethlene bags. *Limnology and Oceanography* 18:155–159.

Townsend, M., Thrush, S. F., and Carbines, M. J. 2011. Simplifying the complex: an "ecosystem principles approach" to goods and services management in marine coastal ecosystems. *Marine Ecology Progress Series* 434:291–301.

Trent, J., Wiley, P., Tozzi, S., McKuin, B., and Reinsch, S. 2012. Research spotlight: the future of biofuels: is it in the bag? *Biofuels* 3:521–524.

Turner, C. H., Ebert, E. E., and Given, R. R. 1968. *The Marine Environment Offshore from Point Loma, San Diego County, California*. Fish Bulletin 146, California Department of Fish and Game, Sacramento, CA.

Turner, N. 1995. *Food Plants of Coastal First Peoples*. University of British Columbia Press, Vancouver, Canada.

Turpen, S., Hunt, J. W., Anderson, B. S., and Pearse, J. S. 1994. Population structure, growth, and fecundity of the kelp forest mysid *Holmesimysis costata* in Monterey Bay, California. *Journal of Crustacean Biology* 14:657–664.

Turrentine, J. W. 1926. *Potash: A Review, Estimate and Forecast*. John Wiley & Sons, New York.

Tutschulte, T. C. and Connell, J. H. 1988. Feeding behavior and algal food of three species of abalones (*Haliotis*) in southern California. *Marine Ecology Progress Series* 49:57–64.

Twain, M. 1897. Quote within a note from Mark Twain to Frank Marshall White. Reprinted in Fishkin, S. F. 1996. *Lighting out for the Territory: Reflections on Mark Twain and American Culture*. Oxford University Press, Oxford, UK.

UCSB (University of Santa Barbara). 2012. *SONGS mitigation monitoring*. http:www.marinemitigation.msi.ucsb.edu/index.html.

Underwood, A. J. 1986. What is a community? Pp. 351–367. in Raup, D. M. and Jablonski, D. (eds), *Patterns and Processes in the History of Life*. Springer-Verlag, Berlin, Germany.

Underwood, A. J. 1999. Physical disturbances and their direct effect on an indirect effect: responses of an intertidal assemblage to a severe storm. *Journal of Experimental Marine Biology and Ecology* 232:125–140.

USDA (United States Department of Agriculture). 2012. *National Nutrient Data Base*. http://www.ndb.nal.usda.gov.

USDE (United States Department of Energy). 2012. *Alternative fuels – ethanol*. www.fueleconomy.gov.

USDT (United States Department of Tranportation). 2011. *Motor Fuel*. US Department of Transporation, Federal Highway Administration, www.fhwa.dot.gov.

Utter, B. D. and Denny, M. W. 1996. Wave-induced forces on the giant kelp *Macrocystis pyrifera* (Agardh): field test of a computational model. *The Journal of Experimental Biology* 199:2645–2654.

Uwai, S. Y., Nelson, W., Neill, K., Wang, W. D., Aguilar-Rosas, L. E., Boo, S. M. et al. 2006. Genetic diversity in *Undaria pinnatifida* (Laminariales, Phaeophyceae) deduced from mitochondria genes origins and succession of introduced populations. *Phycologia* 45:687–695.

Vadas, R. L. 1977. Preferential feeding: optimization strategy in sea urchins. *Ecological Monographs* 47:337–371.

Vadas, R. L., Wright, W. A., and Miller, S. L. 1990. Recruitment of *Ascophyllum nodosum*: wave action as a source of mortality. *Marine Ecology Progress Series* 61:263–272.

Vadas, R. L., Johnson, S., and Norton, T. A. 1992. Recruitment and mortality of early post-settlement stages of benthic algae. *British Phycological Journal* 27:331–351.

Valentine, J. P. and Johnson, C. R. 2003. Establishment of the introduced kelp *Undaria pinnatifida* in Tasmania depends on disturbance to native algal assemblages. *Journal of Experimental Marine Biology and Ecology* 295:63–90.

Valentine, J. P. and Johnson, C. R. 2005. Persistence of the exotic kelp *Undaria pinnatifida* does not depend on sea urchin grazing. *Marine Ecology Progress Series* 285:43–55.

VanBlaricom, G. R. and Estes, J. A. 1988. *The Community Ecology of Sea Otters*. Springer-Verlag, New York.

Vance, R. R. 1979. Effects of grazing by the sea urchin, *Centrostephanus coronatus*, on prey community composition. *Ecology* 60:537–546.

van den Hoek, C. 1982. The distribution of benthic marine algae in relation to the temperature regulation of their life histories. *Biological Journal of the Linnean Society* 18:81–144.

Vanderklift, M. A. and Wernberg, T. 2008. Detached kelps from distant sources are a food subsidy for sea urchins. *Oecologia* 157:327–335.

van Tamelen, P. G. and Woodby, D. 2001. *Macrocystis* biomass, quality, and harvesting effects in relation to the herring spawn-on-kelp fishery in Alaska. *The Alaska Fisheries Research Bulletin* 8:118–131.

Van Tussenbroek, B. I. 1989. Seasonal growth and composition of fronds of *Macrocystis pyrifera* in the Falkland Islands. *Marine Biology* 100:419–430.

VanWagenen, R. F., Foster, M. S., and Bunns, F. 1981. Sea otter predation on birds near Monterey, California. *Journal of Mammalogy* 62:433–434.

Vásquez, J. A. 1993. Effects on the animal community of dislodgement of holdfasts of *Macrocystis pyrifera*. *Pacific Science* 47:180–184.

Vásquez, J. A. 2008. Production, use and fate of Chilean brown seaweeds: resources for a sustainable fishery. *Journal of Applied Phycology* 20:457–467.

Vásquez, J. A. and Buschmann, A. H. 1997. Herbivore-kelp interactions in Chilean subtidal communities: a review. *Revista Chilena De Historia Natural* 70:41–52.

Vásquez, J. A. and McPeak, R. H. 1998. A new tool for kelp restoration. *California Fish and Game* 84:149–158.

Vásquez, J. A., Castilla, J. C., and Santelices, B. 1984. Distribution patterns and diets of four species of sea-urchins in giant kelp forest (*Macrocystis pyrifera*) of Pureto Toro, Navarino Island, Chile. *Marine Ecology Progress Series* 19:55–63.

Vásquez, J. A., Alonso Vega, J. M., and Buschmann, A. H. 2006. Long term variability in the structure of kelp communities in northern Chile and the 1997-98 ENSO. *Journal of Applied Phycology* 18:505–519.

Vásquez, X., Gutiérrez, A., Buschmann, A. H., Flores, R., Farías, D., and Leal, P. 2014. Evaluation of repopulation techniques for the giant kelp Macrocystis pyrifera (Laminariales). *Botanica Marina* 57:123–130.

Velimirov, B. and Griffiths, C. L. 1979. Wave-induced kelp movement and its importance for community structure. *Botanica Marina* 22:169–172.

Verges, A., Steinberg, P. D., Hay, M. E., Poore, A. G. B., Campbell, A. H., Ballesteros, E. et al. 2014. The tropicalization of temperate marine ecosystems: climate-mediated changes in herbivory and community phase shifts. *Proceedings of the Royal Society B: Biological Sciences* 281:20140846.

Vetter, E. W. 1995. Detritus based patches of high secondary production in the nearshore benthos. *Marine Ecology Progress Series* 120:251–262.

Vetter, E. W. 1998. Population dynamics of a dense assemblage of marine detritivores. *Journal of Experimental Marine Biology and Ecology* 226:131–161.

Vetter, E. W. and Dayton, P. K. 1999. Organic enrichment by macrophyte detritus, and abundance patterns of megafaunal populations in submarine canyons. *Marine Ecology Progress Series* 186:137–148.

Villegas, M. J., Laudien, J., Sielfeld, W., and Arntz, W. E. 2007. *Macrocystis integrifolia* and *Lessonia trabeculata* (Laminariales; Phaeophyceae) kelp habitat structures and associated macrobenthic community off northern Chile. *Helgoland Marine Research* 62:33–43.

Villouta, E. and Santelices, B. 1984. Estructura de la comunidad submareal de *Lessonia* (Phaeophyta, Laminariales) en Chile norte y central. *Revista Chilena De Historia Natural* 57:111–122. (in Spanish)

Vogel, S. 1994. *Life in Moving Fluids.* 2nd ed. Princeton University Press, Princeton, NJ.

Voltaire, F. M. A. 1759. *Candide, ou l'optimisme.* Paris, Siréne, France. (in French)

Walker, F. T. 1952. Chromosome number of *Macrocystis integrifolia* Bory. *Annals of Botany* 16:23–27.

Wargacki, A. J., Leonard, E., Win, M. N., Regitsky, D. D., Santos, C. N. S., Kim, P. B. et al. 2012. An engineered microbial platform for direct biofuel production from brown macroalgae. *Science* 335:308–313.

Watanabe, J. M. 1984a. The influence of recruitment, competition and benthic predation on spatial distribution of three species of kelp forest gastropods (Trochidae, Tegula). *Ecology* 65:920–936.

Watanabe, J. M. 1984b. Food preference, food quality and diets of three herbivorous gastropods (Trochidae, Tegula) in a temperate kelp forest habitat. *Oecologia* 62:47–52.

Watanabe, J. M. and Harrold, C. 1991. Destructive grazing by sea urchins *Strongylocentrotus* spp. in a central California kelp forest: potential roles of recruitment, depth and predation. *Marine Ecology Progress Series* 71:125–141.

Watanabe, J. M., Phillips, R. E., Allen, N. H., and Anderson, W. A. 1992. Physiological response of the stipitate understory kelp, *Pterygophora californica* Ruprecht, to shading by the giant kelp *Macrocystis pyrifera* C. Agardh. *Journal of Experimental Marine Biology and Ecology* 159:237–252.

Waters, D. L., Oda, K. T., and Mello, J. 2001. Pacific herring. Pp. 456–459. In Leet, W. S., Dewees, C. M., Klingbeil, R., and Larson, E. J. (eds), *California's Living Marine Resources: A Status Report.* California Department Fish and Game, Sacramento, CA.

Waters, T. 2002. The paddy principle. Pp. 111–114. In Gibson, B. (ed.), *Offshore Saltwater Fishing.* Creative Publishing International, Minnetonka, MN.

Watson, J. 2000. The effects of sea otters (*Enhydra lutris*) on abalone (*Haliotis* spp.) populations. Pp. 123–132. In Campbell, A. (ed.), *Workshop on Rebuilding Abalone Stocks in British Columbia.* Canadian Special Publication on Fisheries and Aquatic Science, NRC Research Press, Canada.

Watson, J. and Estes, J. A. 2011. Stability, resilience, and phase shifts in rocky subtidal communities along the west coast of Vancouver Island, Canada. *Ecological Monographs* 81:215–239.

Weaver, A. M. 1977. *Aspects of the Effects of Particulate Matter on the eCology of a Kelp Forest (Macrocystis pyrifera (L.) C. A. Agardh) Near a Small Domestic Sewer Outfall.* PhD thesis. Stanford University, Stanford, CA.

Wernberg, T., Thomsen, M. S., Tuya, F., Kendrick, G. A., Staehr, P. A., and Toohey, B. D. 2010. Decreasing resilience of kelp beds along a latitudinal temperature gradient: potential implications for a warmer future. *Ecology Letters* 13:685–694.

Wernberg, T., Russell, B. D., Moore, P. J., Ling, S. D., Smale, D. A., Campbell, A. et al. 2011. Impacts of climate change in a global hotspot for temperate marine biodiversity and ocean warming. *Journal of Experimental Marine Biology and Ecology* 400:7–16.

Westermeier, R. and Möller, P. 1990. Population dynamics of *Macrocystis pyrifera* (L) C Agardh in the rocky intertidal of southern Chile. *Botanica Marina* 33:363–367.

Westermeier, R., Patino, D., and Müller, D. G. 2007. Sexual compatibility and hybrid formation between the giant kelp species *Macrocystis pyrifera* and *M. integrifolia* (Laminariales, Phaeophyceae) in Chile. *Journal of Applied Phycology* 19:215–221.

Westermeier, R., Patino, D. J., Murua, P., and Müller, D. G. 2011. *Macrocystis* mariculture in Chile: growth performance of heterosis genotype constructs under field conditions. *Journal of Applied Phycology* 23:819–825.

Westermeier, R., Murua, P., Patino, D. J., Muñoz, L., Ruiz, A., and Mueller, D. G. 2012. Variations of chemical composition and energy content in natural and genetically defined cultivars of *Macrocystis* from Chile. *Journal of Applied Phycology* 24:1191–1201.

Westermeier, R., Murua, P., Patino, D. J., Munoz, L., Ruiz, A., Atero, C. et al. 2013. Utilization of holdfast fragments for vegetative propagation of *Macrocystis integrifolia* in Atacama, Northern Chile. *Journal of Applied Phycology* 25:639–642.

Wheeler, P. A. and North, W. J. 1981. Nitrogen supply, tissue composition and frond growth-rates for *Macrocystis pyrifera* off the coast of southern California. *Marine Biology* 64:59–69.

Wheeler, W. N. 1980a. Effect of boundary layer transport on the fixation of carbon by the giant kelp *Macrocystis pyrifera*. *Marine Biology* 56:103–110.

Wheeler, W. N. 1980b. Pigment content and photosynthetic rate of the fronds of *Macrocystis pyrifera*. *Marine Biology* 56:97–102.

Wheeler, W. N. 1982. Nitrogen nutrition of *Macrocystis*. Pp. 121–135. In Srivastava, L. M. (ed.), *Synthetic and Degradative Processes in Marine Macrophytes*. Walter de Gruyter, New York.

Wheeler, W. N. and Druehl, L. D. 1986. Seasonal growth and productivity of *Macrocystis integrifolia* in British Columbia, Canada. *Marine Biology* 90:181–186.

Wheeler, W. N. and Neushul, M. 1981. The aquatic environment. Pp. 229–247. In Lange, O. L., Nobel, P. S., Osmond, C. B., and Ziegler, H. (eds), *Encyclopedia of Plant Physiology*. Springer-Verlag, Berlin, Germany.

Widdowson, T. B. 1970. A taxonomic revision of the genus *Alaria* Greville. *Syesis* 4:11–49.

Wild, P. W. and Ames, J. A. 1974. *A Report on the Sea Otter, Enhydra lutris L., in California.* California Department of Fish and Game Marine Resources Technical Report 20, California Department of Fish and Game, Sacramento, CA.

Wildish, D. J. and Kristmanson, D. D. 1997. *Benthic Suspension Feeders and Flow*. Cambridge University Press, Cambridge, UK.

Williams, T. D., Allen, D. D., Groff, J. M., and Glass, R. L. 1992. An analysis of California sea otter (*Enhydra lutris*) pelage and integument. *Marine Mammal Science* 8:1–18.

Wilmers, C. C., Estes, J. A., Edwards, M., Laidre, K. L., and Konar, B. 2012. Do trophic cascades affect the storage and flux of atmospheric carbon? An analysis of sea otters and kelp forests. *Frontiers in Ecology and the Environment* 10:409–415.

Wilson, K. C. and McPeak, R. 1983. Kelp restoration. Pp. 199–216. In Bascom, W. (ed.), *The Effects of Waste Disposal on Kelp Communities*. Southern California Coastal Water Research Project, Long Beach, CA.

Wilson, K. C. and Togstad, H. 1983. Storm caused changes in the Palos Verdes kelp forest. Pp. 301–307. In Bascom, W. (ed.), *The Effects of Waste Disposal on Kelp Communities*. Southern California Coastal Water Research Project, Long Beach, CA.

Wing, S. R. and Patterson, M. R. 1993. Effects of wave-induced lightflecks in the intertidal zone on photosynthesis in the macroalgae *Postelsia palmaeformis* and *Hedophyllum sessile* (Phaeophyceae). *Marine Biology* 116:519–525.

Womersley, H. B. S. 1954. The species of *Macrocystis* with special reference to those on southern Australian coasts. *University of California Publications in Botany* 27:109–132.

Wood, H. L., Spicer, J. I., and Widdicombe, S. 2008. Ocean acidification may increase calcification rates, but at a cost. *Proceedings of the Royal Society B: Biological Sciences* 275:1767–1773.

Woodhouse, C. D., Cowen, R. K., and Wilcoxon, L. R. 1977. *A Summary of Knowledge of the Sea Otter, Enhydra lutris L., in California and an Appraisal of the Completeness of Biological Understanding of the Species*. Final Report to U.S. Marine Mammal Commission for Contract MM6AC008. U.S. Marine Mammal Commission, Washington, DC.

Worm, B., Sandow, M., Oschlies, A., Lotze, H. K., and Myers, R. A. 2005. Global patterns of predator diversity in the open oceans. *Science* 309:1365–1369.

Worm, B., Barbier, E. B., Beaumont, N., Duffy, E., Folke, C., Halpern, B. S. et al. 2006. Impacts of biodiversity loss on ocean ecosystem services. *Science* 314:787–790.

Wotton, D. M., O'Brien, C., Stuart, M. D., and Fergus, D. J. 2004. Eradication success down under: heat treatment of a sunken trawler to kill the invasive seaweed *Undaria pinnatifida*. *Marine Pollution Bulletin* 49:844–849.

Wright, S. 1943. Isolation by distance. *Genetics* 28:114–138.

Yabu, H. and Sanbonsuga, Y. 1987. Chromosome count in *Macrocystis integrifolia* Bory. *Bulletin of the Faculty of Fisheries Hokkaido University* 38:339–342.

Yaninek, J. S. 1980. *Beach Wrack: Phenology of an Imported Resource and Utilization by Macroinvertebrates of Sandy Beaches*. MSc thesis. University of California, Berkeley, CA.

Yeates, L. C., Williams, T. M., and Fink, T. L. 2007. Diving and foraging energetics of the smallest marine mammal, the sea otter (*Enhydra lutris*). *Journal of Experimental Biology* 210:1960–1970.

Yellin, M. B., Agegian, C. R., and Pearse, J. S. 1977. *Ecological Benchmarks in the Santa Cruz County Kelp Forests before the Re-Establishment of Sea Otters*. Technical Report Number 6, Center for Coastal Marine Studies (now Institute of Marine Sciences), University of California, Santa Cruz, CA.

Yoklavich, M., Cailliet, G., Lea, R. N., Greene, H. G., Starr, R., De Marignac, J. et al. 2002. Deepwater habitat and fish resources associated with the Big Creek Marine Ecological Reserve. *CalCOFI Reports* 43:120–140.

Yoon, H. S., Hackett, J. D., Ciniglia, C., Pinto, G., and Bhattacharya, D. 2004. A molecular timeline for the origin of photosynthetic eukaryotes. *Molecular Biology and Evolution* 21:809–818.

Yoon, H. S., Lee, J. Y., Boo, S. M., and Bhattacharya, D. 2001. Phylogeny of Alariaceae, Laminariaceae, and Lessoniaceae (Phaeophyceae) based on plastid-encoded RuBisCo spacer and nuclear-encoded ITS sequence comparisons. *Molecular Phylogenetics and Evolution* 21:231–243.

York, R. and Foster, M. S. 2005. *Issues and Environmental Impacts Associated with Once-Through Cooling at California's Coastal Power Plants.* California Energy Commission, Sacramento, CA.

Yorke, C. E., Miller, R. J., Page, H. M., and Reed, D. C. 2013. Importance of kelp detritus as a component of suspended particulate organic matter in giant kelp *Macrocystis pyrifera* forests. *Marine Ecology Progress Series* 493:113–125.

Young, I. R., Zieger, S., and Babanin, A. V. 2011. Global trends in wind speed and wave height. *Science* 332:451–455.

Zaytsev, O., Cervantes-Duarte, R., Montante, O., and Gallegos-Garcia, A. 2003. Coastal upwelling activity on the pacific shelf of the Baja California Peninsula. *Journal of Oceanography* 59:489–502.

Zimmerman, M. B. 2009. Iodine deficiency. *Endocrine Reviews* 30:376–408.

Zimmerman, R. C. and Kremer, J. N. 1984. Episodic nutrient supply to a kelp forest ecosystem in Southern California. *Journal of Marine Research* 42:591–604.

Zimmerman, R. C. and Kremer, J. N. 1986. In situ growth and chemical composition of the giant kelp, *Macrocystis pyrifera*, response to temporal changes in ambient nutrient availability. *Marine Ecology Progress Series* 27:277–285.

Zimmerman, R. C. and Robertson, D. L. 1985. Effects of El Niño on local hydrography and growth of the giant kelp, *Macrocystis pyrifera*, at Santa Catalina Island, California. *Limnology and Oceanography* 30:1298–1302.

ZoBell, C. E. 1971. Drift seaweeds on San Diego County beaches. Pp. 269–314. In North, W. J. (ed.), *The Biology of Giant Kelp Beds (Macrocystis) in California.* Nova Hedwigia 32. Verlag Von J. Cramer, Lehre, Germany.

# INDEX